T0342529

Drought

Titles in the Series

Hydrometeorological Hazards: Interfacing Science and Policy
Edited by Philippe Quevauviller

Coastal Storms: Processes and Impacts
Edited by Paolo Ciavola and Giovanni Coco

Drought: Science and Policy
Edited by Ana Iglesias, Dionysis Assimacopoulos, and Henny A.J. Van Lanen

Forthcoming Titles

Flash Floods Early Warning Systems: Policy and Practice
Edited by Daniel Sempere-Torres

Facing Hydrometeorological Extreme Events: A Governance Issue
Edited by Isabelle La Jeunesse and Corinne Larrue

Drought

Science and Policy

Edited by

Ana Iglesias
Universidad Politécnica de Madrid
Madrid, Spain

Dionysis Assimacopoulos
National Technical University of Athens
Athens, Greece

Henny A.J. Van Lanen
Wageningen University
Wageningen, The Netherlands

This edition first published 2019

Registered Office(s)
John Wiley & Sons, Inc., 111 River Street, Hoboken, NJ 07030, USA
John Wiley & Sons Ltd, The Atrium, Southern Gate, Chichester, West Sussex, PO19 8SQ, UK

Editorial Office
The Atrium, Southern Gate, Chichester, West Sussex, PO19 8SQ, UK

For details of our global editorial offices, customer services and more information about Wiley products, visit us at www.wiley.com.

Wiley also publishes its books in a variety of electronic formats and by print-on-demand. Some content that appears in standard print versions of this book may not be available in other formats.

Library of Congress Cataloging-in-Publication Data

Names: Iglesias, Ana, editor. | Assimacopoulos, Dionysis, 1950- editor. | Lanen, Henny A. J. Van, 1952- editor.
Title: Drought : science and policy / edited by Dr. Ana Iglesias, Professor Dionysis Assimacopoulos, Dr. Henny A. Van Lanen.
Description: First edition. | Hoboken, NJ : John Wiley & Sons, 2018. | Series: Hydrometeorological extreme events | Includes index. |
Identifiers: LCCN 2018015076 (print) | LCCN 2018029390 (ebook) | ISBN 9781119017219 (pdf) | ISBN 9781119017172 (epub) | ISBN 9781119017202 (cloth)
Subjects: LCSH: Droughts.
Classification: LCC QC929.24 (ebook) | LCC QC929.24 .D768 2018 (print) | DDC 363.34/929--dc23
LC record available at https://lccn.loc.gov/2018015076

Cover design: Wiley
Cover image: © Henny Van Lanen

Set in 10/12pt TimesTenLTStd by SPi Global, Chennai, India
Printed in Singapore by C.O.S. Printers Pte Ltd

10 9 8 7 6 5 4 3 2 1

Contents

Series Preface xi

The Series Editor – Philippe Quevauviller xiii

List of Contributors xv

PART ONE: UNDERSTANDING DROUGHT AS A NATURAL HAZARD 1

1.1 Diagnosis of Drought-Generating Processes 3
Henny A.J. Van Lanen, Anne F. Van Loon, and Lena M. Tallaksen

1.1.1 Introduction 3
1.1.2 Background 4
1.1.3 Climate drivers of drought 7
1.1.3.1 Atmospheric and oceanic drivers 7
1.1.3.2 Summer drought of 2015 8
1.1.3.3 Influence of humans (Climate change) 9
1.1.4 Soil moisture drought processes 11
1.1.4.1 Processes 11
1.1.4.2 Human influences 13
1.1.5 Hydrological drought processes (Groundwater and streamflow) 14
1.1.5.1 Groundwater 14
1.1.5.2 Streamflow 17
1.1.6 Drought propagation 19
1.1.6.1 Climate–hydrology links 19
1.1.6.2 Hydrological drought typology 20
1.1.6.3 Human influences 20
1.1.7 Concluding remarks and outlook 21
1.1.7.1 Conclusions 21
1.1.7.2 Outlook 22
Acknowledgements 23
References 23

1.2 Recent Trends in Historical Drought **29**
Kerstin Stahl, Lena M. Tallaksen, and Jamie Hannaford

1.2.1 Introduction 29
1.2.2 Trend analysis and data 30
1.2.2.1 Methodology 30
1.2.2.2 Data 31
1.2.3 Trends in river flow across Europe 32
1.2.3.1 Trends in observed river flow 32
1.2.3.2 Trends in modelled runoff 34
1.2.3.3 Influence of decadal-scale variability on trends in long streamflow records 36
1.2.4 Discussion 38
1.2.5 Conclusions – Future needs 40
1.2.5.1 Conclusions 40
1.2.5.2 Future needs 41
Acknowledgements 41
References 41

1.3 Historic Drought from Archives: Beyond the Instrumental Record **45**
Emmanuel Garnier

1.3.1 Introduction 45
1.3.2 Methodology 45
1.3.2.1 The historical material 45
1.3.2.2 Rebuild the droughts of the past 47
1.3.3 United Kingdom 48
1.3.3.1 Chronological variation and severity of the UK droughts 49
1.3.3.2 The most extreme events in a 500-year period 50
1.3.4 France: Ile-de-France 55
1.3.4.1 Chronological variation and severity of French droughts 55
1.3.4.2 Two examples of very severe droughts 58
1.3.5 Valley of the Upper Rhine (Germany, Switzerland, France) 59
1.3.5.1 Chronological variation and severity of Rhenish droughts 60
1.3.5.2 The drought, a subject of collective memory, the Laufenstein 61
1.3.6 Conclusions 64
Acknowledgements 65
References 65

1.4 Future Drought **69**
Henny A.J. Van Lanen, Christel Prudhomme, Niko Wanders, and Marjolein H.J. Van Huijgevoort

1.4.1 Introduction 69
1.4.2 Overview of studies 70
1.4.3 Assessment of future hydrological drought 71
1.4.3.1 Future drought across climate regions 71
1.4.3.2 Future streamflow drought in Europe 72
1.4.3.3 Hotspots of future drought 75
1.4.3.4 Future low flows 76
1.4.4 Human influences on future drought 78
1.4.4.1 Impact of reservoirs on future drought across the globe 78
1.4.4.2 Impact of water use on streamflow drought in Europe 80
1.4.4.3 Impact of gradual change of hydrological regime on future drought 81

1.4.5 Uncertainties in future drought 82
1.4.5.1 Uncertainty assessments 82
1.4.5.2 Impact sources of uncertainty 84
1.4.6 Conclusions – Future needs 86
1.4.6.1 Conclusions 86
1.4.6.2 Future needs 87
Acknowledgements 87
References 88

PART TWO: VULNERABILITY, RISK, AND POLICY 93

2.5 On the Institutional Framework for Drought Planning and Early Action 95
Ana Iglesias, Luis Garrote, and Alfredo Granados

2.5.1 Introduction 95
2.5.2 Drought planning and water resources planning 96
2.5.3 A code for best practices: early action and risk management plans 98
2.5.4 Institutions involved in drought planning 99
2.5.4.1 Main issues and guidelines for the institutional analysis of drought planning 99
2.5.4.2 The legal framework and complexity of institutional systems 100
2.5.4.3 Examples of legal provisions in Spain 103
2.5.4.4 Key aspects of drought management in mediterranean countries 106
2.5.5 Conclusions 108
Acknowledgements 108
References 109

2.6 Indicators of Social Vulnerability to Drought 111
Gustavo Naumann, Hugo Carrão, and Paulo Barbosa

2.6.1 Introduction 111
2.6.2 Theoretical framework 112
2.6.3 Selection of policy-relevant variables 115
2.6.3.1 Normalisation of variables to a common baseline 117
2.6.3.2 Composite indicator of drought vulnerability (weighting and aggregation) 118
2.6.3.3 Sensitivity analysis and model validation 122
2.6.4 Application: Drought risk assessment in Latin America 123
References 124

2.7 Drought Vulnerability Under Climate Change: A Case Study in La Plata Basin 127
Alvaro Sordo-Ward, María D. Bejarano, Luis Garrote, Victor Asenjo, and Paola Bianucci

2.7.1 Introduction 127
2.7.2 Methods 128
2.7.2.1 Framework 128
2.7.2.2 Data and regional extent 129
2.7.2.3 Data analysis 130
2.7.2.4 Drought indicator used 132
2.7.2.5 Limitations of the methodology 132
2.7.3 Results and discussion 133

2.7.3.1 Drought characterisation: Changes in rainfall 133
2.7.3.2 Drought characterisation: Changes in SPEI drought indicator 134
2.7.3.3 Graphical user interface as a tool for drought management 140
2.7.4 Conclusions 140
Acknowledgements 144
References 144

2.8 Drought Insurance **147**
Teresa Maestro, Alberto Garrido, and María Bielza

2.8.1 Introduction 147
2.8.2 Main difficulties and challenges in developing drought insurance 148
2.8.3 Types of drought insurance 149
2.8.4 Drought indemnity-based insurance 150
2.8.5 Drought index-based insurance 150
2.8.5.1 Meteorological-based index insurance 154
2.8.5.2 Remote sensing-based index insurance 155
2.8.5.3 Hydrological drought index insurance 155
2.8.6 Conclusions 157
Acknowledgements 157
References 157

PART THREE: DROUGHT MANAGEMENT EXPERIENCES AND LINKS TO STAKEHOLDERS **163**

3.9 Drought and Water Management in The Netherlands **165**
Wouter Wolters, Henny A.J. Van Lanen, and Francien van Luijn

3.9.1 General context 165
3.9.1.1 Physical and socioeconomic systems 165
3.9.1.2 Scenarios 167
3.9.1.3 Drought characteristics: Frequency and severity 168
3.9.1.4 Current management framework 169
3.9.2 Drought risk and mitigation 170
3.9.2.1 Vulnerability to drought 170
3.9.2.2 Existing framework for drought management 171
3.9.2.3 Stakeholder involvement in drought management 174
3.9.2.4 Example of the 2015 drought monitoring and management 175
3.9.2.5 Comparison of the 2015 drought with 2003 177
3.9.2.6 Responses to past drought events and assessment of their effects on drought impact mitigation 178
3.9.3 Conclusions – Future needs 179
3.9.3.1 Conclusions 179
3.9.3.2 Preparing for the future through establishing ‘service levels’ 180
3.9.3.3 International dimension 180
References 181

3.10 Improving Drought Preparedness in Portugal **183**
Susana Dias, Vanda Acácio, Carlo Bifulco, and Francisco Rego

3.10.1 Local context 183
3.10.1.1 Climate and land use 183
3.10.1.2 Water resources – use and consumption 184

3.10.1.3 Past drought events, impacts, and forecasted trends 186
3.10.1.4 Lessons learned from the 2004–2006 drought event 188
3.10.2 Current approach to drought monitoring and management 189
3.10.2.1 Drought monitoring systems 189
3.10.2.2 Existing framework for drought management 189
3.10.3 Improving drought preparedness and drought management 190
3.10.3.1 Stakeholder involvement in drought management 190
3.10.3.2 Vulnerability to drought: Analysis of SPIs as indicators of future drought impacts 191
3.10.3.3 Strengthening national drought information systems 193
3.10.3.4 Policy gaps and measures to improve drought preparedness and management 193
3.10.4 Conclusions 195
Acknowledgements 195
References 196

3.11 Drought Management in the Po River Basin, Italy 201
Dario Musolino, Claudia Vezzani, and Antonio Massarutto

3.11.1 General context 201
3.11.1.1 Physical and socioeconomic system 201
3.11.1.2 Drought characteristics and water availability 203
3.11.2 Drought risk and mitigation 206
3.11.2.1 Vulnerability to drought 206
3.11.2.2 The framework for drought management: Current situation and on-going changes 208
3.11.2.3 Policy responses to the 2003 drought event: A qualitative assessment 210
3.11.3 Conclusions 213
References 214

3.12 Experiences in Proactive and Participatory Drought Planning and Management in the Jucar River Basin, Spain 217
Joaquin Andreu, David Haro, Abel Solera, and Javier Paredes

3.12.1 Introduction 217
3.12.2 Droughts characterisation 219
3.12.2.1 Past droughts 219
3.12.2.2 Future droughts 223
3.12.3 Methods for drought vulnerability and risk assessment 223
3.12.3.1 Assessment of vulnerability during the planning phase 224
3.12.3.2 Assessment of vulnerability during the management phase (Real time) 224
3.12.3.3 Use of DSSs for drought management in real time 226
3.12.4 Proactive and participatory drought management 228
3.12.4.1 Culture of adaptation to droughts in the JRBD 228
3.12.4.2 Institutional, legal, and normative framework for drought planning and management 230
3.12.4.3 Measures included in the SDP 232
3.12.4.4 Responses to past drought events and assessment of their effects on drought impact mitigation 232
3.12.5 Conclusions 234
Acknowledgements 235
References 235

3.13 Drought Risk and Management in Syros, Greece **239**
Dionysis Assimacopoulos and Eleni Kampragou

3.13.1 Introduction 239
3.13.2 Droughts in Syros 240
3.13.2.1 Past droughts 240
3.13.2.2 Future droughts 242
3.13.2.3 Key messages 242
3.13.3 Drought risk and mitigation 243
3.13.4 Lessons learnt – the need for participatory drought management 246
Acknowledgements 247
References 248

Index **251**

Series Preface

The rising frequency and severity of hydrometeorological extreme events have been reported in many studies and surveys, including the IPCC Fifth Assessment Report. This report and other sources highlight the increasing probability that these events are partly driven by climate change, while other causes are linked to the increased exposure and vulnerability of societies in exposed areas (which are not only due to climate change but also to mismanagement of risks and 'lost memories' about them). Efforts are ongoing to enhance today's forecasting, prediction and early warning capabilities in order to improve the assessment of vulnerability and risks and to develop adequate prevention, mitigation and preparedness measures.

This book series, titled 'Hydrometeorological Extreme Events', has the ambition to gather available knowledge in this area, taking stock of research and policy developments at the international level. While individual publications exist on specific hazards, the proposed series is the first of its kind to propose an enlarged coverage of various extreme events that are generally studied by different (not necessarily interconnected) research teams.

The series comprises several volumes dealing with the various aspects of hydrometeorological extreme events, primarily discussing science–policy interfacing issues, and developing specific discussions about floods, coastal storms (including storm surges), droughts, resilience and adaptation, and governance. While the books examine the crisis management cycle as a whole, the focus of the discussions is generally oriented towards the knowledge base of the different events; prevention and preparedness; and improved early warning and prediction systems.

The involvement of internationally renowned scientists (from different horizons and disciplines) behind the knowledge base of hydrometeorological events makes this series unique in this respect. The overall series will provide a multidisciplinary description of various scientific and policy features concerning hydrometeorological extreme events, as written by authors from different countries, making it a truly international book series.

Following a first volume introducing the series and a second volume on coastal storms, the 'drought' volume is the third book of this series. This book has been written by renowned experts in the field, covering various horizons and (policy and scientific) views. It offers the reader an overview of scientific knowledge about droughts (understanding the natural hazard, vulnerability, risks and policy, and management experiences). The forthcoming volumes of the series will focus on floods, governance and climate and health aspects.

Philippe Quevauviller
Series Editor

The Series Editor – Philippe Quevauviller

Philippe Quevauviller began his research career in 1983 at the University of Bordeaux I, France, studying lake geochemistry. Between 1984 and 1987, he was associate researcher in the Institut de Géologie du Bassin d'Aquitaine, in the framework of a scientific cooperation between the office of the Portuguese Environment State Secretary and the University of Bordeaux I. Here, he performed a multidisciplinary study (sedimentology, geomorphology, and geochemistry) of the coastal environment of the Galé coastline and the Sado Estuary, which was the topic of his PhD in oceanography at the University of Bordeaux I in 1987. In 1988, he became associate researcher in the framework of a contract between the University of Bordeaux I and the Dutch Ministry for Public Works (Rijkswaterstaat), in which he investigated the organotin contamination levels in Dutch coastal environments and waterways. For this research work, he was awarded a PhD in chemistry from the University of Bordeaux I in 1990. From 1989 to 2002, he worked at the European Commission (EC) Directorate-General for Research in Brussels, where he managed various research and technological development (RTD) projects in the field of quality assurance, analytical method development, and pre-normative research for environmental analyses in the framework of the EC Standards, Measurements and Testing Programme. In 1999, he obtained an HDR (*habilitation à diriger des recherches*) in chemistry from the University of Pau, France, for a study of the quality assurance of chemical species' determination in the environment.

In 2002, he left the research arena to move to the policy sector at the EC Directorate-General for Environment, where he developed a new EU directive on groundwater protection against pollution, and chaired European science–policy expert groups on groundwater and chemical monitoring in support of implementing the EU Water Framework Directive. He moved back to the EC Directorate-General for Research and Innovation in 2008, where he functioned as a research programme officer and managed research projects about climate change impacts on the aquatic environment and on hydrometeorological hazards, while ensuring strong links with policy networks. In April 2013, he moved to another area of work – security research – at the EC Directorate-General for Enterprise and Industry, and then at EC Direction-General for Migration and Home Affairs, where he is a research programming and policy officer in the fields of crisis management and chemical, biological, radiological, and nuclear (CBRN) risk mitigation.

Besides his EC career, Philippe Quevauviller has also remained active in the academic and scientific realms. He is associate professor at Vrije Universiteit Brussel, and is a promoter of a master's thesis in the Interuniversity Programme on Water Resource Engineering (IUPWARE), which is a collaboration between Vrije Universiteit Brussel and KU Leuven. It is under this role that he is functioning as the series editor of this 'Hydrometeorological Extreme Events' series for Wiley. He also teaches integrated water management issues and their links to EU water science and policies to students of the 'Master EuroAquae – Erasmus Mundus' programme at the Polytech Nice-Sophia (University of Nice-Sophia Antipolis, France).

Philippe Quevauviller has authored and co-authored more than 220 scientific and policy publications internationally, in addition to 54 book chapters, 80 reports, and six books. He has functioned as the editor and co-editor for 26 special issues of scientific journals and 15 books. He also coordinated a book series for Wiley titled 'Water Quality Measurements', which resulted in 10 books published between 2000 and 2011.

List of Contributors

Vanda Acácio
Centro de Ecologia Aplicada "Prof. Baeta Neves" (CEABN), Instituto Superior de Agronomia, Universidade de Lisboa, Lisboa, Portugal

Joaquin Andreu
Instituto Universitario de Investigación de Ingenieria del Agua y Medio Ambiente, Universitat Politècnica de València, València, Spain

Victor Asenjo
GIS and Environmental Sciences Consultant, Spain

Dionysis Assimacopoulos
National Technical University of Athens, Athens, Greece

Paulo Barbosa
European Commission, Joint Research Centre (JRC), Ispra, Italy

María D. Bejarano
Universidad Politécnica de Madrid, Madrid, Spain

Paola Bianucci
AQUATEC (Suez Group), Dep. Basin Management, Madrid, Spain

María Bielza
Universidad Politécnica de Madrid, Madrid, Spain

Carlo Bifulco
Centro de Ecologia Aplicada "Prof. Baeta Neves" (CEABN), Instituto Superior de Agronomia, Universidade de Lisboa, Lisboa, Portugal

Hugo Carrão
Space4Environment, Niederanven, Luxemburg

Susana Dias
Centro de Ecologia Aplicada "Prof. Baeta Neves" (CEABN), Instituto Superior de Agronomia, Universidade de Lisboa, Lisboa, Portugal

Emmanuel Garnier
French National Centre for Scientific Research, Besancon, France

Luis Garrote
Universidad Politécnica de Madrid, Madrid, Spain

Alberto Garrido
Universidad Politécnica de Madrid, Madrid, Spain

Alfredo Granados
Universidad Politécnica de Madrid, Madrid, Spain

Jamie Hannaford
Centre for Ecology and Hydrology, Wallingford, United Kingdom

David Haro
School of Water, Energy and Environment, Cranfield University, Bedfordshire, United Kingdom

Ana Iglesias
Universidad Politécnica de Madrid, Madrid, Spain

Eleni Kampragou
Hellenic Ministry of Environment and Energy, Athens, Greece

Teresa Maestro
Universidad Politécnica de Madrid, Madrid, Spain

Antonio Massarutto
Dipartimento di Scienze Economiche e Statistiche, Università di Udine, Udine, Italy

Dario Musolino
Università Bocconi – Centro di Economia Regionale, dei Trasporti e del Turismo, Milan, Italy

Gustavo Naumann
European Commission, Joint Research Centre (JRC), Ispra, Italy

Javier Paredes
Instituto Universitario de Investigación de Ingenieria del Agua y Medio Ambiente, Universitat Politècnica de València, València, Spain

Christel Prudhomme
European Centre for Medium-Range Weather Forecasts, Reading, United Kingdom

Francisco Rego
Centro de Ecologia Aplicada "Prof. Baeta Neves" (CEABN), Instituto Superior de Agronomia, Universidade de Lisboa, Lisboa, Portugal

Alvaro Sordo-Ward
Universidad Politécnica de Madrid, Madrid, Spain

Abel Solera
Instituto Universitario de Investigación de Ingenieria del Agua y Medio Ambiente, Universitat Politècnica de València, València, Spain

Kerstin Stahl
University of Freiburg, Freiburg, Germany

Lena M. Tallaksen
University of Oslo, Oslo, Norway

Marjolein H.J. Van Huijgevoort
KWR Watercycle Research Institute, Nieuwegein, The Netherlands

Henny A.J. Van Lanen
Wageningen University, Wageningen, The Netherlands

Anne F. Van Loon
University of Birmingham, Birmingham, United Kingdom

Francien van Luijn
Rijkswaterstaat Water, Traffic and Environment, Lelystad, The Netherlands

Claudia Vezzani
Autorità di Bacino Distrettuale del Fiume Po, Parma, Italy

Niko Wanders
University of Utrecht, Utrecht, The Netherlands

Wouter Wolters
Wageningen Environmental Research (Alterra), Wageningen, The Netherlands

Part One

Understanding Drought as a Natural Hazard

1.1

Diagnosis of Drought-Generating Processes

Henny A.J. Van Lanen[1], Anne F. Van Loon[2], and Lena M. Tallaksen[3]

[1]*Wageningen University, Wageningen, The Netherlands*
[2]*University of Birmingham, Birmingham, United Kingdom*
[3]*University of Oslo, Oslo, Norway*

1.1.1 Introduction

It is well known that precipitation deficits initiate drought development. In cold climates (snow and glaciers), temperature anomalies (both high and low) too may contribute to drought generation. An in-depth understanding, that is, drought diagnosis, of how climate drives precipitation and temperature anomalies, and subsequently how these anomalies propagate into soil moisture, groundwater, and streamflow deficits (catchment control), is a prerequisite for reducing the immense socioeconomic and environment impacts of droughts (e.g. Stahl et al., 2016). Comprehensive overviews on drought are provided by Wilhite (2000); Tallaksen and Van Lanen (2004); Mishra and Singh (2010); Sheffield and Wood (2011); and Van Loon (2015). This chapter builds upon these overviews and complements them by synthesising knowledge from recently finished EU projects.

The chapter starts with an introduction about key hydroclimatological processes controlling drought generation – that is, how precipitation and temperature drive drought, and which stores and fluxes are affected. Different drought types are explained, and the main drought indices that are used in this chapter are briefly described (Section 1.1.2). Section 1.1.3 gives an overview of the main atmospheric and oceanic drivers for meteorological drought (deficit in precipitation, temperature anomalies), supplemented with a detailed description of the drivers of the 2015 summer drought in Europe, and a discussion of the influence of climate change on the meteorological drought in Europe. Section 1.1.4 continues with a more comprehensive description of the influence of meteorological drought on soil water, followed by the influence on groundwater and streamflow (Section 1.1.5). Both sections focus on key processes controlling the development of droughts in the hydrological system – that is, soil moisture, streamflow, and groundwater drought – followed by the role of human influence in

Drought: Science and Policy, First Edition.
Edited by Ana Iglesias, Dionysis Assimacopoulos, and Henny A.J. Van Lanen.

modifying the drought signal. Section 1.1.6 addresses how these different droughts propagate in the hydrological system – that is, how precipitation deficits and temperature anomalies affect snow accumulation and smelt, propagating into soil water, groundwater, and streamflow (drought propagation). A recently developed hydrological drought typology explains how climate and catchment controls determine drought propagation (Section 1.1.6). The focus in this chapter is on natural processes; however, at the end of each section, human interferences and their feedbacks are briefly touched upon (Sections 1.1.3–1.1.6). Finally, the concluding remarks are given in Section 1.1.7.

1.1.2 Background

The climate in the centre and north of Europe is influenced by the westerlies of the mid-latitudes during the whole year, bringing moisture from the Atlantic Ocean. The Mediterranean region lies in a transitional climate zone, influenced by the Subtropical High-Pressure Belt during summer and the mid-latitude westerlies during winter. Hence, two main climate regions can be distinguished: a temperate climate with a dry summer season in the Mediterranean; and a temperate climate and a cold climate without any dry season in the centre and north of Europe, respectively. Within these regions, climate is modified by numerous other permanent or temporally variable global, regional, or local factors, such as soil moisture, oceanic currents, and topography. Blocking situations disturb the common eastwards movement of the mid-latitude pressure systems, that is, the westerlies. During a blocking phase, an extended, persistent, high-pressure system develops in the eastern Atlantic Ocean at the mid-latitudes that does not move eastwards or moves only very slowly (Stahl and Hisdal, 2004). As a consequence, the moisture-bringing pressure systems divert moisture to Northern Africa and Northern Fennoscandia, causing an extended dry period (precipitation deficits) in mainland Europe (Section 1.1.3). During a dry and warm summer, feedbacks between the land surface and the atmosphere may amplify the drought signal. As the soil dries out, less energy is used for evapotranspiration (latent heat flux), and the partitioning of incident solar energy changes as more energy is used for heating the air (sensible heat flux). Heat waves thus frequently accompany major droughts, as reported for Europe by Ionita et al. (2017).

The lower-than-normal precipitation, usually combined with higher temperature and associated larger potential evapotranspiration (PET), leads to a decreased net precipitation (gross precipitation minus evaporated interception water), and hence infiltration into the topsoil of the vegetated surfaces (Figure 1.1.1). In places without vegetation, the lower gross precipitation directly results in lower infiltration. Evaporation of intercepted water, especially in forests, and overland flow on sloping land is also lower.

The reduced soil infiltration, together with the often-higher atmospheric water demand (increased PET), causes a larger depletion of the soil moisture storage than normal. Consequently, lowered soil evaporation and lessened soil water uptake by vegetation results in reduced evapotranspiration in many cases. Another important effect of the more depleted soil moisture store is the lower recharge to the underlying aquifer. In catchments where interflow takes place (e.g. soils with contrasting hydraulic conductivities, slopes), the aquifer also receives less water. This leads to lower groundwater levels, and hence reduced groundwater discharge to streams and lakes. Deeper, regional

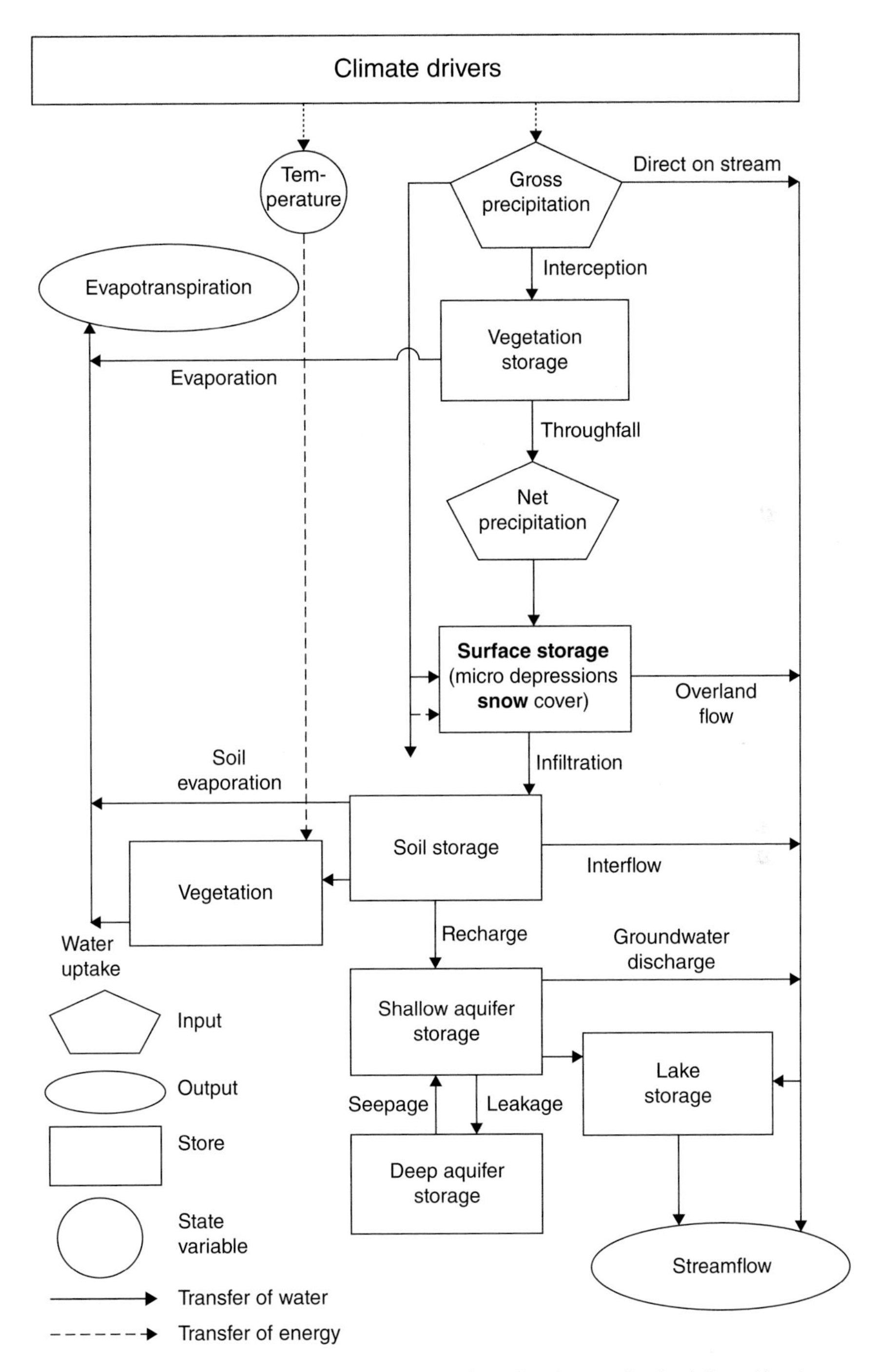

Figure 1.1.1 Water stores and fluxes affected by drought. *Source*: Derived from Van Lanen et al. (2004a).

aquifers also receive less water input (lower leakage), which may have long-lasting impacts in a wide area. Lakes can mitigate the effects of a downstream drought, because their natural role is to store surface water during wet periods and to release it during a dry period (upstream–downstream differences in streamflow). The changes in hydrological processes, which also have different time delays (i.e. response times), are more elaborated in the following sections (Sections 1.1.4 and 1.1.5).

In cold climates, irrespective of precipitation amounts, below-normal temperatures can result in earlier snow accumulation at the start of the cold season, which might lead to lower inflow to the streams. This also takes place when the cold season is longer than normal (delayed snow melt peak). In glaciated regions, cold temperature anomalies also lead to below-normal streamflow. Higher-than-normal winter temperatures may also cause below-normal summer flow, which is further elaborated in Section 1.1.6.

In many places, people try to intervene in natural processes, as described in the preceding text, to reduce drought impacts. For instance, deep soil tillage is applied to increase soil moisture supply capacity, or irrigation is applied to increase soil moisture content. Reservoirs are built to retain streamflow in a certain area to supply irrigation or water supply in general during a drought. However, this may enhance the drought downstream. On the other hand, the reservoirs can also be managed to maintain ecologically minimum flow during a drought. Van Loon et al. (2016a; 2016b) discuss the different implications that human interventions may have in a catchment, and introduce the following terms: (i) *climate-induced drought*, (ii) *human-modified drought*, and (iii) *human-induced drought*. Climate-induced drought is caused by natural climate variability, and is the focal area of this chapter; human-modified drought reflects a situation where humans have enhanced or alleviated the effects of climate-induced drought; and human-induced drought is caused entirely by people's measures (i.e. the drought should not have occurred under natural conditions). In Sections 1.1.3–1.1.6, human influences are further described.

Droughts happen in different domains of the hydrological cycle, and it is important to make distinctions between different drought types and their associated impacts (e.g. Van Lanen et al., 2016). Precipitation deficits and temperature anomalies cause meteorological drought, reduced soil infiltration results in soil water drought, and reduced recharge leads to groundwater drought. The combined effect of lower precipitation on the stream (and reduced overland flow, interflow, and groundwater discharge) is the origin of streamflow drought. Groundwater drought and streamflow drought are both referred to as 'hydrological drought' (Tallaksen and Van Lanen, 2004).

Various indices have been introduced to describe the different drought types, including their onset, duration, severity, and intensity (total deficit divided by duration). Two principal approaches are commonly used: (i) *standardised approaches* and (ii) *threshold approaches*. The *Standardised Precipitation Index* (SPI, McKee et al., 1993) and the *Standardised Precipitation–Evapotranspiration Index* (SPEI, Vicente-Serrano et al., 2010) are the most well-known standardised indices to describe meteorological drought. The SPI is a probabilistic measure that describes the number of standard deviations by which the dry event (accumulated precipitation over a given period) deviates from the median precipitation total over the same period. It can be calculated for different rainfall accumulation periods (e.g. 1–48 months, SPI-1 to SPI-48). The SPEI adds the PET and reflects the *climatic water deficit* (P-PET) over a given period. Similar standardised indices have been developed for groundwater (GRI, Bloomfield and Marchant, 2013) and runoff/streamflow (SRI, Shukla and Wood, 2008). Threshold approaches are better suited to quantify the water needed (volume) to manage and recover from a drought. It is based on defining a threshold, below which, for instance, the precipitation or streamflow is considered as a drought (Yevjevich, 1967). The threshold level method was first introduced for daily time series by Zelenhasic and Salvai (1987), and further explored by Tallaksen et al. (2009) for two catchments with varying storage properties. Both fixed and variable thresholds are used (Hisdal et al., 2004). Van Loon et al. (2010) and Van Loon (2015) describe the implementation of a daily smoothed monthly variable threshold that frequently is used. Heudorfer and Stahl (2017) comprehensively set out differences in outcome between fixed and variable thresholds when using the same time series (e.g. streamflow).

1.1.3 Climate drivers of drought

Regional drought-causing atmospheric situations are characterised by: (i) the anomalous timing of a seasonal phenomenon; (ii) the anomalous location of pressure centres and tracks of cyclones; and/or (iii) the anomalous persistence or persistent recurrence of dry weather patterns (Stahl and Hisdal, 2004). In the Mediterranean region, with its seasonal climate (Section 1.1.2), severe droughts can, for instance, be caused by longer-than-usual influence of the Subtropical High-Pressure Belt. Droughts can accordingly last for several weeks or even months. In the more humid mid-latitudes of western and northern Europe, 'atmospheric blockings' are the major atmospheric anomalies causing extended dry weather periods. Here, a few weeks or months with low rainfall may constitute a severe drought.

1.1.3.1 Atmospheric and oceanic drivers

Quantification of large-scale climate drivers of drought is important to understand and better manage spatially extensive and often prolonged natural hazards such as droughts, particularly their triggering mechanisms and persistence (e.g. Fleig et al., 2010, 2011). Primary drought-controlling mechanisms at the continental scale for Europe were explored by Kingston et al. (2015). Drought events were identified using SPI-6 and SPEI-6 (Section 1.1.2), both calculated using the gridded Water and Global Change forcing dataset (WATCH-WFD) for 1958–2001. Based on correlations between monthly time series of the percentage area in drought and 500 hPa geopotential height, a weakening of the prevailing westerly circulation was found to be associated with drought onset. Such conditions can be linked to variations in the East Atlantic/Western Russia (EA/WR) and North Atlantic Oscillation (NAO) atmospheric circulation patterns (Figure 1.1.2). Kingston et al. (2015) also performed an event-based composite analysis of the most widespread European droughts. It revealed that a higher number of droughts were identified by SPEI-6 than by SPI-6, with SPEI-6 drought events showing a greater variety of locations and start dates. They further concluded that differences between the atmospheric drivers of SPI-6 (associated with the NAO) and SPEI-6 events (associated with the EA/WR) reflected the sensitivity of these indices to the underlying drought type (precipitation versus climatic water balance), as well as sensitivity to the associated differences in their timing and location (northern Europe in winter vs. Europe-wide and year-round).

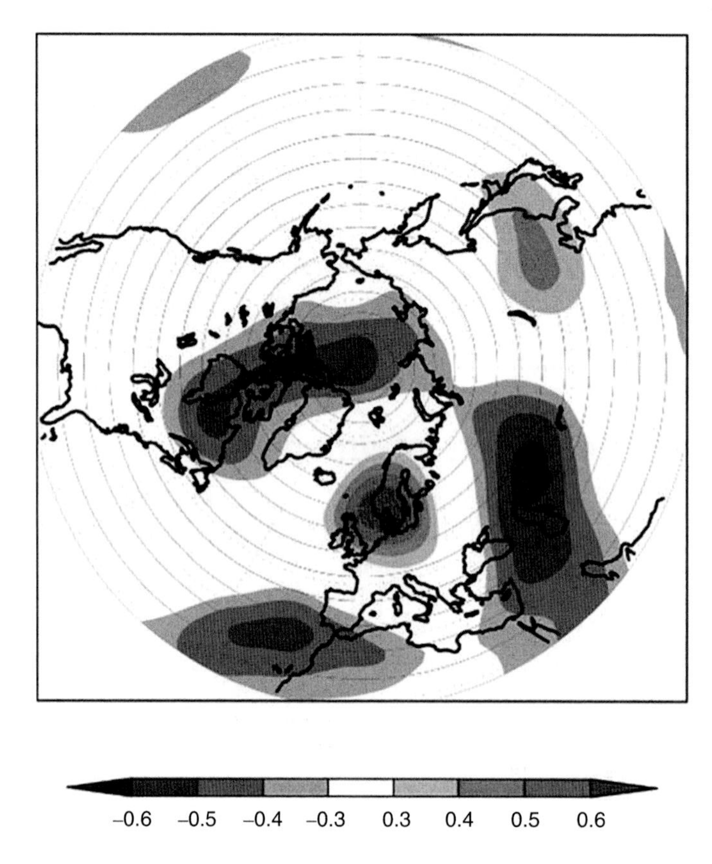

Figure 1.1.2 Correlation of May SPI-6 with December–May mean 500 hPa geopotential height. *Source*: From Kingston et al. (2015). (*See colour plate section for the colour representation of this figure.*)

Persistent dry conditions are associated with anticyclonic circulation, as reported in the preceding text for Europe, but oceanic factors, such as sea surface temperatures (SSTs), too can play a role through interaction with large-scale climatic or oceanic modes of variability, such as the North Atlantic oscillation (NAO) (Ionita et al., 2017; Kingston et al., 2013; 2015; Schubert et al., 2014).

The work of Kingston et al. (2013) illustrates the impact of antecedent sea surface temperatures and atmospheric circulation patterns on summer drought in Great Britain. However, the atmospheric bridge linking North Atlantic SST to drought development was identified as being too complex to be described solely by indices of the NAO. In Ionita et al. (2017), the key drivers of the 2015 drought event in Europe were analysed, with special emphasis on the role played by the SST and large-scale (atmospheric) circulation modes of variability, as described in the following section.

1.1.3.2 Summer drought of 2015

The summer drought of 2015 affected a large portion of continental Europe and was one of the most severe droughts since the summer of 2003, with record high temperatures in many parts of central and eastern Europe (Ionita et al., 2017). Over the summer period, there were four heat wave episodes, all associated with persistent blocking events. Upper-level atmospheric circulation was characterised by positive 500 hPa geopotential height anomalies bordered by a large negative anomaly to the north and west (i.e. over the central North Atlantic Ocean extending to northern Fennoscandia), and another centre of positive geopotential height anomalies over Greenland and northern Canada. At the same time, summer SSTs were characterised by large negative anomalies in the central North Atlantic Ocean and large positive anomalies in the Mediterranean basin (Figure 1.1.3). Ionita et al. (2017) conclude that the lagged relationship between the Mediterranean SST and summer drought conditions, especially over the eastern part of Europe, as identified in their study, holds potential for the prediction of drought conditions over Europe on seasonal to decadal timescales.

1.1.3.3 Influence of humans (Climate change)

Global warming has been most pronounced after 1970 (Hartmann et al., 2013), and Europe has warmed faster than the global mean land trend (Christensen et al., 2013). The most pronounced warming is seen in summer, notably in August (Nilsen et al., 2016). To understand the causes of regional temperature changes, it is common to separate the factors inducing the changes into *changes in atmospheric synoptic circulation* vs. *other factors*, including the so-called 'within-type changes'. Nilsen et al. (2016) revealed that changes in synoptic circulation could not account for all the observed (WATCH Forcing Dataset ERA-Interim, or 'WFDEI') warming in Europe over the period 1981–2010. Significant warming in specific months and regions, such as the large-scale warming in April, June, July, August, and November, must also be caused by other factors. Within-type changes may be caused by positive feedbacks between the land surface and the atmosphere, by forcing from greenhouse gases, or from other potential climate factors. Warming in regions with a seasonal snow cover, such as Scandinavia (Rizzi et al., 2017), is likely influenced by snow albedo feedbacks related to changes in snow cover, particular in spring when incoming radiation is high. Warming in water-limited regions, such as southern Europe during the summer, has been documented to be influenced by soil moisture–temperature feedbacks (Section 1.1.4). A general increase in temperature, and thus potential evapotranspiration, likely will enhance drought, regardless of changes in precipitation.

The frequency of meteorological droughts in Europe has increased since 1950 in parts of southern and central Europe, whereas droughts have become less frequent in northern Europe and parts of eastern Europe (EEA, 2017; Stagge et al., 2017), consistent with climate change projections (e.g. Stagge et al., 2015). Trends in drought severity (based on a combination of indices, including SPI and SPEI) also show significant increases in the Mediterranean region and parts of central and southeast Europe, and decreases in northern and parts of eastern Europe (EEA, 2017; Spinoni, et al., 2015; 2016). Stagge et al. (2017) document an increasing deviation in European drought frequency using SPI and SPEI, derived from instrumental records (1958–2014). Notably, they conclude that increases in temperature and reference evapotranspiration have enhanced droughts in southern Europe while counteracting increased precipitation in northern Europe.

Based on an ensemble from the EURO-CORDEX community project, Stagge et al. (2015) project that the frequency and duration of extreme meteorological droughts

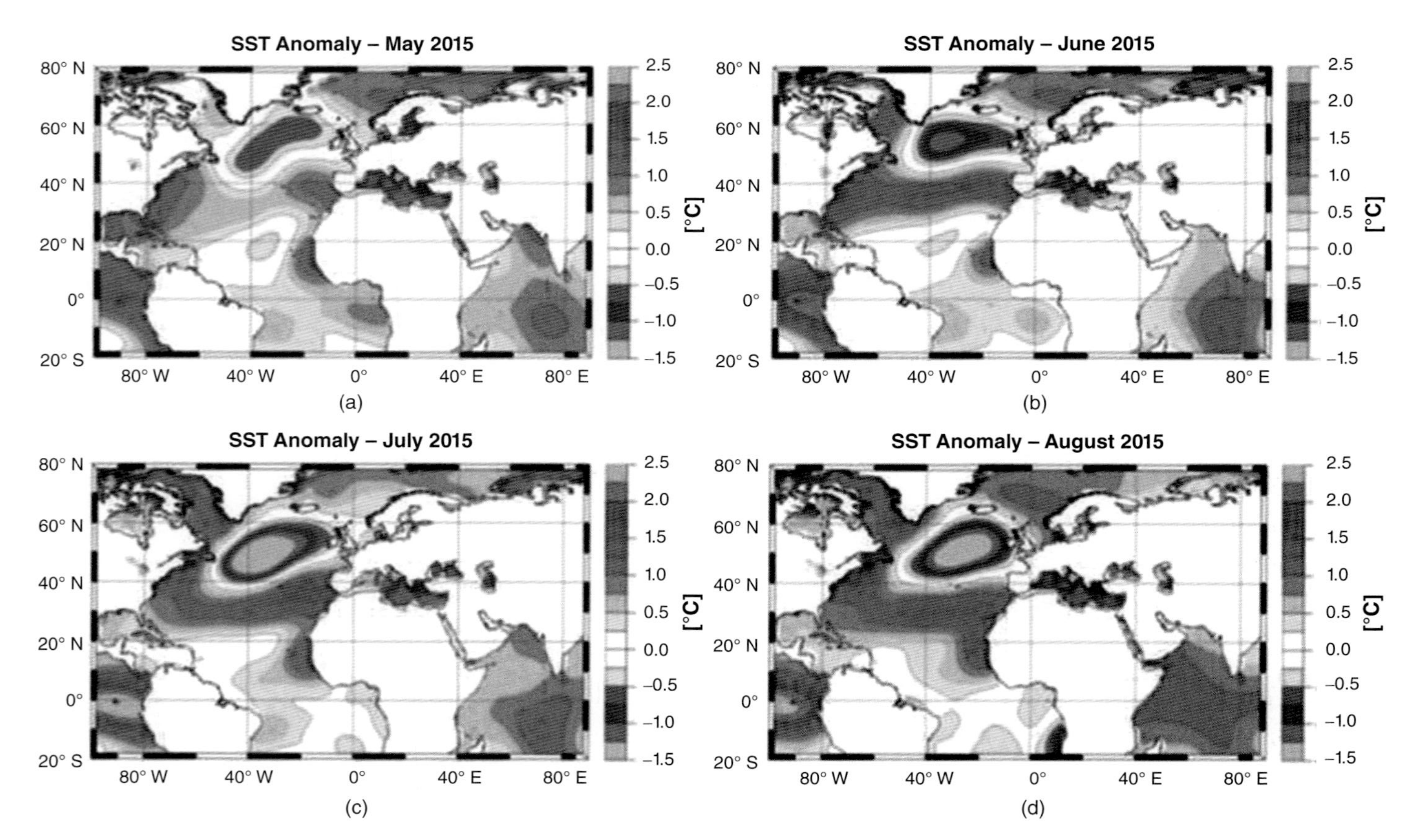

Figure 1.1.3 Monthly SST anomalies: (a) May 2015; (b) June 2015; (c) July 2015; and (d) August 2015; computed relative to the 1971–2000 period. *Source*: From Ionita et al. (2017). (*See colour plate section for the colour representation of this figure.*)

(SPI-6 < –2) will significantly increase in the future with respect to the baseline period (1971–2000). These projections show the largest increases in frequency for extreme droughts in parts of the Iberian Peninsula, southern Italy, and the eastern Mediterranean region, especially towards the end of the century (Stagge et al., 2015). Drought projections that also consider PET (e.g. SPEI) showed substantially more severe increases in the areas affected by drought than those based on precipitation alone (e.g. SPI) (EEA, 2017).

1.1.4 Soil moisture drought processes

The influence of a lower-than-normal infiltration into the soil surface is described in the following text, including the effects on evapotranspiration and recharge to the underlying aquifer. An example illustrates the change in processes, that is, the development of a soil water drought. The section concludes with how human interferences affect soil water drought. For a comprehensive description of drought-relevant flow-generating processes in the unsaturated zone, readers also are referred to Van Lanen et al. (2004a) and Sheffield and Wood (2011).

1.1.4.1 Processes

Lower-soil infiltration (Figure 1.1.1) influences *actual evapotranspiration* (ET_a) through a reduction in soil moisture. *Soil water storage* (SM) depletes quicker than normal, particularly because PET usually is higher. Several cases can be distinguished:

(i) Soils with high *soil moisture supply capacity* (SMSC) in humid climates
(ii) Soils with low SMSC in humid climates
(iii) Soils with high SMSC in dry climates
(iv) Soils with low SMSC in dry climates

In case (i), soils are more depleted, because of the high SMSC. SM does not reach the *critical soil moisture* (SM_c) level implying that the ET_a equals PET. Water losses to the atmosphere are higher during a drought than normal, at least in the first phase of the drought, because of the increased PET. Teuling et al. (2013) illustrate, for four catchments in west and central Europe, that ET_a increases, contributing to the aggravation of the drought. In case (ii), in the early phase of the drought, evapotranspiration might be higher – see case (i) – but this is followed by a situation where soil moisture becomes lower than the critical level ($SM < SM_c$), which leads to an earlier reduction of the PET ($ET_a < PET$). This causes vegetation stress, resulting in biomass reduction (e.g. reduced crop yield). Only rarely does the soil moisture level *not* reach the wilting point for plants (SM_w) – that is, soil conditions that the plants fully dry out. Case (iii) is probably similar to case (ii): the higher PET first results in higher losses to the atmosphere, which are then followed by a situation where $ET_a < PET$. In case (iv), $ET_a < PET$ is the dominant situation that eventually leads to an earlier full depletion of soil moisture ($SM = SM_w$; SMSC fully used).

The above-mentioned anomalies in precipitation and evapotranspiration affect soil moisture change (SM'). In case (i), the lower precipitation and higher PET have to be fully compensated by SM'. This results in a soil water drought. Later, in a wet period, when the rainfall is higher than normal, SM' has to be fully balanced by a reduction in recharge to the underlying aquifer, RCH' = SM'. In cases (ii) and (iii), the earlier reduction of the PET results in a situation where SM' is smaller than the sum of the anomalies in precipitation and PET. The soil water drought that develops is relatively smaller than in case (i). In case (iv), there is an impact on the ET_a that is lower due to the full depletion of SMSC earlier. A soil water drought develops, but finishes at the time of full depletion. SM at the start of a following wet period does not deviate from normal conditions (SM' = 0). This implies that, in this case, the recharge is not affected. In other words, a meteorological drought leads to a temporary soil water drought, but it does not result in a hydrological drought.

Soil water drought is driven by precipitation and PET anomalies, but it is constrained by the SMSC (the soil moisture determines to what level the anomalies in precipitation and PET can be compensated). Furthermore soil water drought is conditioned by the climate. In humid climates, soils are more likely to be fully replenished than in dry climates. In cold climates , temperature anomalies play a role too. Earlier snow accumulation or later snow melt than normal results into below-normal soil infiltration. This may imply the development of a soil water drought or the continuation of an ongoing drought. Likewise, seasonality, as a climate characteristic, contributes to how a soil water drought develops and recovers, and how it affects drought characteristics such as drought duration (DD) and deficit of soil moisture drought (DSMD) (Van Loon et al., 2014). For drought management, it is relevant to know whether a deficit quickly increases or not during a drought event. Long-lasting droughts can have either a small or large deficit. In a modelling experiment, Van Loon et al. (2014) investigated the relation between DD and DSMD for over 1000 grid cells distributed across 27 Köppen–Geiger climate types for the period 1958–2001. They used the variable threshold approach. For each grid cell, the correlation (R^2) between DD and DSMD of the soil moisture drought events was calculated (Figure 1.1.4).

Climate types with significant precipitation in all seasons (denoted by 'f' in the Köppen–Geiger climate type acronyms) are characterised by high correlations – for example, temperate and cold climates without a dry season and warm summer, or tropical rainforests (e.g. Af, Cfb, Dfb) that cover large parts of Europe and Southeast United States. Soil moisture droughts can occur in all seasons as a response to precipitation anomalies, and drought duration and deficit have a strong linear relation ($R^2 > 0.88$). Climate types with a dry summer or winter ('s' or 'w' in the climate type acronyms) have strong seasonality. These are characterised by a divergent density field, and the associated lower R^2 (<0.68). A typical example is the tropical savanna climate (Aw) – for example, in South Brazil, Southeast Asia, North Australia, and large areas around the equator in Africa. Drought events occur both in the wet and dry season. In the wet season, the variable threshold is rather big, and soil moisture does not reach wilting point ($SM > SM_w$), which allows large deficits (SM') to develop. On the contrary, in the dry period, the SM is already close to wilting point under normal conditions (variable threshold is small). During a drought, SM reaches wilting points and the change in soil moisture is minor, which means that only small deficits develop. Hence, in climate types such as Aw, droughts of the same duration can either have a large deficit (wet season drought) or a small deficit (dry season drought) leading to low R^2.

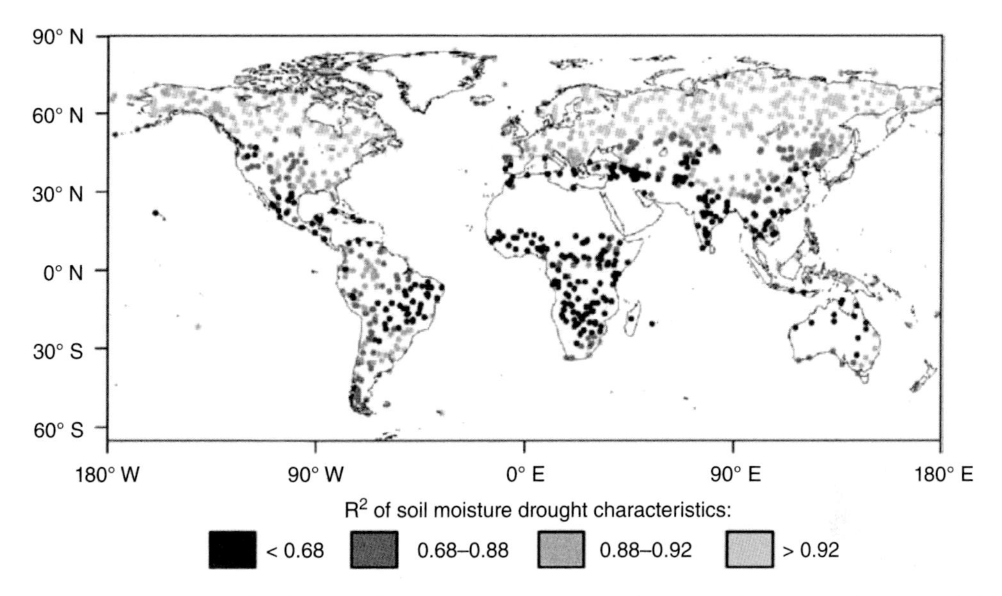

Figure 1.1.4 Spatial distribution of the correlation between DD and DSMD events for the period 1958–2001. *Source*: From Van Loon et al. (2014). (*See colour plate section for the colour representation of this figure.*)

1.1.4.2 Human influences

There are several studies that have investigated the impact of global warming on soil water moisture – in other words, whether soil water drought has changed over time. Long time series of soil moisture are required, but observed series with a continental or global coverage are lacking. Time series derived from satellite products are still rather short (De Jeu and Dorigo, 2016). Hence, large-scale changes of soil moisture have mainly been investigated using modelled and reanalysis data (e.g. Dai, 2012; Sheffield et al., 2012). These studies, based on the Palmer drought severity index (PDSI, Palmer, 1965), report apparently conflicting results on how drought is changing under climate change. Trenberth et al. (2014) explain that these disparities are caused by: (i) the different ways of calculating PET – that is, the more physically based Penman–Monteith concept (PDSI_PM) versus the only-temperature-driven Thornthwaite approach (PDSI_Th) (Van der Schrier et al., 2011); (ii) the weather data from global datasets (PDSI_PM uses more weather data than PSDI-Th, but some of these data are less reliable); and (iii) the differences in the global precipitation datasets. They conclude that the use of PDSI to assess the impact of global warming on soil water drought should be treated with care. Orlowsky and Seneviratne (2013) used the threshold-based soil moisture anomaly (SMA), which is a more comprehensive approach than PDSI, to explore the impact of global warming on soil water drought. The outcome of about 30 Climate Model Intercomparison Project Phase 5 (CMIP5) Global Climate Models (GCMs) for the period 1979–2009 was used, which was tested against three global monthly precipitation datasets. They found no significant changes in SMA in 12 major regions across the world for this historic period.

Irrigation is an important and direct human activity that considerably influences soil water drought. Over 70% of human need for water is intended for irrigation purposes, aimed at raising biomass production. The abstracted volume increased from ~500 km^3/year to ~4000 km^3/year over the last 100 years (e.g. Oki and Kanae, 2006). In the Northern Hemisphere, extensive irrigated areas are present in several countries between 20° and 50° latitudes, and similar regions are found in the Southern Hemisphere in South America, South Africa, and Australia (FAO, 2015). In these irrigated areas, soil water drought is substantially reduced or even eliminated.

1.1.5 Hydrological drought processes (Groundwater and streamflow)

The influence of a lower-than-normal recharge from the soil (Section 1.1.4) affects groundwater storage and, consequently, groundwater flow to the streams. Some examples are presented in the following text to illustrate the change in processes – that is, the development of a hydrological drought in response to a deficit in recharge. How human interferences affect hydrological drought is described towards the end. For a comprehensive description of drought-relevant flow-generating processes in water-saturated stores (such as groundwater, lakes, and wetlands), readers are also referred to Van Lanen et al. (2004a).

1.1.5.1 Groundwater

Processes A below-normal recharge (Section 1.1.4.1) leads to a more rapid decline of the groundwater heads, which causes a lower-than-normal groundwater storage in the shallow aquifer (Figure 1.1.1). Groundwater table response to a meteorological drought may occur months later (termed 'delay'; see Section 1.1.6) if the water table is deep – that is, several tens of meters or more below the soil surface (Van Lanen, 2004a).

In case a deep aquifer occurs, the faster-than-normal water table decline of the shallow aquifer is counteracted by either a lower leakage or higher seepage, which implies that the groundwater drought develops slower than in an area with only a shallow aquifer. The change in leakage or seepage, however, also affects the groundwater storage in the deep aquifer, where also a groundwater drought is induced. In multiple-aquifer systems, groundwater drought develops slowly, and recovery can last long. The effect of responsiveness of groundwater systems on drought has been explored by Van Lanen et al. (2013) in a modelling experiment. Responsiveness depends on aquifer properties, such as the thickness of the aquifer, hydraulic conductivity, and storativity. The distance between the streams to which the groundwater flows also plays a role; the larger the distance, the lower the responsiveness. In a modelling experiment, 1495 grid cells across the world that were proportionally distributed over the major climates of the globe (Köppen–Geiger) were selected. Hydrological characteristics for each grid cell were derived from the time series of simulated groundwater discharge obtained with a hydrological model that was forced with 44 years of observed weather data. Summary statistics of the number of droughts are given in Figure 1.1.5. Fast-responding groundwater systems have a substantially higher number of droughts as

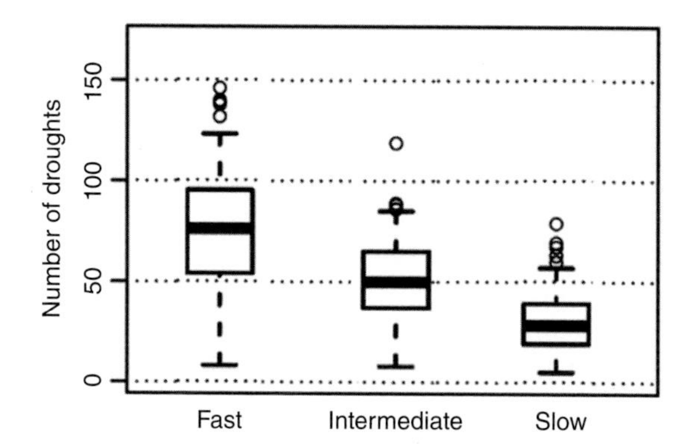

Figure 1.1.5 Summary statistics (box: 25, 50 and 75 deciles; whiskers: 5 and 95 deciles) of the number of hydrological droughts of all drought events (1495 grid cells) for three different groundwater systems. The circles represent extreme events beyond 95 decile. *Source*: Derived from Van Lanen et al. (2013).

compared to the slow-responding ones. Of the major climates (polar climates excluded), (semi-)arid climates have the lowest number of droughts. However, the droughts last longer there.

Bloomfield and Marchant (2013) also point at the importance of hydrogeology for groundwater drought development. They studied the long time series of a groundwater drought index from 14 different observation wells across the United Kingdom. They concluded that, in granular aquifers, drought duration is mainly determined by the effects of intrinsic aquifer properties on saturated groundwater flow, which can lead to long droughts (drought duration and frequency are negatively correlated in general). However, in fractured aquifers, drought duration is usually shorter and more connected to the temporal distribution of the recharge. In general, fractured aquifers respond faster than granular aquifers, implying that these are more affected by the recharge pattern. The role of *hydrogeology*, that is, how groundwater drought affects streamflow drought development, is explained in the next section (Section 1.1.5.2).

Human influences Groundwater abstraction is one of the best examples of human influence on the groundwater system. Abstraction usually leads to the enhancement of natural groundwater drought; exceptions occur under irrigated fields with leakage losses. The effect of abstraction on groundwater drought also depends on whether it is a permanent extraction (e.g. public water supply) or non-permanent (supplementary irrigation), and clearly on the distance to the well field. Van Lanen et al. (2004b) elaborated on the effects of both permanent and non-permanent abstractions on groundwater drought in a Dutch lowland river basin. The total abstraction of the non-permanent abstraction is half of the abstraction of the permanent well, and occurs only in the wet season. It is obvious that the permanent abstraction leads to more severe drought (expressed as the cumulative deviation between the groundwater table and the threshold). However, there are non-linear effects making the severity of the permanent abstraction more than twice as extreme as the non-permanent one. The main reason for this is that, during a drought, even under natural conditions, a large part of the drainage system (e.g. ditches, small streams) in a lowland area does not carry water. Under these conditions, abstracted groundwater cannot come from reduced groundwater flow to the surface water, but it comes from a decrease of groundwater storage, which means larger drawdowns of the water table than in normal or wet periods. The influence of this partly dry drainage network is larger for permanent abstractions than for non-permanent ones.

Van Loon and Van Lanen (2013) investigated the impact of groundwater abstraction for irrigation on groundwater drought in the Upper Guadiana river basin in Spain. They compared the observed groundwater levels, in the period during which the hydrological system is affected by the abstractions, with naturalised groundwater levels (Figure 1.1.6). The naturalised groundwater levels were obtained by running a hydrological model without groundwater abstractions. The model was calibrated with data from the period before irrigation expanded. Further, both time series of groundwater levels were compared against the variable threshold (Section 1.1.2) to obtain anomalies. The anomalies were used to distinguish between climate-induced drought versus human-modified and human-induced droughts (Section 1.1.2). The variable threshold curve (Figure 1.1.6) is identical for every year, and has been derived from monthly cumulative frequency distributions of the groundwater level before 1980 (almost no disturbance).

Climate-induced droughts occur when the naturalised and observed groundwater levels more or less coincide, and are below the threshold (e.g. 1991–1993, Figure 1.1.6), whereas human-modified droughts happen when both the observed and naturalised groundwater levels are below the threshold (e.g. 1992–1995). Human-induced droughts are rarer, and take place when the naturalised groundwater level is above the threshold and the observed level below (e.g. 1986 and 1996). Until 1983, the observed and naturalised groundwater levels more or less coincided, meaning that there was almost no human disturbance, and only minor climate-induced drought (1981–1982). However, this changed from 1984 onwards, and particularly since 1986. There is a clear increase of the human impact (groundwater abstraction) leading to the enhancement of groundwater drought and the development of human-modified drought. Van Loon and Van Lanen (2013) report that the number of groundwater

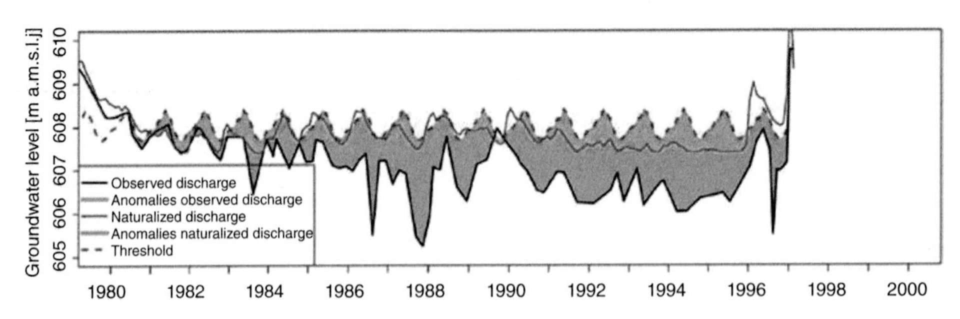

Figure 1.1.6 Anomalies in groundwater level in Upper Guadiana during the period 1980–1997 (disturbed period), using a variable 80% monthly threshold, and derived from the observed and naturalised groundwater levels. Blue areas can be attributed to climate, and grey areas to humans. *Source*: Derived from Van Loon and Van Lanen (2013). (*See colour plate section for the colour representation of this figure.*)

droughts has decreased by a factor of four, and that the duration of groundwater droughts has increased by a factor of six, owing to the pooling of droughts associated with groundwater abstraction.

1.1.5.2 Streamflow

Processes A below-normal groundwater discharge due to groundwater drought (see the subsection titled 'Processes' under Section 1.1.5.1) leads to lower stream levels and associated streamflow than normal (hydrological drought). In addition to groundwater discharge, overland flow and interflow (Figure 1.1.1), and other stores (such as lakes and wetlands), feed a stream. Substantial interflow is rather rare, but overland flow occurs in many catchments after intensive rainfall or snowmelt on frozen soil, particularly in areas with slopes and soils with a low infiltration capacity. During a meteorological drought, overland flow is lower than normal or absent, meaning that less rainwater reaches the stream quickly. During a drought, a stream is fully dependent on the groundwater discharge, and drainage from lakes and wetlands, if occurring in a region, which gradually becomes lower than normal. Overland flow can play a role in a temporary recovery from a drought; long-lasting droughts are interrupted by flow peaks. Streamflow drought does not recover unless the groundwater discharge (and lake or wetland drainage) becomes normal or above normal.

Response of streams to meteorological drought has much in common with groundwater response (see the subsection titled 'Processes' under Section 1.1.5.1), as long as overland flow does not play a dominant role. The number of streamflow droughts decreases, going from flashy to slowly responding catchments. The duration shows an opposite signal (Van Lanen et al., 2004a; Tallaksen et al., 2009).

The link between recharge deficit and streamflow response was investigated by Stoelzle et al. (2014) in several almost-natural catchments in Southwest Germany with different hydrogeological frameworks (hydraulic conductivities, storage capacities). Similar to Bloomfield and Marchant (2013), they found that hydrological drought in karstic and fractured aquifers rather quickly responds to drought in recharge. Specific recharge events could be traced back in the hydrological drought. Hydrological drought in granular aquifers responds more slowly to recharge deficits. Patterns in hydrological drought are more dominated by subsurface characteristics than by specific events in recharge.

Streamflow drought in Austria was studied using data from 44 catchments covering a wide spectrum of geo-climatic settings (Van Loon and Laaha, 2015). Drought duration is primarily driven by storage (accumulation and release) in the catchment. Large accumulation and slow release mean slow responsiveness, expressed by a high *Base Flow Index* (BFI, part of annual volume in the river coming from stores, e.g. groundwater). Clearly, drought duration in streamflow is also affected by the length of dry spells (in rainfall) or cold spells (in temperature). The streamflow deficit volume is more connected with seasonal storage in the snow pack and glaciers, which in Austria is linked to the elevation and the associated mean annual precipitation. Haslinger et al. (2014) add a dynamic component to the discussion on streamflow drought development in not-too-large catchments (<650 km^2). Under still rather wet conditions, droughts are largely determined by climate forcing, and can be predicted reasonably well by a simple meteorological index. However, when more dry conditions develop, the predictive skill decreases, and underground storage becomes more dominant. In the United Kingdom, research has shown that streamflow drought termination characteristics are also related to elevation and catchment annual precipitation, resulting in shorter drought terminations in wet upland catchments (Parry et al., 2016).

Lakes also impact downstream drought by storing upstream surface water in wet periods (rise of lake level) and releasing it (decline of level) during dry periods. The more the lake volume can change, the more upstream droughts are pooled, leading to less droughts downstream of the lake, but these can become longer during long dry periods (Van Lanen et al., 2004a). Wetlands have a similar role in alleviating and enhancing downstream streamflow drought.

Human influences In many parts of the world, humans have influenced streamflow drought (e.g. surface water abstraction, groundwater abstraction, water transfers, land use change; Van Lanen et al., 2004b). Similar to the influence on groundwater levels (see the subsection titled 'Human Influences' under Section 1.1.5.1), Van Loon and Van Lanen (2013) investigated the impact of groundwater abstraction on streamflow drought in the Upper Guadiana river basin. The observed flow, which was strongly impacted after 1980, was compared against the naturalised flow and the variable threshold (similar to Figure 1.1.6). The latter two were derived from the flow in the undisturbed period. The human-modified streamflow drought due to groundwater abstraction appeared to be four times larger than the natural, that is, the climate-induced drought.

In the Bilina catchment (Czech Republic), large-scale mining activities triggered water transfer from an adjacent catchment, implying that, in the disturbed period, the observed flow was usually higher than under natural conditions. Van Loon and Van Lanen (2015) show that streamflow drought was substantially alleviated. The frequency dropped by 85%, and the mean deficit volume by almost 90%. The streamflow drought in the catchment that supplied the water was not analysed, but due to the lower flow there likely the drought has been enhanced (depending on the ratio of water supply relative to the total discharge).

Surface water reservoirs impact discharge downstream of the dam. Usually, the high flows decrease substantially, whereas the low flows are higher. However, if water is abstracted directly from the reservoir downstream, low flows cannot be sustained during drought. López-Moreno et al. (2009) carried out a comprehensive drought analysis of the impact of the Alcántara reservoir, which was built on the Tagus River in the late 1960s. The reservoir is one of the largest in Europe, and it affected the river's flow in Spain and downstream in Portugal. Since the construction of the dam, the streamflow drought became more severe, particularly in the Portuguese part of the river, before tributaries enter the main river. Rangecroft et al. (2016) analysed the influence of the Santa Juana dam (a reservoir in northern Chile built for downstream water supply) on streamflow drought. They compared the upstream–downstream relation in the pre- and post-dam periods using a set of drought identification methods, including standardised and threshold approaches (Section 1.1.2). The dam caused a reduction in the average duration and deficit volume downstream (–42 and –86%, respectively). Hence, the dam alleviated downstream drought, although the study also showed that the capacity was insufficient to fully alleviate the severe multi-year droughts.

1.1.6 Drought propagation

1.1.6.1 Climate–hydrology links

Meteorological, soil moisture, and hydrological drought cannot be regarded separately. They are part of the same interconnected system in which one drought type influences the other by changing evapotranspiration, infiltration, and runoff processes (Figure 1.1.1). This transfer of drought through the terrestrial part of the hydrological cycle is called *propagation* (Peters et al., 2003; Van Loon, 2015).

Because a catchment can be regarded as a low-pass filter, the drought signal will undergo changes when it propagates from meteorological to hydrological drought (Figure 1.1.7). A soil moisture drought or hydrological drought starts later than a meteorological drought ('delay'), has a longer duration ('lengthening') and is of lower intensity ('attenuation'). Multiple meteorological droughts can also grow together into one hydrological drought ('pooling'). These four propagation processes are dependent on catchment characteristics and climate – that is, more storage in the catchment and more seasonality in climate lead to more delay, lengthening, attenuation, and pooling. The result is fewer but longer droughts in discharge than in precipitation (Peters et al., 2003; 2006; Tallaksen et al., 2009; Vidal et al., 2010; Van Loon and Van Lanen, 2012; Fendeková and Fendek, 2012; Van Loon and Laaha, 2015).

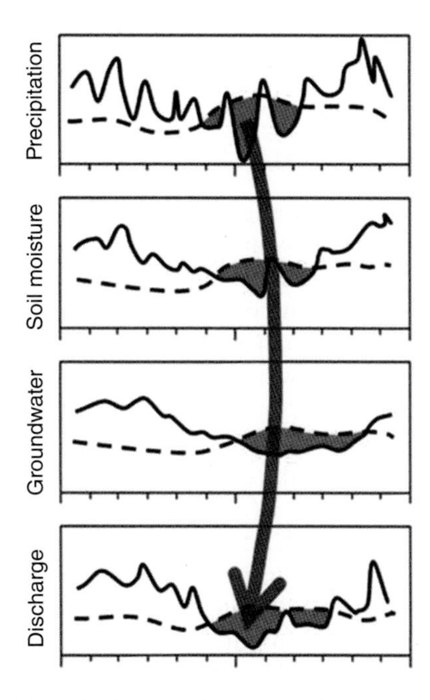

Figure 1.1.7 Propagation of drought through the terrestrial hydrological cycle, from meteorological drought to soil moisture drought to hydrological drought, showing pooling, delay, lengthening, and attenuation. (*See colour plate section for the colour representation of this figure.*)

1.1.6.2 Hydrological drought typology

Different climate seasonality and catchment storage result in clearly distinguishable hydrological drought types with different causing factors, development, and termination processes (Van Loon and Van Lanen, 2012; Van Loon et al., 2015). The most common hydrological drought type that occurs around the world is the *classical rainfall deficit drought*, in which a meteorological drought propagates through the hydrological system to form a groundwater drought or streamflow drought (Figure 1.1.7). In extremely seasonal climates, such a classical rainfall deficit drought developing in the high-flow season (summer in seasonally snow-covered catchments and winter in seasonally dry catchments) might continue into the low-flow season because all precipitation falls as snow or is lost to evapotranspiration. This results in *rain-to-snow-season drought* and *wet-to-dry-season drought*, respectively, which tend to have long durations (Van Loon and Van Lanen, 2012).

In snow- and ice-dominated catchments, anomalies in temperature play a role in drought development besides precipitation deficits (Van Loon and Van Lanen, 2012; Haslinger et al., 2014; Van Loon et al., 2015). In these studies, several drought types have been distinguished – for example, *glaciermelt drought*, *snowmelt drought*, *warm snow season drought*, and *cold snow season drought*. The variety in hydrological drought types related to snow and seasonality in cold climates makes it very important to consider snow and ice in drought monitoring and management (e.g. Staudinger et al., 2014; Haslinger et al., 2014; Harpold et al., 2017).

Finally, *composite droughts* are multi-year hydrological droughts that undergo extreme pooling, as a result of which different drought types in different seasons and years are combined into one long drought event (Van Loon and Van Lanen, 2012).

1.1.6.3 Human influences

Besides natural factors, each of the drought propagation characteristics and hydrological drought types is also influenced by anthropogenic factors. Human activities, intentionally or unintentionally, influence catchment storage and transfer processes, thereby enhancing or alleviating drought (Van Loon et al., 2016a). For example, reservoir building results in more pooling and attenuation of drought (e.g. Rangecroft et al., 2016), and groundwater abstraction leads to growing together of droughts into composite droughts (e.g. Van Loon and Van Lanen, 2013; see also the subsection titled 'Human Influences' under Section 1.1.5.2).

In most drought research till date, the climate system is seen as the driver of drought, and the societal system as the receiver (Figure 1.1.8, top). In reality, these systems are intertwined, and feedbacks occur with both the climate system and the societal system (Figure 1.1.8, bottom). For example, during the *Millennium Drought* in Australia, river regulation and abstraction influenced drought propagation and made the streamflow drought almost twice as severe as in natural circumstances (Van Dijk et al., 2013). In the recent multi-year drought in California, the presence and management of reservoirs decreased streamflow drought deficit by 50% in some areas, but irrigation water use increased streamflow drought deficit by 50–100% in others (He et al., 2017). In the future, the impact of human water use on changes in streamflow drought severity is expected to be as important as the effects of climate change in many regions around the world (Wanders and Wada, 2015). By including these feedbacks in drought analysis, we

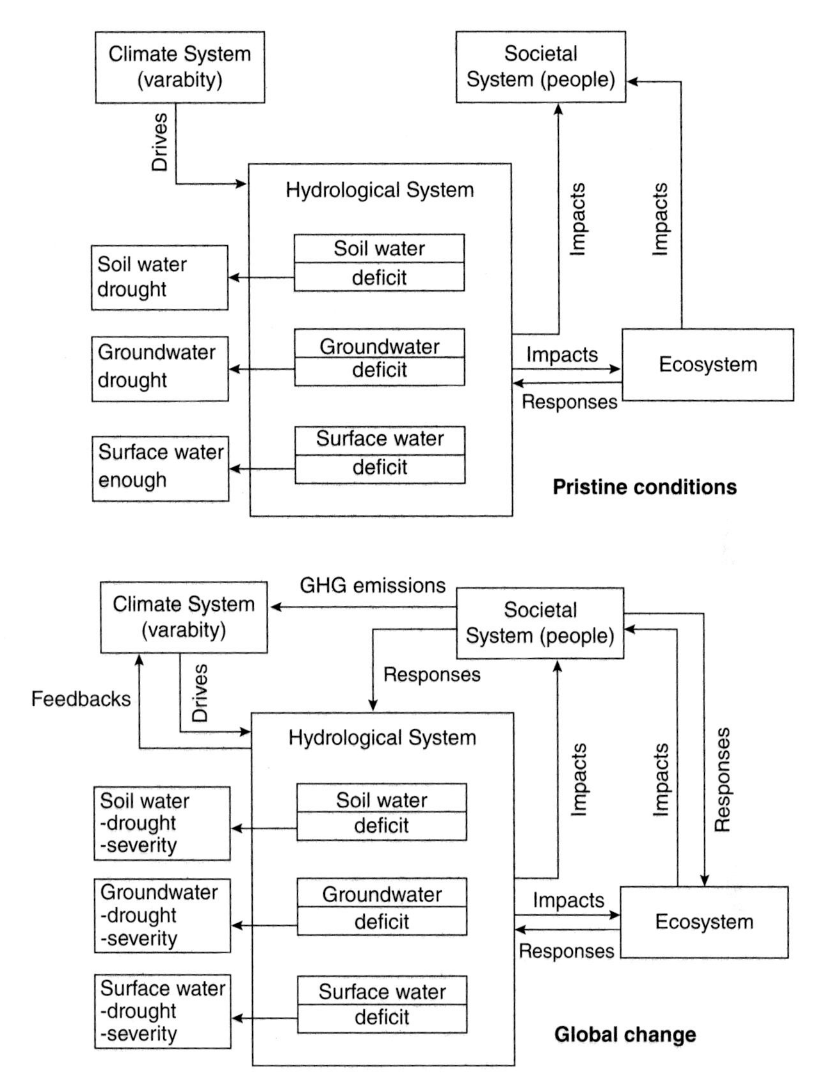

Figure 1.1.8 Relationships between the climate system, hydrological system, ecosystem, and societal system. Top: pristine conditions (representing a unidirectional drought propagation). Bottom: global change conditions (representing a multi-directional drought propagation characterised by feedbacks).

will not only increase our understanding of drought, but also be able to manage drought more effectively (Van Loon et al., 2016b).

1.1.7 Concluding remarks and outlook

1.1.7.1 Conclusions

In the centre and north of Europe, blocking situations that disturb the common eastwards movement of the mid-latitudes pressure systems, that is, the westerlies, are the main cause for meteorological drought development. In the Mediterranean region, severe droughts can be caused by the longer-than-usual influence of the Subtropical High-Pressure Belt. SSTs also play a role through their interaction with large-scale climatic or oceanic modes of variability.

Documented trends of drought over Europe show apparently conflicting outcomes (both drying and wetting trends are reported; see also Chapter 1.2 in this book, titled 'Recent Trends in Historical Drought'). Improved knowledge regarding the diagnosis of drought-generating processes can expose the underlying reasons – for example, regional geophysical settings, drought types, drought characteristics, and drought index.

Soil water drought has disappeared or been reduced in irrigated areas, but it is largely unclear what happened with the groundwater drought and surface drought in the irrigated area itself and beyond (downstream areas).

Hydrogeology (e.g. aquifer characteristics) plays an important role in the development of groundwater drought, and consequently streamflow drought is affected through drought propagation. Catchments with slowly responding groundwater systems, such as granular aquifers, have a lower number of droughts as compared to the fast-responding ones (e.g. fractured or consolidated karstic aquifers).

The development of the observation-modelling framework enabled us to better distinguish between climate-induced and human-modified/human-induced drought. So far, the framework has mainly been used for studying the effects of groundwater abstraction and reservoirs in a limited number of geo-climatic settings.

Climate seasonality and catchment storage trigger various drought generation processes, including termination, which lead to clearly distinguishable hydrological drought types. Similar to the flood typology, drought types can help us better understand the impacts associated with drought.

1.1.7.2 Outlook

Research on drought-generating processes should move from a focus on only natural processes to studies that integrate the physical system (climate–hydrology) and the societal system, including feedbacks. By intertwining the two systems in drought analysis, our understanding improves, from which drought management will benefit.

Improved knowledge on the diagnosis of drought-generating processes may enhance our abilities for monthly and seasonal drought forecasting. Teleconnections, including the lagged relationship between the SSTs and drought conditions, holds the potential for drought prediction on seasonal to decadal timescales over Europe. In addition to further improvements of weather/climate models to forecast drought, statistical models that use teleconnections need to be investigated further.

Improved knowledge on the processes underlying drought should be used to better explain the reasons for the apparently contradicting trends in drought, and may help to attribute to underlying causes for the trends (e.g. decadal climate variability, land use change, abstractions).

Although groundwater is an important water resource that is often intensively exploited during drought, the development of groundwater drought is still not extensively studied. More focus is required on the role of hydrogeology in drought generation.

Hydrological drought (groundwater, streamflow) as a response to water use for irrigation needs to be further investigated in the irrigated and downstream areas.

Application of the observation-modelling framework to distinguish between climate-induced and human-modified/human-induced drought needs more attention – for example, for better coverage of geo-climatic settings and human interferences that affect drought generation.

The established series of drought types (drought typology) should be further tested in other geo-climatic settings. Furthermore, robustness of the typology needs investigation – for example, do drought types change due to humans interferes, such as abstractions, reservoirs, global warming.

Acknowledgements

Funding was provided by the EU WATCH project (WATer and global CHange), contract 036946, and the EU project DROUGHT–R&SPI (Fostering European Drought Research and Science–Policy Interfacing), contract 282769. In the last phase, support came from the EU H2020 ANYWHERE (contract no. 700099). The work was partly funded by the Dutch NWO Rubicon project 'Adding the human dimension to drought' (reference number: 2004/08338/ALW). The study is also a contribution to the UNESCO IHP-VII programme (Euro FRIEND-Water project) and the Panta Rhei Initiative of the International Association of Hydrological Sciences (IAHS).

References

Bloomfield, J.P. and Marchant, B.P. (2013). Analysis of groundwater drought building on the standardised precipitation index approach. *Hydrology and Earth System Sciences* 17: 4769–4787. doi: 10.5194/hess-17-4769-2013.

Christensen, J., Krishna Kumar, K., Aldrian, E. An S.I., Cavalcanti, I., De Castro, M., Dong, W., Goswami, P., Hall, A., Kanyanga, J., Kitoh, A., Kossin, J., Lau, N.C., Renwick, J., Stephenson, D., Xie, S.P., and Zhou, T. (2013). *Climate Phenomena and Their Relevance for Future Regional Climate Change*. Book section 14. Cambridge: Cambridge University Press, ISBN 978-1-107-66182-0: Cambridge, UK and New York, NY, 1217–1308. doi: 10.1017/CBO9781107415324.028.

Dai, A. (2012). Increasing drought under global warming in observations and models. *Nature Climate Change* 3: 52–58. doi: 10.1038/nclimate1633.

De Jeu, R. and Dorigo, W. (2016). On the importance of satellite observed soil moisture. *International Journal of Applied Earth Observation and Geoinformation* 45: 107–109. doi: 10.1016/j.jag.2015.10.007.

EEA (European Environmental Agency) (2017). Climate Change, Impacts and Vulnerability in Europe 2016. An Indicator-based Report. *EEA Report No. 1/2017, European Environmental Agency*, Copenhagen. doi: 10.2800/534806.

FAO (2015). AQUASTAT Main Database, Food and Agriculture Organization of the United Nations (FAO). Available from: http://www.fao.org/nr/water/aquastat/main/index.stm, website accessed on 22/03/2018.

Fendeková, M. and Fendek, M. (2012). Groundwater drought in the Nitra River Basin –identification and classification. *Journal of Hydrology and Hydromechanics* 60 (3): 185–193.

Fleig, A.K., Tallaksen, L.M., Hisdal, H., Stahl, K., and Hannah, D.M. (2010). Inter-comparison of weather and circulation type classifications for hydrological drought development. *Physics and Chemistry of the Earth* 35: 507–515. doi: 10.1016/j.pce.2009.11.005, 2010.

Fleig, A.K., Tallaksen, L.M., Hisdal, H., and Hannah, D.M. (2011). Regional hydrological droughts and associations with the objective Grosswetterlagen in north-western Europe. *Hydrological Processes* 25: 1163–1179. doi: 10.1002/hyp.7644, 2011.

Harpold, A.A., Dettinger, M., and Rajagopal, S. (2017). Defining snow drought and why it matters. *Eos* 98. doi: 10.1029/2017EO068775.

Hartmann, D.L., Klein Tank, A.M.G., Rusticucci, M., Alexander, L.V., Brönnimann, S., Charabi, Y., Dentener, F.J., Dlugokencky, E.J., Easterling, D.R., Kaplan, A., Soden, B.J., Thorne, P.W., Wild, M., and Zhai, P.M. (2013). Observations: atmosphere and surface. In: *Climate Change 2013: The Physical Science Basis. Contribution of Working Group I to the Fifth Assessment Report of the Intergovernmental Panel on Climate Change* (ed. T.F. Stocker, D. Qin, G.-K. Plattner, M. Tignor, S.K. Allen, J. Boschung, A. Nauels, Y. Xia, V. Bex, and P.M. Midgley). Cambridge, UK and New York, NY, USA: Cambridge University Press.

Haslinger, K., Koffler, D., Schöner, W., and Laaha, G. (2014). Exploring the link between meteorological drought and streamflow: effects of climate–catchment interaction. *Water Resources Research* 50: 2468–2487. doi: 10.1002/2013wr015051.

He, X., Wada, Y., Wanders, N., and Sheffield, J. (2017). Human water management intensifies hydrological drought in California. *Geophysical Research Letters* 44: 1777–1785. doi: 10.1002/2016GL071665.

Heudorfer, B. and Stahl, K. (2017). Comparison of different threshold level methods for drought propagation analysis in Germany. *Hydrology Research* 48 (5): 1311–1326. doi: 10.2166/nh.2016.258.

Hisdal, H., Tallaksen, L.M., Clausen, B., Peters, E., and Gustard, A. (2004). Drought characteristics. In: *Hydrological Drought: Processes and Estimation Methods for Streamflow and Groundwater; Developments in Water Science, Volume 48* (ed. L.M. Tallaksen and H. Van Lanen), 139–198. Amsterdam: Elsevier.

Ionita, M., Tallaksen, L.M., Kingston, D.G., Stagge, J.H., Laaha, G., Van Lanen, H.A.J., Chelcea, S.M., and Haslinger, K. (2017). The European 2015 drought from a climatological perspective. *Hydrology and Earth System Sciences* 21: 1397–1419. doi: 10.5194/hess-21-1397-2017.

Kingston, D.G., Fleig, A.K., Tallaksen, L.M., and Hannah, D.M. (2013). Ocean–atmosphere forcing of summer streamflow drought in Great Britain. *Journal of Hydrometeorology* 14: 331–344. doi: 10.1175/JHM-D-11-0100.1.

Kingston, D.G., Stagge, J.H., Tallaksen, L.M., and Hannah, D.M. (2015). European-scale drought: understanding connections between atmospheric circulation and meteorological drought indices. *Journal of Climate* 28: 505–516. doi: 10.1175/JCLI-D-14-00001.1.

López-Moreno, J.I., Vicente-Serrano, S.M., Beguería, S., García-Ruiz, J.M., Portela, M.M., and Almeida, A.B. (2009). Dam effects on droughts magnitude and duration in a transboundary basin: the Lower River Tagus, Spain and Portugal. *Water Resources Research* 45: W02405. doi: 10.1029/2008WR007198.

Mckee, T.B., Doesken, N.J., and Kleist, J. (1993). The relationship of drought frequency and duration to time scale. In: *Proceedings of 8th Conference on Applied Climatology*, Anaheim, California, 17–22 January 1993. Boston: American Meteorological Society, 179–184.

Mishra, K.K. and Singh, V.P. (2010). A review of drought concepts. *Journal of Hydrology* 391: 202–216.

Nilsen, I.B., Stagge, J.H., and Tallaksen, L.M. (2016). A probabilistic approach for attributing temperature changes to synoptic type frequency. *International Journal of Climatology* 37 (6): 2990–3002. doi: 10.1002/joc.4894.

Oki, T. and Kanae, S. (2006). Global hydrological cycles and world water resources. *Science* 313: 1068–1072. doi: 10.1126/science.1128845.

Orlowsky, B. and Seneviratne, S.I. (2013). Elusive drought: uncertainty in observed trends and short- and long-term CMIP5 projections. *Hydrology and Earth System Sciences* 17: 1765–1781. doi: 10.5194/hess-17-1765-2013.

Palmer, W.C. (1965). Meteorological drought, *Weather Bureau, Research Paper No. 45,* U.S. Department of Commerce, Washington, DC.

Parry, S., Wilby, R.L., Prudhomme, C., and Wood, P.J. (2016). A systematic assessment of drought termination in the United Kingdom. *Hydrology and Earth System Sciences* 20: 4265–4281. doi: 10.5194/hess-20-4265-2016.

Peters, E., Torfs, P.J.J.F., Van Lanen, H.A.J., and Bier, G. (2003). Propagation of drought through groundwater – a new approach using linear reservoir theory. *Hydrological Processes* 17(15): 3023–3040.

Peters, E., Bier, G., Van Lanen, H.A.J., and Torfs, P.J.J.F. (2006). Propagation and spatial distribution of drought in a groundwater catchment. *Journal of Hydrology* 321(1): 257–275.

Rangecroft, S., Van Loon, A.F., Maureira, H., Verbist, K., and Hannah, D.M. (2016). Multi-method assessment of reservoir effects on hydrological droughts in an arid region. *Earth System Dynamics Discussions.* doi: 10.5194/esd-2016-57.

Schubert, S.D., Wang, H., Koster, R.D., Suarez, M.J., and Groisman, P. (2014). Northern Eurasian heat waves and droughts. *Journal of Climate* 27: 3169–3207. doi:10.1175/JCLI-D-13-00360.1.

Rizzi, J., Nilsen, I.B., Stagge, J.H., Grini, K., and Tallaksen, L.M. (2017). Five decades of warming: impacts on snow cover in Norway. *Hydrological Research* (Submitted).

Sheffield, J. and Wood, E.F. (2011). *Drought: Past Problems and Future Scenarios*. London: Earthscan.

Sheffield, J., Wood, E.F., and Roderick, M.L. (2012). Little change in global drought over the past 60 years. *Nature*, 491: 435–438. doi: 10.1038/nature11575.

Shukla, S. and Wood, A.W. (2008). Use of a standardized runoff index for characterizing hydrologic drought. *Geophysical Research Letters* 35 (2): L02,405.

Spinoni, J., Naumann, G., Vogt, J., and Barbosa, P. (2015). European drought climatologies and trends based on a multi-indicator approach. *Global and Planetary Change* 127: 50–57.

Spinoni, J., Naumann, G., Vogt, J., and Barbosa, P. (2016). *Meteorological Droughts in Europe. Events and Impacts Past Trends and Future Projections*. Luxembourg, EUR 27748 EN: Publications Office of the European Union. doi: 10.2788/450449.

Stagge, J.H., Rizzi, J., Tallaksen, L.M., and Stahl, K. (2015). Future Meteorological Drought: Projections of Regional Climate Models for Europe. *DROUGHT R&SPI Technical Report No. 25*, University of Oslo, Norway.

Stagge, J.H., Kingston, D., Tallaksen, L.M., and Hannah, D. (2017). Observed drought indices show increasing divergence across Europe. *Proceedings of the National Academy of Sciences* (Submitted).

Stahl, K. and Hisdal, H. (2004). Hydroclimatology (chapter 2). In: *Hydrological Drought. Processes and Estimation Methods for Streamflow and Groundwater; Developments in Water Science, Volume 48* (ed. L.M. Tallaksen and H.A.J. Van Lanen), 19–51. Amsterdam: Elsevier Science B.V.

Stahl, K., Kohn, I., Blauhut, V., Urquijo, J., de Stefano, L., Acacio, V., Dias, S., Stagge, J.H., Tallaksen, L.M., Kampragou, E., Van Loon, A.F., Barker, L.J., Melsen, L.A., Bifulco, C., Musolino, D., de Carli, A., Massarutto, A., Assimacopoulos, D., and Van Lanen, H.A.J. (2016). Impacts of European drought events: insights from an international database of text-based reports. *Natural Hazards and Earth System Sciences* 16: 801–819. doi:10.5194/nhess-16-801-2016.

Staudinger, M., Stahl, K., and Seibert, J. (2014). A drought index accounting for snow. *Water Resources Research* 50 (10): 7861–7872. doi: 10.1002/2013WR015143.

Stoelzle, M., Stahl, K., Morhard, A., and Weiler, M. (2014). Streamflow sensitivity to drought scenarios in catchments with different geology. *Geophysical Research Letters* 41 (17): 6174–6183. doi: 10.1002/2014GL061344.

Tallaksen, L.M. and Van Lanen, H.A.J. (2004). *Hydrological Drought: Processes and Estimation Methods for Streamflow and Groundwater*. No. 48 in Development in Water Science, Amsterdam: Elsevier Science B.V.

Tallaksen, L.M., Hisdal, H., and Van Lanen, H.A.J. (2009). Space–time modelling of catchment scale drought characteristics. *Journal of Hydrology* 375 (3): 363–372.

Teuling, A.J., Van Loon, A.F., Seneviratne, S.I., Lehner, I., Aubinet, M., Heinesch, B., Bernhofer, C., Grünwald, T., Prasse, H., and Spank, U. (2013). Evapotranspiration amplifies European summer drought. *Geophysical Research Letters* 40: 2071–2075. doi:10.1002/grl.50495.

Trenberth, K.E., Dai, A., Van der Schrier, G., Jones, P.D., Barichivich, J., Briffa, K.R., and Sheffield, J. (2014). Global warming and changes in drought. *Nature Climate Change* 4: 17–22. doi: 10.1038/nclimate2067.

Van der Schrier, G., Jones, P.D., and Briffa, K.R. (2011). The sensitivity of the PDSI to the Thornthwaite and Penman–Monteith parameterizations for potential evapotranspiration. *Journal of Geophysical Research* 116: D03106.

Van Dijk, A.I.J.M., Beck, H.E., Crosbie, R.S., De Jeu, R.A.M., Liu, Y.Y., Podger, G.M., Timbal, B., and Viney, N.R. (2013). (2001–2009): Natural and human causes and implications for water resources, ecosystems, economy, and society. *Water Resources Research* 49 (2): 1040–1057. doi:10.1002/wrcr.20123.

Van Lanen, H.A.J., Fendeková, M., Kupczyk, E., Kasprzyk, A., and Pokojski, W. (2004a). Flow generating processes (chapter 3). In: *Hydrological Drought. Processes and Estimation Methods for Streamflow and Groundwater; Developments in Water Science, Volume 48* (ed. L.M. Tallaksen and H.A.J. Van Lanen), 53–96. Amsterdam: Elsevier Science B.V.

Van Lanen, H.A.J., Kašpárek, L., Novický, O., Querner, E.P., Fendeková, M., and Kupczyk, E. (2004b). Human influences (chapter 9). In: *Hydrological Drought. Processes and Estimation Methods for Streamflow and Groundwater; Developments in Water Science, Volume 48* (ed. L.M. Tallaksen and H.A.J. Van Lanen), 347–410. Amsterdam: Elsevier Science B.V.

Van Lanen, H.A.J., Wanders, N., Tallaksen, L.M., and Van Loon, A.F. (2013). Hydrological drought across the world: impact of climate and physical catchment structure. *Hydrology and Earth System Sciences* 17: 1715–1732. doi: 10.5194/hess-17-1715-2013.

Van Lanen, H.A.J., Laaha, G., Kingston, D.G., Gauster, T., Ionita, M., Vidal, J.-P., Vlnas, R., Tallaksen, L.M., Stahl, K., Hannaford, J., Delus, C., Fendekova, M., Mediero, L., Prudhomme, C., Rets, E., Romanowicz, R.J., Gailliez, S., Wong, W.K., Adler, M.-J., Blauhut, V., Caillouet, L., Chelcea, S., Frolova, N., Gudmundsson, L., Hanel, M., Haslinger, K., Kireeva, M., Osuch, M., Sauquet, E., Stagge, J.H. and Van Loon, A.F. (2016). Hydrology needed to manage droughts: the 2015 European case. *Hydrological Processes* 30: 3097–3104. doi: 10.1002/hyp.10838.

Van Loon, A.F. (2015). Hydrological drought explained. *WIREs Water*. doi: 10.1002/wat2.1085.

Van Loon, A.F. and Van Lanen, H.A.J. (2012). A process-based typology of hydrological drought. *Hydrology and Earth System Sciences* 16: 1915–1946. doi: 10.5194/hess-16-1915-2012.

Van Loon, A.F. and Van Lanen, H.A.J. (2013). Making the distinction between water scarcity and drought using an observation-modeling framework. *Water Resources Research* 49: 1483–1502. doi:10.1002/wrcr.20147.

Van Loon, A.F., and Laaha, G. (2015). Hydrological drought severity explained by climate and catchment characteristics. *Journal of Hydrology* 526: 3–14. doi:10.1016/j.jhydrol.2014.10.059.

Van Loon, A.F. and Van Lanen, H.A.J. (2015): Testing the observation-modelling framework to distinguish between hydrological drought and water scarcity in case studies around Europe. *European Water* 49: 65–75, www.ewra.net/ew/issue_49.htm.

Van Loon, A.F., Van Lanen, H.A.J., Hisdal, H., Tallaksen, L.M., Fendeková, M., Oosterwijk, J., Horvát, O., and Machlica, A. (2010). Understanding hydrological winter drought in Europe. In: *Global Change: Facing Risks and Threats to Water Resources* (ed. E. Servat, S. Demuth, A. Dezetter, T. Daniell, E. Ferrari, M. Ijjaali, R. Jabrane, H. Van Lanen, and Y. Huang), 189–197. Wallingford: IAHS Publ. No. 340.

Van Loon, A.F., Tijdeman, E., Wanders, N., Van Lanen, H.A.J., Teuling, A.J., and Uijlenhoet, R. (2014). How climate seasonality modifies drought duration and deficit. *Journal of Geophysical Research – Atmospheres* 119 (8): 4640–4656. doi: 10.1002/2013JD020383.

Van Loon, A.F., Ploum, S.W., Parajka, J., Fleig, A.K., Garnier, E., Laaha, G., and Van Lanen, H.A.J. (2015). Hydrological drought types in cold climates: quantitative analysis of causing factors and qualitative survey of impacts. *Hydrology and Earth System Sciences* 19: 1993–2016. doi: 10.5194/hess-19-1993-2015.

Van Loon, A.F., Gleeson, T., Clark, J., Van Dijk, A., Stahl, K., Hannaford, J., Di Baldassarre, G., Teuling, A., Tallaksen, L.M., Uijlenhoet, R., Hannah, D.M., Sheffield, J., Svoboda, M., Verbeiren, B., Wagener, T., Rangecroft, S., Wanders, N., and Van Lanen, H.A.J. (2016a). Drought in the Anthropocene. *Nature Geoscience* 9 (2): 89–91.

Van Loon, A. F., Gleeson, T., Clark, J., Van Dijk, A.I.J.M., Stahl, K., Hannaford, J., Di Baldassarre, G., Teuling, A.J., Tallaksen, L.M., Uijlenhoet, R., Hannah, D.M., Sheffield, J., Svoboda, M., Verbeiren, B., Wagener, T., Rangecroft, S., Wanders, N., and Van Lanen, H.A.J. (2016b). Drought in a human-modified world: reframing drought definitions, understanding and analysis approaches. *Hydrology and Earth System Sciences* 20 (9): 3631–3650. doi: 10.5194/hess-20-3631-2016.

Vicente-Serrano, S.M., Beguería, S., and López-Moreno, J.I. (2010). A multi-scalar drought index sensitive to global warming: the Standardized Precipitation Evapotranspiration Index – SPEI. *Journal of Climate* 23 (7): 1696–1718. doi: 10.1175/2009JCLI2909.1.

Vidal, J. P., Martin, E., Franchistéguy, L., Habets, F., Soubeyroux, J.M., Blanchard, M., and Baillon, M. (2010). Multilevel and multiscale drought reanalysis over France with the Safran-Isba-Modcou hydrometeorological suite. *Hydrology and Earth System Sciences* 14 (3): 459–478.

Wanders, N. and Wada, Y. (2015). Human and climate impacts on the 21st century hydrological drought. *Journal of Hydrology* 526: 208–220.

Wilhite, D.A. (Ed.) (2000). Droughts as a natural hazard: concepts and definitions. In: *Drought, A Global Assessment, Vol I and II, Routledge Hazards and Disasters Series*. London: Routledge.

Yevjevich, V. (1967). An objective approach to definition and investigations of continental hydrologic droughts. *Hydrology Papers 23*, Colorado State University, Fort Collins, USA.

Zelenhasic, E. and Salvai, A. (1987). A method of streamflow drought analysis. *Water Resources Research* 23 (1): 156–168.

1.2
Recent Trends in Historical Drought

Kerstin Stahl[1], Lena M. Tallaksen[2], and Jamie Hannaford[3]

[1]*University of Freiburg, Freiburg, Germany*
[2]*University of Oslo, Oslo, Norway*
[3]*Centre for Ecology and Hydrology, Wallingford, United Kingdom*

1.2.1 Introduction

Drought is caused by climate anomalies. Combined with low antecedent storage in surface and subsurface systems, these may lead to water deficit in the hydrological cycle, and thus to hydrological drought (e.g. Tallaksen and Van Lanen, 2004; Tallaksen et al., 2015). Trends in hydrological drought – as typically characterised by deficits in streamflow or groundwater – are thus important indicators of changes in water availability. For policy development and water management at international scales, these large-scale trends in water availability can provide crucial information. Recent years have witnessed a number of severe droughts in different regions across Europe. These events have prompted numerous national and regional studies on past trends, and projected future changes, in drought – as summarised by Stahl et al. (2014). Particular concerns are the impacts of extremely low streamflow on aquatic ecology and a variety of human water uses, such as for water supply, energy production, waterborne transportation, industrial use, etc. (Stahl et al., 2016). In Europe, an exacerbation of drought-associated low flows would make it more difficult for countries to meet their obligation to improve the ecological status of water bodies according to the EU Water Framework Directive. In times of global warming and expected increases in the occurrence and severity of hydrological extremes, there is considerable concern over potential changes in hydrological drought. The focus in this chapter is on trends in streamflow drought.

Any large-scale assessment of trends in streamflow (as well as other fluxes and state variables in the hydrological cycle), as a response to climatic change, presents a number of challenges. Common methods to assess past changes include time series analysis of observed records, and hydrological model experiments or model chain experiments – from Global Climate Models (GCMs), to Regional Climate Models (RCMs),

Drought: Science and Policy, First Edition.
Edited by Ana Iglesias, Dionysis Assimacopoulos, and Henny A.J. Van Lanen.

to hydrological models – for simulation of the recent past. All approaches have their strengths and weaknesses, and the use of a diversity of observational datasets, methods (see Section 1.2.2) and models, across a range of temporal and spatial scales, complicates any comparative assessments between studies.

The key aim of a number of continental-scale studies on trends in Europe's hydrology and drought, in particular, has been to harmonise the previously fragmented knowledge on past streamflow trends for the European domain (Stahl et al., 2014). Two studies contributed to this harmonisation through: (i) an empirical analysis of trends in a specially assembled network of near-natural streamflow records across Europe, the most comprehensive dataset of this type yet assembled (Stahl et al., 2010); and (ii) thorough in-depth analysis of the influence of interdecadal river flow variability on trends, based on long time series from a subsample of that near-natural network (Hannaford et al., 2013). Expanding on these papers, one study tested the use of runoff simulated by a multi-model ensemble of global land surface and hydrological models to 'fill in the white space' between the observations on the European map (Stahl et al., 2012). The following section summarises the findings of these studies on trends in hydrological variables relevant to drought. The final section puts the studies into perspective with a wider range of studies on drought trends in multiple variables.

1.2.2 Trend analysis and data

1.2.2.1 Methodology

Temporal trends can be analysed in many ways (e.g. Chandler and Scott, 2011). Often, simple linear regression with time as a predictor is applied under the assumption of monotonic increase or decrease in the variable in question. A statistical test for fitted slope (change in time) being different from zero is then employed to quantify the significance of a trend, whereas the slope represents the trend magnitude. Trends in environmental variables are often analysed by a non-parametric, rank-based variant of this regression approach: the Mann–Kendall test (Mann, 1945; Kendall, 1975), and an estimate of trend direction and magnitude from the slope of the Kendall–Theil robust line (Theil, 1950). The studies summarised in Section 1.2.3 used trends that were calculated by the slopes of the Kendall–Theil robust line for annual and monthly streamflow, as well as for summer low flow magnitude and timing.

Other approaches do not imply monotonic or linear trends, and either assume a different trend behaviour or simply employ time-variable smoothing methods to visualise non-linear trends. For example, LOESS is a robust and widely used smoothing method (e.g. Chandler and Scott, 2011) for which a span parameter regulates the level of smoothing by controlling the proportion of the dataset that is used in the 'local' smoothing window. This method was employed in the study discussed in Section 1.2.3. For hypotheses on decadal and multi-decadal variability, a variable 'multi-temporal' approach with shifting time periods may be used, and the co-variability with hypothesised drivers such as large-scale modes of ocean–atmosphere variability (e.g. the North Atlantic Oscillation, NAO) are often analysed.

1.2.2.2 Data

Observational hydrological datasets have some constraints with regard to station density, and the gauging network coverage is also shrinking in many countries. In addition, centralised access to data is often limited due to political, administrative, and technical constraints (Hannah et al., 2011), as well as economic barriers (Viglione et al., 2010). Hannah et al. (2011) argue that large-scale river flow archives hold vital data to identify and understand the changing water cycle, to underpin the modelling of future regional and global hydrology, and to inform water resource assessment and decision-making. They describe examples of international datasets, notably those held by the WMO Global Runoff Data Centre (GRDC) and the UNESCO-FRIEND-Water program's European Water Archive (EWA). With a large number of sovereign countries and an even larger number of authorities responsible for streamflow gauging and collection and archiving of hydrometric data, establishing these international databases has been a major challenge. The challenge thereby is not only to get access to the data itself, but to ensure high data quality and regular updates.

In addition, many regions in Europe are densely populated, with a long history of settlement. A majority of rivers, particularly the larger ones, have been regulated for over a century. River regulation, particularly dams and abstractions, but also river training and indirect human influences within the basin, can change the streamflow regime and potentially compromise the climate sensitivity of the streamflow signal. Several countries have therefore established the so-called 'reference' or 'benchmark' networks (Whitfield et al., 2012). Streamflow records from these networks represent near-natural river flow regimes and allow observation of how hydrological processes respond to climatic changes. Taken to be representative of wider regions, they thus provide a basis for investigating the predominant climate and catchment processes that govern changes in regional hydrology.

Stahl et al. (2010) determined the streamflow trends in a dataset of near-natural streamflow records from 441 small catchments in 15 countries across the European continent. To do so, they updated the European Water Archive (EWA). The update included homogeneous, quality-controlled records of daily mean flow suitable for low flow analysis, with a key criteria selection being the lack of appreciable direct human influences on river flow during low flow (e.g. through abstractions, reservoir storage). Catchments were relatively small (mostly <1000 km^2), and time series covered 40 years or longer and included data at least to the year 2004. Four time periods were considered: 42, 52, 62, and 72 years (starting, respectively, in 1962, 1952, 1942, and 1932, and all ending in 2004). The period 1962–2004 provides the best spatial coverage, but analyses were also carried out for the three longer time periods (with fewer stations).

Based on the same archive, Hannaford et al. (2013) analysed trends in hydrometric records from 132 catchments from northern and central Europe with long time series (1932–2004). In a multi-temporal approach, trends were calculated for every possible combination of start and end years in a record. This approach reveals the influence of interdecadal variability, and thus the co-variability with the NAO Index was also investigated as a first step towards trend attribution.

Observed trend patterns provide a valuable benchmark for continental-scale model simulations. However, observational records have spatial gaps and represent a selection of rather specific catchment types – that is, relatively small near-natural headwater catchments with long records. Stahl et al. (2012) tested whether trends derived from model simulations of European runoff could fill the white space between the observations on the maps of change.

Modelled hydrological data are available from a number of global hydrological models and the many land surface schemes developed for coupling to global or regional climate models. The use of model ensembles in large-scale hydrology either describes an ensemble derived from one model with different parameters, or different models (developed by different groups). In analogy to the World Climate Research Programme's (WCRP) Coupled Model Intercomparison Projects (CMIPs), which have provided the multi-model dataset referenced in the Intergovernmental Panel on Climate Change Assessment Reports, a number of hydrology model intercomparison projects (MIPs) have been launched. They include WaterMIP, supported by the multi-model comparison within the EU-funded WATer and global CHange (WATCH) project (www.eu-watch.org). Many (8–12) large-scale hydrological models were run for the period 1958–2000 on a global 0.5° grid and forced by the WATCH Forcing Dataset. Details of the particular models included in the studies summarised here (i.e. GWAVA, HTESSEL, JULES, LPJml, MATSIRO, MPI-HM, Orchidee, and Water-GAP) and their performance can be found in Haddeland et al. (2011) and Gudmundsson et al. (2012a, 2012b).

In contrast with the two trend studies based on observational data, Stahl et al. (2012) assessed the WATCH project's large-scale hydrological model ensemble simulation for trends in similar variables on the entire European continent. The study summarised in Section 1.2.3 used daily total runoff as the sum of fast and slow components, simulated for each grid cell in Europe (4425 land cells). The same catchments as in Stahl et al. (2010) were used for model verification. The mostly small catchments with observational data are subscale or of about the same scale as a grid cell of the WATCH Forcing Dataset, and thus used the model grid cell's value closest to the river basin centroid for comparison. The derived trends were thus compared for 293 grid cells across the European domain with observation-based trend estimates from the EURO-FRIEND near-natural catchments.

1.2.3 Trends in river flow across Europe

1.2.3.1 Trends in observed river flow

Overall, a regionally coherent picture of annual streamflow trends emerged from the analyses of observed streamflow (Stahl et al., 2010), with negative trends in southern and eastern regions, and generally positive trends elsewhere. Trends in monthly streamflow for 1962–2004 elucidated the potential causes for these changes, as well as for changes in hydrological regimes across Europe. Positive trends were found in the winter months in most catchments. A marked shift towards negative trends was observed in April, gradually spreading across Europe to reach a maximum extent in August.

Most relevant for assessing potential trends in the drought hazard, however, were the analyses of trends in the minimum of the hydrological regime – that is, the monthly mean flow of the month with the (on average) minimum flow; the trend in 7-day low flows, AM(7), of the summer half-year; and the trend in the timing of these summer

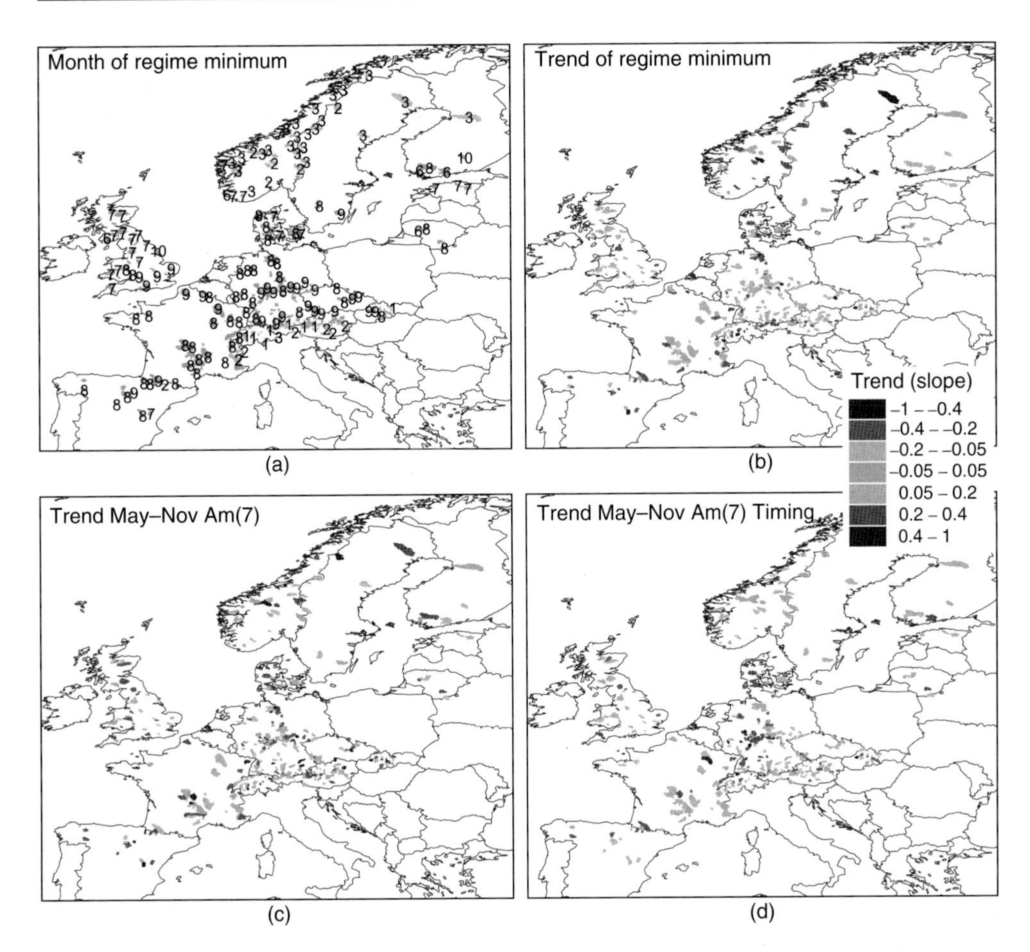

Figure 1.2.1 (a) The month of the regime minimum; (b) trend for mean monthly streamflow for the month of the regime minimum; (c) trend for AM(7) during May-November; (d) trend for the timing of AM(7) during May-November. All trends are for the period 1962–2004 and given in standard deviations per year. *Source*: From Stahl et al. (2010). (*See colour plate section for the colour representation of this figure.*)

AM(7) occurrences (Figure 1.2.1). The results show that monthly flows have increased in most regions that have winter low-flow regimes and decreased in most regions with summer low-flows (Figure 1.2.1b). However, the decreasing trends in the areas with summer low flow regimes are often weak and exhibit high spatial variability, particularly for regions with only secondary flow minima in summer. The increasing trends dominate in the area around the Baltic Sea region, where the lowest mean monthly flow occurs in summer.

Trends in 7-day low flows for only the summer season revealed a higher number of stations with negative trends, and these trends also often show a larger magnitude (e.g. in the United Kingdom and central Germany). Whereas the lowest summer flows have decreased in several Norwegian catchments that exhibit an absolute winter minimum, this is not the case in the Alps, Switzerland, Germany, and the western part of Austria.

Here, many positive AM(7) trends were found. In many of these catchments, the timing of the low flow has shifted to an earlier date.

The pan-European study largely confirmed findings from national- and regional-scale trend analyses for annual flows in similar time periods. However, for summer low flows, many national-scale studies did not confirm the negative trends, and some even reported increasing summer low flows (Stahl et al., 2014). One reason may be that the pan-European studies ended with widespread summer droughts, whereas national studies covered different periods. Since the studies discussed in detail herein were published, several other national-scale studies have been produced, which add to the picture of decreasing low flow trends in southern Europe, particularly Iberia. Decreasing low flows were observed in near-natural catchments in Spain for the 1949–2009 period (Coch and Medeiro, 2016). Nevertheless, pan-European empirical studies clearly added to the collection of national studies by confirming that these tendencies are part of coherent patterns of change that cover a much larger region. Despite large gaps between the catchments with near-natural streamflow records, a broad continental-scale pattern of change was found that appeared congruent with the hydrological responses expected from future climatic changes, as projected by climate models.

1.2.3.2 Trends in modelled runoff

The simulated trends in annual runoff analysed by Stahl et al. (2012) more clearly demonstrated the pronounced continental dipole pattern – of positive trends in western and northern Europe and negative trends in southern and parts of Eastern Europe, as already suggested by the trends in the observations. Overall, positive trends in annual streamflow appeared to reflect the marked wetting trends of the winter months, whereas negative annual trends resulted primarily from a widespread decrease in streamflow in the spring and summer months, consistent with the decrease in summer low flow in large parts of Europe.

Figure 1.2.2 shows the trends in summer climate (temperature and precipitation), and the summer low flows derived from the modelled runoff over the modelling period 1962–2000. According to the simulations, low flows decreased in southern Europe, parts of central and eastern Europe, Denmark, southern Norway, Sweden, and in some areas of the United Kingdom. July and August temperatures have increased in all these regions, suggesting an increase in evapotranspiration as a contributing cause. Some of the positive summer low flow trends coincide with regions that have seen precipitation increases in June and July.

The different models agreed on the predominant continental-scale pattern of trends, but disagreed on the magnitude and even the direction of trends in some areas – particularly in the transition zones between regions with increasing and decreasing runoff trends, in complex terrain with a high spatial variability and in snow-dominated regimes. Model estimates are most reliable in reproducing observed trends in annual runoff and winter runoff. Modelled trends in runoff during the summer months, spring (for snow-influenced regions) and autumn, as well as trends in summer low flow, were more variable – both among models and in the spatial patterns of agreements between models and observations. The ensemble mean of these trends compared best to the observations (Figure 1.2.3).

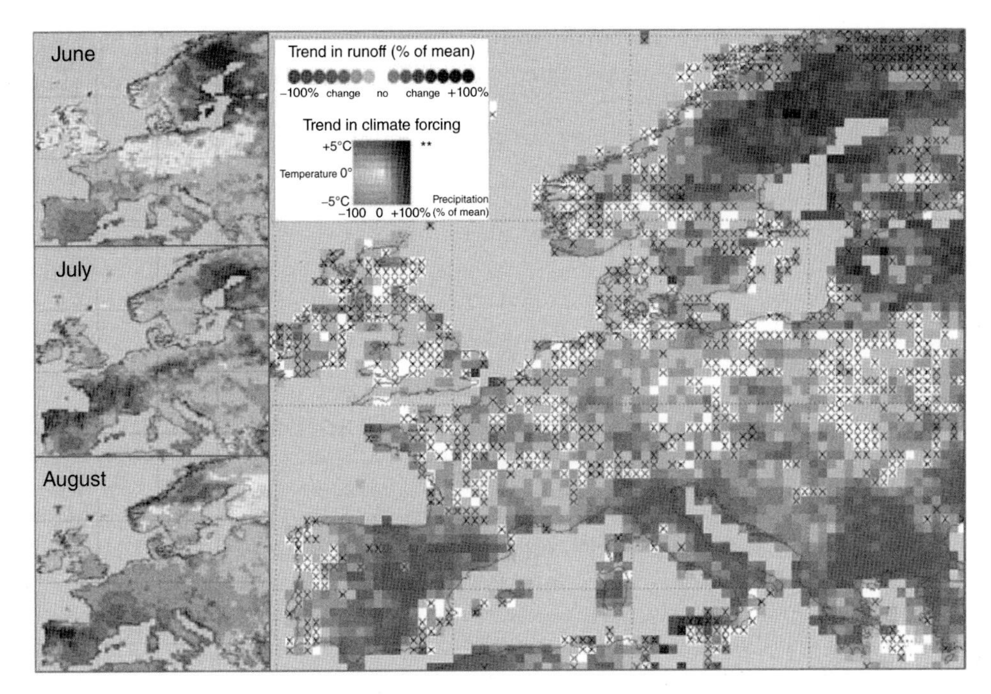

Figure 1.2.2 Trends in summer hydroclimate and modelled low flows in Europe over the period 1962–2000. Left: precipitation and temperature trends from the WATCH Forcing Dataset (www.eu-watch.org/data_availability) for June (upper), July (middle) and August (lower). Right: Low flow (AM(7)) trends for the summer half-year from the WATCH multi-model ensemble (from Stahl et al., 2012). Crosses: Less than three-quarters of the models agree on the sign of the trend. *Source*: Reproduced from Stahl et al. (2014). (*See colour plate section for the colour representation of this figure.*)

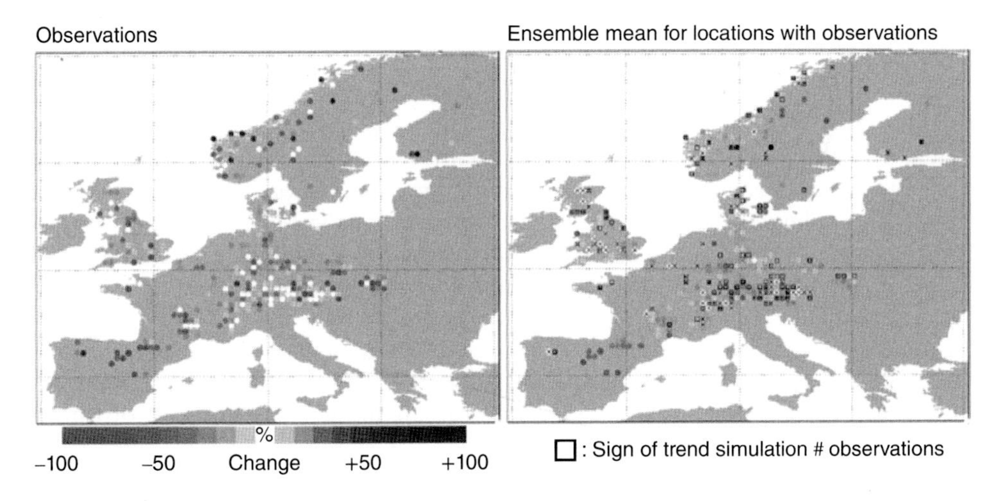

Figure 1.2.3 Spatial distribution of 7-day summer low flow trends from observed streamflow records (left) and from the model ensemble mean (right). As with Figure 1.2.2, crosses denote that less than three-quarters of the models agree on the sign of the trend. *Source*: From Stahl et al. (2012). (*See colour plate section for the colour representation of this figure.*)

In summary, it appears that the multi-model ensemble mean agreed best with observations in northwest Europe, where advective rain from the North Atlantic is the driving meteorology. The findings of increasing flows in this area agree with observed changes in the meteorology. However, trends from models and observations agree less well in places where changes in hydrological characteristics are due to changing snow processes and storage depletion, or where catchment characteristics at smaller scales matter for the timing and magnitude of events. Hence, in these cases, models generally less reliably represent hydrological droughts, which strongly depend on these catchment hydrological processes.

1.2.3.3 Influence of decadal-scale variability on trends in long streamflow records

It is well known that derived trends are sensitive to the chosen period. Thus, illustrating these sensitivities to the period of record on a regional scale can guide the conclusions to be drawn. The study by Hannaford et al. (2013) provides such a long-term context for the trend analyses presented in Section 1.2.3.1. In this study, records were first clustered into five regions that were relatively homogenous in terms of interdecadal variability of annual mean flow (Figure 1.2.4). The LOESS-smoothed cluster time series illustrate that the annual mean flow series are fairly homogeneous within each region and show variability on a broadly decadal scale. As the clustering was performed on annual flows, the annual 7-day minimum flow series are somewhat less homogeneous, but are still indicative of decadal variability for the majority of catchments within each cluster.

The cluster averages (unsmoothed) were then subjected to multi-temporal trend analysis with varying start and end dates. The resulting trend patterns differ by region (Figure 1.2.5). Some parts of the pattern for annual low flows resemble those of annual mean flows, but others show considerable differences.

Negative low flow trends appear to be more common for shorter, late-starting time periods across most of the domain, except for Northern Coastal. In the Northern region, low flow trends differ from the annual flow trends in particular. Decreasing low flows in northern and central East Europe are rather stable for different starting years when ending in the most recent decade of the analyses. For other regions low flow trends are more sensitive to the starting period. Monthly flow trends (not shown here) provide an even more complex and differentiated trend pattern for different start and end dates.

Hannaford et al. (2013) hence posed the question as to how representative the previously reported trends – for example, from Stahl et al. (2010; 2012) (Section 1.3.2) – were. They caution that the dipole trend pattern of drying in the south and east of Europe and wetting in the north and west may reverse for trends of other time periods. The general drying trend for the east and south, however, appears to be rather stable when including post-1990 records. This is not the case for other regions, particularly western Europe. Here, trends are less clear when starting with the drier 1940s. In addition, several studies have found a link of streamflow to the decadal variability component of the

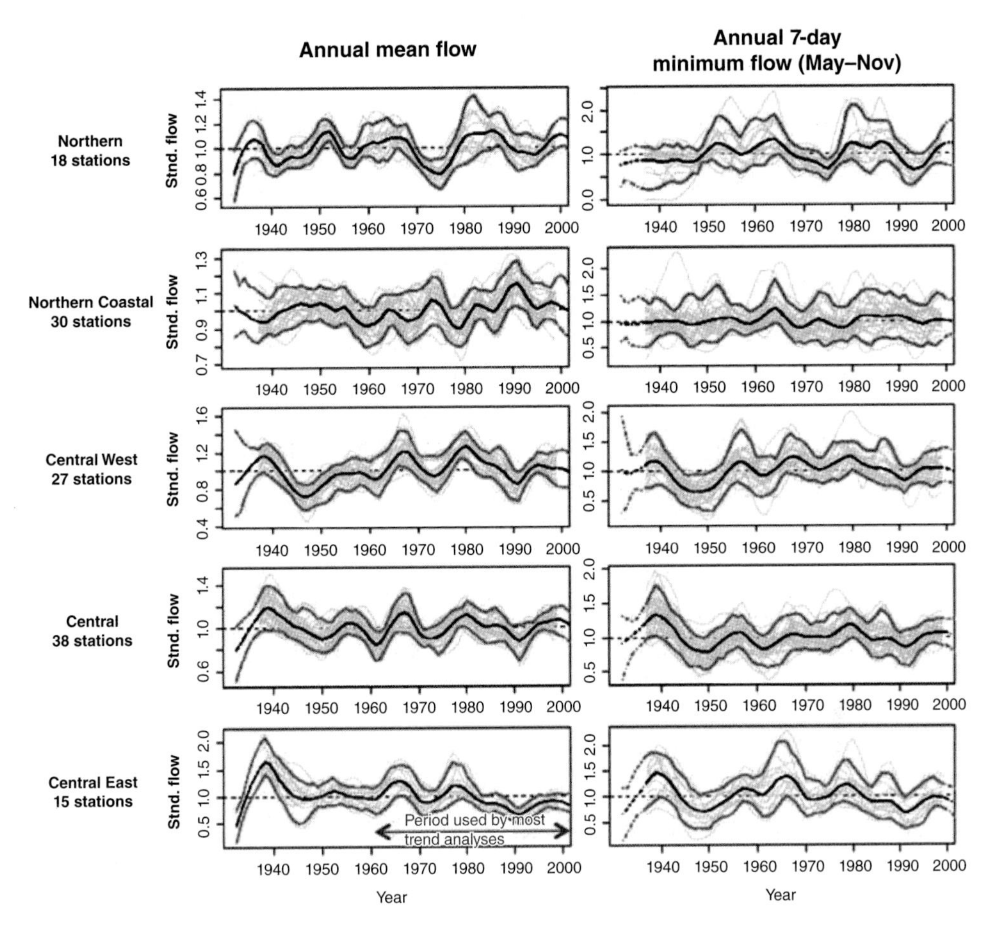

Figure 1.2.4 Interdecadal variability of streamflow records using LOESS smoothing. Grey lines show individual station series; black line shows the cluster average smoothed series for each cluster (region). Red lines show the 5th and 95th percentiles of the combined cluster member series. *Source*: From Hannaford et al. (2013). (*See colour plate section for the colour representation of this figure.*)

North Atlantic Oscillation (NAO) – for example, with higher correlation to the NAO index than to time (Giuntoli et al., 2013).

Overall, the multi-temporal analysis cautioned that trends for any fixed period should not be extrapolated. Long-term fluctuations may be driven by multiple competing influences. The general climate change signal is one of them, overlain by decadal ocean–atmosphere oscillation signals. The work provided some preliminary attribution of these drivers, via the NAO, but called for more detailed work. Other studies have subsequently revealed the many complex, interacting, large-scale drivers on European drought – both proximal North Atlantic drivers and more global-scale phenomena (e.g. Ionita et al., 2015; Folland et al., 2015; Kingston et al., 2015).

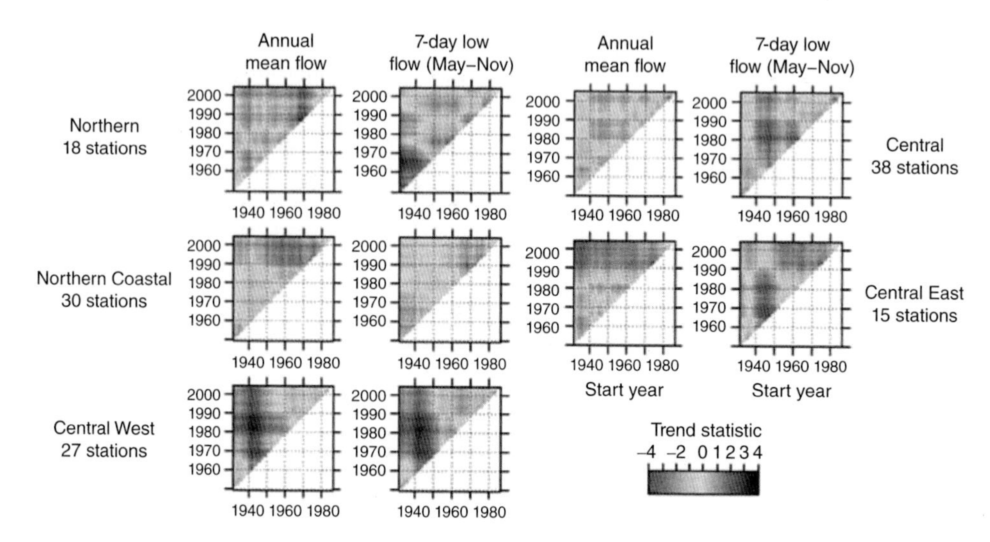

Figure 1.2.5 Multi-temporal trend analysis for annual mean and annual 7-day minimum flow, with x-axis showing start year and y-axis showing end year of the period for which the displayed trend statistic is shown. The Mann-Kendall test applied to all start and end years, and the corresponding pixel is coloured according to the resulting Z statistic, with red representing negative trends and blue representing positive trends. *Source*: Modified from Hannaford et al. (2013). (*See colour plate section for the colour representation of this figure.*)

1.2.4 Discussion

A number of trend studies have been carried out globally, at the European scale and at smaller (national or regional) scales, analysing drought trends in indices other than those of hydrological drought, most often the *Standardised Precipitation Index* (SPI) and *Standardised Precipitation–Evapotranspiration Index* (SPEI). Generally, the trends of hydro-meteorological variables conclude on the increased severity and frequency of droughts in southern and southeastern Europe. Figure 1.2.6 shows drought trends in a combination of indices presented by Spinoni et al. (2015) that illustrate this pattern, but also demonstrate a large spatial variability (see also Spinoni et al., 2017). Gudmundsson and Seneviratne (2015) found a tendency for decreased drought frequency in northern Europe, and a likely increase in drought frequency in selected southern regions to be more pronounced for longer accumulation periods in the standardised indices. It is worth noting that more widespread drying trends are seen for the SPEI due to the broadly increasing air temperature, which has increased potential evapotranspiration (Stagge et al., 2016; Spinoni et al., 2017).

Trends in drought indices that include estimates of evapotranspiration, such as the Palmer drought severity index (PDSI), have been providing apparently conflicting results of how drought is changing under climate change (Sheffield et al., 2012; Dai, 2013). In these and other global studies, the Mediterranean region stands out as a drought hotspot undergoing significant changes (Greve et al., 2014; Orlowsky and Seneviratne, 2013). Also, attendant increases in hydrological drought severity were attributed to be driven primarily by temperature increases (e.g. Vicente-Serrano et al., 2014). However, even the severities of drying trends in Southern Europe differ among

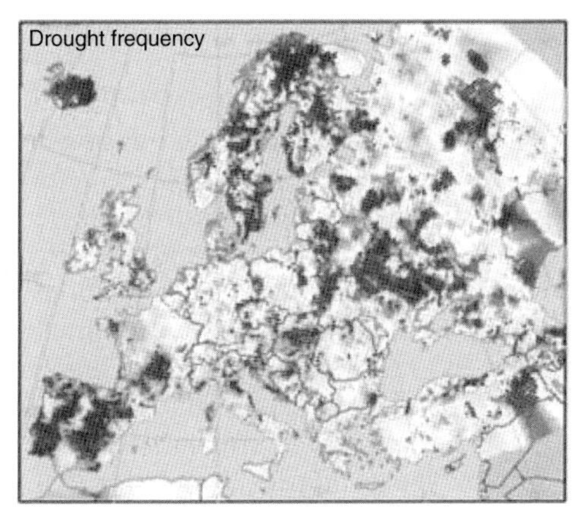

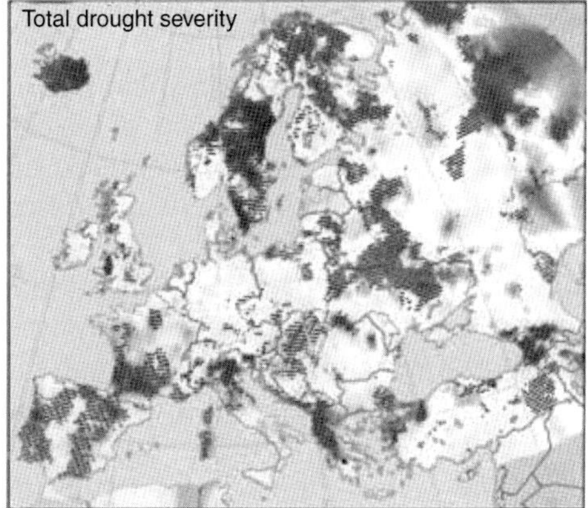

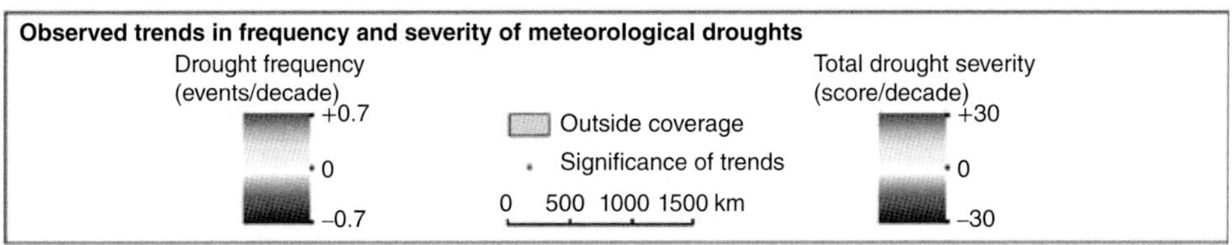

Note: This map shows the trends in drought frequency (number of events per decade; left) and severity (score per decade; right) of meteorological droughts between 1950 and 2012. The severity score is the sum of absolute values of three different drought indices (SPI, SPEI and RDI) accumulated over 12-month periods. Dots show trends significant at the 5% level.

Source: Adapted from Spinoni, Naumann, Vogt et al., 2015.

Figure 1.2.6 Trends in frequency and severity of meteorological droughts. *Source*: From Spinoni et al. (2015). Reproduced from EEA (2017). (*See colour plate section for the colour representation of this figure.*)

the studies. Reasons include, in particular, the exaggeration of the temperature influence on evapotranspiration estimates, considerable discrepancies among precipitation datasets at the continental and global scale as well as the previously discussed question of long-term variability for trend analyses (Trenberth et al., 2014). Similarly, as previously discussed, precipitation has a strong linkage to decadal modes of atmosphere–ocean variability that influence the findings.

Clear continental patterns that mostly result from individual meteorological variables or modelled hydrology are somewhat in contrast with the variability of observed trends in derived indices and hydrological drought indices such as summer low flows, as reviewed in Section 1.2.3. Low flow trends are much less consistent and show both decreases and increases with rare statistical significance. Many studies cover a period that starts in the 1960s or 1970s, consistent with the timing of the widespread expansion of many national hydrometric networks. The impact of decadal variability articulated by Hannaford et al. (2013) means that, in some countries, positive trends in low flows are 'hard coded', as a majority of records start in the late 1960s or early 1970s during a drier period, followed by increases that coincide with a shift towards a more prominent, positive NAO – as discussed for the United Kingdom (Hannaford and Marsh, 2006) and Ireland (Murphy et al., 2013; Nasr and Bruen, 2017). This network expansion is particularly the case for smaller basins that tend to be less influenced by regulation, and hence more suited for use in climate-sensitivity studies.

Another important factor explaining the variability in trends in low flows is that low flow is controlled by the capacity of the catchment to store and release water during dry weather. The spatial diversities of these storage-and-release processes – that are in turn strongly linked to heterogeneity in landscape properties, most notably hydrogeological characteristics – influence the response of streamflow to prolonged meteorological drought. The results of studies suggest that particularly the headwater catchments that are highly variably regarding topography, vegetation, soils, and geology may not directly follow seasonal climate trends. In high mountains and northern latitudes, the relative influence of meteorological drivers and the role of snow and ice assume prominent importance. A study of trends by Bard et al. (2015) in a range of streamflow indices across the entire Alpine Region confirmed this by separating the trend signals by hydrological regimes. Decreasing trends in the severity of winter droughts were mainly found for glacial- and snowmelt-dominated regimes, whereas trends differed for mixed snowmelt–rainfall regimes.

River regulation is poorly documented in many countries, and therefore human influences present another challenge for low flow assessments, both with models and observations. In addition, gradual changes in land use (e.g. afforestation) may influence the water balance, and thus low flow. Some observation-based studies have used data from near-natural reference networks, but most modelling, particularly for large rivers, needs to incorporate flow management and regulation. While reference networks are vital for discerning climate signals, it is for managed and impacted rivers that the greatest utility is to be gained from projections. Hydrometric networks and datasets need to be improved with respect to the metadata on these influences that may confound climate change attribution.

1.2.5 Conclusions – Future needs

1.2.5.1 Conclusions

Studies on past trends in drought indices, and in particular those of hydrological drought, have shed light on the changing nature of the drought hazard in Europe. In particular, they have revealed considerable local variability within the general large-scale pattern of drying trends most consistently found for southern and eastern Europe. The studies reviewed here also illustrate the many methodological aspects that affect trend assessments: data and catchment selection criteria, considered season, selected time period, measure of change, and the index analysed. Stahl et al. (2014) suggest that, among these aspects, the choice of methodology for analysing changes and trends, such as the low flow index or trend test, appears to matter less than the initial data choices, thus highlighting the need for better data documentation, availability, and consistent observation networks.

For model-based assessments of future changes, the knowledge base is rapidly increasing, but ensemble approaches have only recently started to quantify the different sources of error and uncertainty. With considerable differences between GCMs in projected climatic changes (particularly for precipitation), further up in the model chain, changes in low flows and hydrological drought also vary considerably. In addition, comparisons of multiple hydrological models suggest that the uncertainty from the use of different hydrological models can be high as well, especially for low flows – the part of the regime on which catchment storage and release characteristics tend to exert a major control, and where estimation of losses by evapotranspiration can have a strong influence on future change.

1.2.5.2 Future needs

A close monitoring is needed of the trends in all climatic and hydrological variables and derived indices relevant to drought. This monitoring requires that existing gauges with long, climate-sensitive records be maintained and ideally extended via the use of data recovery or reconstruction approaches where appropriate. They need to be provided freely to science to improve trend assessment methodologies. In addition to the climate-sensitive, near-natural records analysed mainly so far, trends in regulated rivers should also be analysed to investigate the degree of mitigation or exacerbation that human impacts have on streamflow droughts. Future research needs to address these aspects, and also aim for better integration across different model types and spatial scales.

Acknowledgements

Funding was provided by the EU WATCH project (WATer and global CHange), contract 036946, and the EU project DROUGHT–R&SPI (Fostering European Drought Research and Science–Policy Interfacing), contract 282769. The study is also a contribution to UNESCO IHP-VII programme (Euro FRIEND-Water project). The authors thank all data providers and the national authorities for the observed streamflow records.

References

Bard, A., Renard, B., Lang, M., Giuntoli, I., Korck, J., Koboltschnig, G., Janza, M., D'Amico, M., and Volken, D. (2015). Trends in the hydrologic regime of Alpine rivers. *Journal of Hydrology* 529: 1823–1837. doi: 10.1016/j.jhydrol.2015.07.052.

Chandler, R. and Scott, M. (2011). *Statistical Methods for Trend Detection and Analysis in the Environmental Sciences*. Chichester, UK: Wiley.

Coch, A. and Mediero, L. (2016). Trends in low flows in Spain in the period 1949–2009. *Hydrological Sciences Journal* 21: 568–584. doi: 10.1080/02626667.2015.1081202.

Dai, A. (2013). Increasing drought under global warming in observations and models. *Nature Climate Change* 3: 52–58.

EEA (European Environment Agency). (2017). Climate change, impacts and vulnerability in Europe 2016. An indicator-based report. *EEA Report No 1/2017*. doi: 10.2800/534806.

Folland, C.K., Hannaford, J., Bloomfield, J.P., Kendon, M., Svensson, C., Marchant, B.P., Prior, J., and Wallace, E. (2015). Multi-annual droughts in the English lowlands: a review of their characteristics and climate drivers in the winter half-year. *Hydrology and Earth System Sciences* 19: 2353–2375. doi: 10.5194/hess-19-2353-2015.

Greve, P., Orlowsky, B., Mueller, B., Sheffield, J., Rechstein, M., and Seneviratne, S.I. (2014). Global assessment of trends in wetting and drying over land. *Nature Geoscience* 7: 716–721.

Gudmundsson, L. and Seneviratne, S.I. (2015) European drought trends. *IAHS* 369: 75–79.

Gudmundsson, L., Tallaksen, L.M., Stahl, K., Clark, D.B., Dumont, E., Hagemann, S., Bertrand, N., Gerten, D., Heinke, J., Hanasaki, N., Voss, F., and Koirala, S. (2012a). Comparing large-scale hydrological model simulations to observed runoff percentiles in Europe. *Journal of Hydrometeorology* 13: 604–620. doi:10.1175/JHM-D-11-083.1.

Gudmundsson, L., Wagener, T., Tallaksen, L.M., and Engeland, K. (2012b). Evaluation of nine large-scale hydrological models with respect to the seasonal runoff climatology in Europe. *Water Resources Research* 48: W11504. doi:10.1029/2011WR010911.

Giuntoli, I., Renard, N., Vidal, J.P., and Bard, A. (2013). Low flows in France and their relationship to large-scale climate indices. *Journal of Hydrology* 482: 105–118.

Haddeland, I., Clark, D.B., Franssen, W., Ludwig, F., Voss, F., Arnell, N.W., Bertrand, N., Best, M., Folwell, S., Gerten, D., Gomes, S., Gosling, S.N., Hagemann, S., Hanasaki, N., Harding, R., Heinke, J., Kabat, P., Koirala, S., Oki, T., Polcher, J., Stacke, T., Viterbo, P., Weedon, G.P., and Yeh, P. (2011). Multi-model estimate of the global terrestrial water balance: Setup and first results. *Journal of Hydrometeorology* 12 (5): 869–884. doi:10.1175/2011JHM1324.1.

Hannaford, J. and Marsh, T.J. (2006). An assessment of trends in UK runoff and low flows using a network of undisturbed catchments. *International Journal of Climatology* 26: 1237–1253. doi: 10.1002/JOC.1303.

Hannaford, J., Buys, G., Stahl, K., and Tallaksen, L.M. (2013). The influence of decadal-scale variability on trends in long European streamflow records. *Hydrology and Earth System Sciences* 17: 2717–2733. doi: 10.5194/HESS-17-2717-2013.

Hannah, D.M., Demuth, S., Van Lanen, H.A.J., Looser, U., Prudhomme, C., Rees, G., Stahl, K., and Tallaksen, L.M. (2011). Large-scale river flow archives: importance, current status and future needs. *Hydrological Processes* 25: 1191–1200.

Ionita, M., Boroneanţ, C., and Chelcea, S. (2015). Seasonal modes of dryness and wetness variability over Europe and their connections with large scale atmospheric circulation and global sea surface temperature. *Climate Dynamics* 45: 2803. doi: 10.1007/S00382-015-2508-2.

Kendall, M.G. (1975). *Rank Correlation Methods*. London: Griffin, 202 pp.

Kingston, D.G., Stagge, J.H., Tallaksen, L.M., and Hannah, D.M. (2015). European-scale drought: understanding connections between atmospheric circulation and meteorological drought indices. *Journal of Climate* 28: 505–516. doi: 10.1175/JCLI-D-14-00001.1.

Mann, H.B. (1945). Nonparametric tests against trend. *Econometrica* 13: 245–259.

Murphy, C., Harrigan, S., Hall, J., and Wilby, R.L. (2013). Hydrodetect: The Identification and Assessment of Climate Change Indicators for an Irish Reference Network of River Flow Stations. *Climate Change Research Programme (Ccrp) 2007–2013 Report Series No. 27. ISBN 978-1-84095-507-1. Technical Report*. Environmental Protection Agency, Co. Wexford.

Nasr, A. and Bruen, M. (2017). Detection of trends in the 7- ay sustained low-flow time series of Irish rivers. *Hydrological Sciences Journal* 62 (6): 947–959. doi: 10.1080/02626667.2016.1266361.

Orlowsky, B. and Seneviratne, S.I. (2013). Elusive drought: uncertainty in observed trends and short- and long-term CMIP5 projections. *Hydrology and Earth System Sciences* 17: 1765–1781.

Sheffield, J., Wood, E.F., and Roderick, M.L. (2012). Little change in global drought over the past 60 years. *Nature* 491: 435–438.

Spinoni, J., Naumann, G., and Vogt, J. (2015). Spatial patterns of European droughts under a moderate emission scenario. *Advances in Science and Research* 12: 179–186.

Spinoni, J., Naumann, G., and Vogt, J.V. (2017). Pan-European seasonal trends and recent changes of drought frequency and severity. *Global and Planetary Change* 148: 113–130. doi: 10.1016/J.GLOPLACHA.2016.11.013.

Stagge, J.H., Kingston, D., Tallaksen, L.M., and Hannah, D. (2016). Diverging trends between meteorological drought indices (SPI and SPEI). *Geophysical Research Abstracts* 18: EGU2016-10703-1.

Stahl, K., Hisdal, H., Hannaford, J., Tallaksen, L.M., Van Lanen, H.A.J., Sauquet, E., Demuth, S., Fendekova, M., and Jodar, J. (2010). Streamflow trends in Europe: evidence from a dataset of near-natural catchments. *Hydrology and Earth System Sciences* 14: 2367–2382. doi:10.5194/hess-14-2367-2010.

Stahl, K., Tallaksen, L.M., Hannaford, J., and Van Lanen, H.A.J. (2012). Filling the white space on maps of European runoff trends: estimates from a multi-model ensemble. *Hydrology and Earth System Sciences* 16: 2035–2047. doi:10.5194/hess-16-2035-2012.

Stahl, K., Vidal, J.P., Hannaford, J., Prudhomme, C., Laaha, G., and Tallaksen, L. (2014). Synthesizing changes in low flows from observations and models across scales. In: *Hydrology in a Changing World: Environmental and Human Dimensions* (ed. T.M. Daniell, H.A.J. Van Lanen, S. Demuth, G. Laaha, E. Servat, G. Mahe, J.-F. Boyer, J.-E., Paturel, A. Dezetter, and D. Ruelland), 30–35. Wallingford: IAHS Publ. No. 363.

Stahl, K., Kohn, I., Blauhut, V., Urquijo, J., De Stefano, L., Acácio, V., Dias, S., Stagge, J.H., Tallaksen, L.M., Kampragou, E., Van Loon, A.F., Barker, L.J., Melsen, L.A., Bifulco, C., Musolino, D., de Carli, A., Massarutto, A., Assimacopoulos, D., and Van Lanen, H.A.J. (2016). Impacts of European drought events: insights from an international database of text-based reports. *Natural Hazards and Earth System Sciences* 16: 801–819.

Tallaksen, L.M. and Van Lanen, H.A.J. (2004). *Hydrological Drought: Processes and Estimation Methods for Streamflow and Groundwater*. No. 48 in Development in Water Science. Amsterdam: Elsevier Science B.V.

Tallaksen, L.M., Stagge, J.H., Stahl, K., Gudmundsson, L., Orth, R., Seneviratne, S.I., Van Loon, A.F., and Van Lanen, H.A.J. (2015). Characteristics and drivers of drought in Europe – a summary of the DROUGHT–R&SPI project. In: *Drought: Research and Science–Policy Interfacing* (ed. J. Andreu, A. Solera, J. Paredes-Arquiola, D. Haro-Monteagudo, and H.A.J. Van Lanen). Boca, Raton, London, New York, and Leiden: CRC/Balkema Publishers.

Theil, H. (1950). A rank-invariant method of linear and polynomial regression analysis. *Indagationes Mathematicae* 12: 85–91.

Trenberth, K.E., Dai, A., van der Schrier, G., Jones, P.D., Barichivich, J., Briffa, K.R., and Justin Sheffield, J. (2014). Global warming and changes in drought. *Nature Climate Change* 4: 17–22. doi: 10.1038/NCLIMATE2067.

Vicente-Serrano, S.M., Lopez-Moreno, J.I., Begueria, S., Lorenzo-Lacruz, J., Sanchez-Lorenzo, A., Garcia-Ruiz, J.M., Azorin-Molina, C., Moran-Tejeda, E., Revuelto, J., and Trigo, R. (2014). Evidence of increasing drought severity caused by temperature rise in southern Europe. *Environmental Research Letters* 9: 044001. doi:10.1088/1748-9326/9/4/044001.

Viglione, A., Borga, M., Balanbanis, P., and Bloschl, G. (2010). Barriers to the exchange of hydrometeorological data in Europe: results from a survey and implications for data policy. *Journal of Hydrology* 394: 63–77. doi:10.1016/j.jhydrol.2010.03.023.

Whitfield, P., Burn, D., Hannaford, J., Higgins, H., Hodgkins, G.A., Marsh, T.J., and Looser, U. (2012). Hydrologic reference networks I. The status of national reference hydrologic networks for detecting trends and future directions. *Hydrological Sciences Journal* 57: 1562–1579.

1.3 Historic Drought from Archives: Beyond the Instrumental Record

Emmanuel Garnier
French National Centre for Scientific Research, Besancon, France

1.3.1 Introduction

Before our current instrumental record of the second part of the twentieth century, historic droughts, because of their impacts on society, left multiple indications in the archives of the last 500 years. For the record, it is necessary to remind ourselves that the general term 'drought' covered different notions. In the most frequent meaning, the term was synonymic with a pluviometric deficit and an extreme climate event.

It is thus important to understand that, for the historian, droughts are viewed through the 'social signature' of these extreme events, as recorded over the centuries in the European archives. They can thus be appreciably different from the definitions used by hydrologists or climatologists, and need to be assessed and categorised according to another metric – for example, the HSDS. This to obtain comparative data series (Garnier, 2015).

In this chapter, we will explore how the combination of textual and rare instrumental data recorded in the archives since the sixteenth century can improve our knowledge of European droughts. We will report on several cases from the United Kingdom, France, and Upper Rhine valley to better understand the variations in these climatic extreme events during the last 500 years, and their social and economic impacts on European ancient societies.

1.3.2 Methodology

1.3.2.1 The historical material

Because of the unpredictable character and the absence of civil services specially dedicated to the study of these extreme events before the middle of the nineteenth century,

Drought: Science and Policy, First Edition.
Edited by Ana Iglesias, Dionysis Assimacopoulos, and Henny A.J. Van Lanen.

historians have to make maximum use of the entire corpus of sources. The information we need is often hidden at random in the margins of some documentation, and we cannot afford to neglect any type of archive if we hope to reconstruct long and relatively reliable chronologies (Garnier, 2010a).

Particularly useful are diaries drafted by private persons (priests, middle-class persons, aristocrats) and municipal chronicles. Often very sensitive to the extreme events that engender a disaster, they provide an integrated approach to drought by combining visual observations (water heights at staff gauges on bridges), the phenology (state of the vegetation, the fires), the prices on markets, and even its social expression (scarcity, religious processions, riots). From the 1750s, some diaries became real meteorological journals because of the new passion for meteorology in Europe. These diaries include daily meteorological data (Figure 1.3.1).

Up to the eighteenth century, droughts were often considered to be demonstrations of the 'god's wrath' (*ira dei* in Latin), and the Roman Catholic Church is hence a reliable source of information. Thanks to records of the religious processions organised by the Catholic Church, historians have been able to glean some relatively homogeneous series on archival and historic plans, because they emanate from the same lay or religious institution that registered them over long periods. These religious ceremonies allow the reconstruction of historic series generally covering the period between 1500 and 1800, and sometimes even beyond in case of Spain. The Roman Catholic Church or the municipal authorities ordered these qualified ceremonies of rogations (*rogativas*)

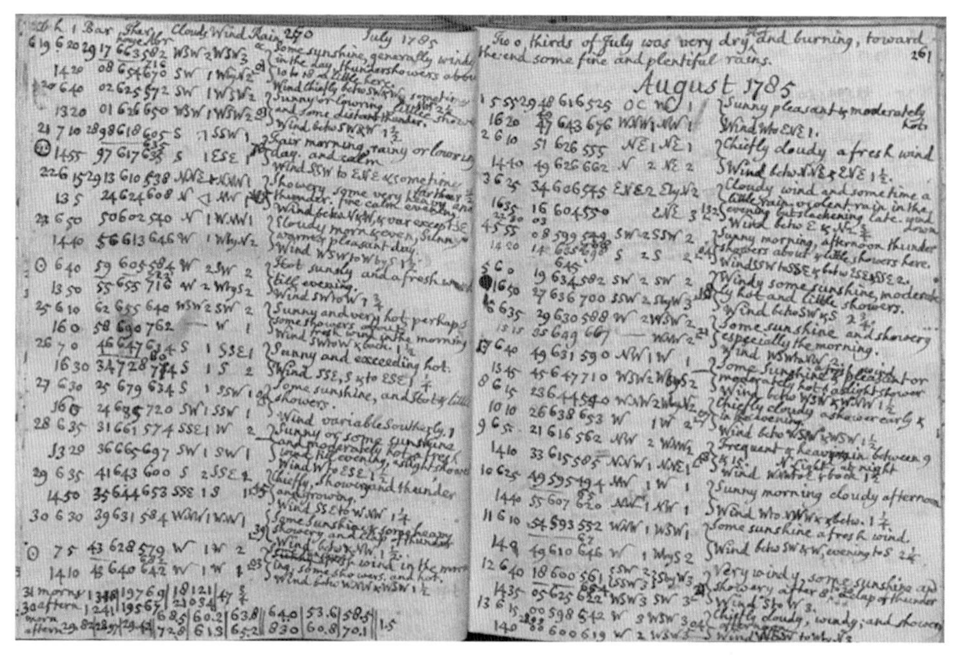

Figure 1.3.1 Extract of Thomas Barker's meteorological journal (Cambridgeshire, United Kingdom) for the summer of the big European drought of 1785. The page details the temperatures and precipitation, as well as the state of the sky, during August. *Source*: National Meteorological Library and Archives b1488027, Met Office, Exeter.

in Spain, or processions in Portugal and France, which were meant to maintain and not to endanger the established order or socioeconomic balance. In the case of droughts, processions were organised *pro pluvia* – literally, 'for the rain'.

The municipal archives contain the registers of the municipal deliberations and the accounts. These documents frequently begin by the end of the fifteenth century. Deliberations and municipal accounts constitute an inexhaustible deposit of climatic data. Meteorological information is omnipresent in these registers and arises from the understandable desire to anticipate the risks of breaks in supplies, of diseases, and riots. Hence, any sustained drought sparks off a discussion within the government of the city. That is why the state or municipal authorities intervened by using diverse mechanisms such as processions, price controls, requisitions of wheat, and imports of wheat.

Unfortunately, it is necessary to wait for the beginning of the nineteenth century to have the first instrumental data on pluviometry or water flows. They result from the creation of scientific societies – such as the Royal Society of London, the Royal Academy of Sciences of Paris or the *Societas Meteorologica Palatina* of Mannheim in Germany (Figure 1.3.2) – or come from private companies – such as the Bedford Level Company in the English Fens or the French 'Canal du Midi'. Both hired engineers especially for the task of monitoring rivers. For this purpose, they also produced the early data series.

1.3.2.2 Rebuild the droughts of the past

To address the lack of reliable instrumental data before 1750, the contents of archives offer two methodological solutions to estimate these natural events, for which we have only textual descriptions. The first solution is to use all the chronological mentions of a drought appearing in archives. Concretely, it is noting, for example, the first mention of a religious procession *pro pluvia*, then the municipal acts, which evokes the drying up of the public fountains, the ban on drawing water from certain places, the lay-off of wheat mills and, in the most extreme cases, the problems of supply of wheat and wood via the waterways. Naturally, this list is not exhaustive. However, the time of mention of these indications in the archives allows proposing the duration in days for the very great majority of the droughts of the past.

Another methodological choice that can complete the evaluation of the duration consists of creating an indexed scale of severity directly built according to the descriptive contents of the drought. Naturally, it results from a systematic inventory of the impacts engendered by the extreme event on societies. Thus, the historian can observe the chronology of an event, which is well recorded in archives. An HSDS, which is an index that varies between 1 and 5, as shown in Table 1.3.1, can be derived from this inventory.

At the level HSDS=1, the absence of precipitation (atmospheric drought) begins to be felt. If this continues, agriculture is affected, and a fall in the levels of water is observed in records (HSDS = 2). When HSDS = 3 or 4, the limitation of resources becomes important. The situation deteriorates as the absence of precipitation affects societies, resulting in high price of farm products, lay-off of wheat mills, and degradation of ecosystems (HSDS = 4). The paroxysm of the social crisis is reached with HSDS = 5, when the drought becomes exceptional, with a very clear deterioration in living conditions and an increase of the social tensions centred around who has access to water.

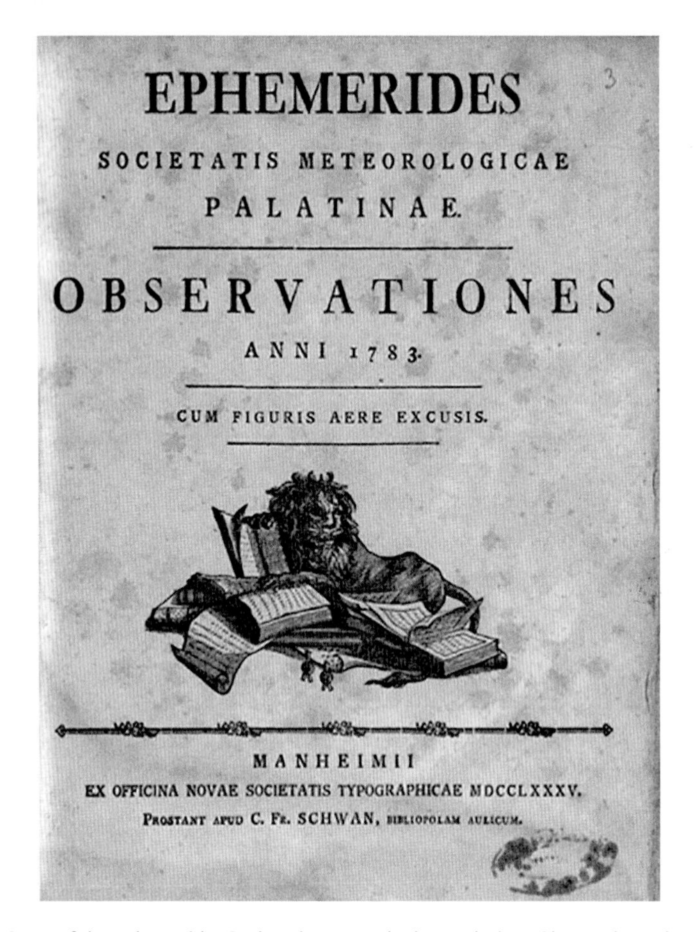
EPHEMERIDES
SOCIETATIS METEOROLOGICAE
PALATINAE.
OBSERVATIONES
ANNI 1783.
CUM FIGURIS AERE EXCUSIS.
MANHEIMII
EX OFFICINA NOVAE SOCIETATIS TYPOGRAPHICAE MDCCLXXXV.
PROSTANT APUD C. FR. SCHWAN, BIBLIOPOLAM AULICUM.

Figure 1.3.2 Cover of the *Ephemerides Societatis Meteorologicae Palatinae Observationes* (1783), Mannheim. *Source*: Bayerische Staatsbibliothek München, Bl., 424 S., (9) Bl.

Table 1.3.1 Description of HSDS (sixteenth–eighteenth centuries).

Index	Descriptions
5	Exceptional drought: no supply possible, shortage, sanitary problems, very high prices of wheat, forest fires
4	Severe low-water marks: navigation impossible, lay-off of wheat mills, search for new springs, forest fires, death of cattle
3	General low waters (navigation difficult) and low water reserves
2	Local low waters of rivers, first effects on vegetation
1	Absence of rainfall: rogations, prayers, evidences in texts
−1	Insufficient qualitative and quantitative information, but the event is kept in the chronological reconstruction

1.3.3 United Kingdom

In the case of the droughts in the United Kingdom, multiple archives and printed sources were exploited in Cambridge and London, particularly the remarkable diary of

Samuel Pepys, Secretary of the Admiralty under the reign of Charles II and James II (Cooper, 1827; Smith, 1825).

1.3.3.1 Chronological variation and severity of the UK droughts

The available UK archives allowed us to count 42 droughts of variable severity between 1500 and 2014 according to HSDS. As shown in Figure 1.3.3, the chronology shows very strong disparities from one century to the next in terms of frequencies and severity.

The distribution of these extreme events over a period of 50 years (Figure 1.3.4) reveals the big dry phases in the history of the United Kingdom since the 1500s. We observe a rather clear break between the period prior to 1700 and the last three centuries. The first period has only 11 events, while we count 31 for the last 300 years. This low number before 1700 can be explained in two ways. The first reason would be the less-rich historic documentation during the period, including the lack of archives. The second reason, undoubtedly the most relevant, is connected to the climatic context of the period. Indeed, the sixteenth and seventeenth centuries correspond in Europe to a very severe phase of the Little Ice Age (1300–1850), which is characterised by much wetter and cooler seasons, particularly between 1540–1640 and 1683–1693 (Le Roy Ladurie, 1972; Garnier, 2010b).

After the 1700s, the drought phenomenon seems clearly to become more frequent. The first dry episode corresponds to the years 1700–1750, during which continental Europe experienced different weather conditions. From 1705 to the end of 1730s, the climate was hotter and drier, even if 1709 remains until today as the year of the 'Great Winter' in Europe. Moreover, several droughts, such as those of 1714, 1715 and 1740, began during very cold and very dry winters. The second turning point occurs after 1800 with a greater frequency of droughts, very likely the beginning of the global warming from around 1830–1850. From then on, in every period of 50 years, six–seven dry events can be found, although an increase in the phenomenon might have occurred from the 1950s onwards.

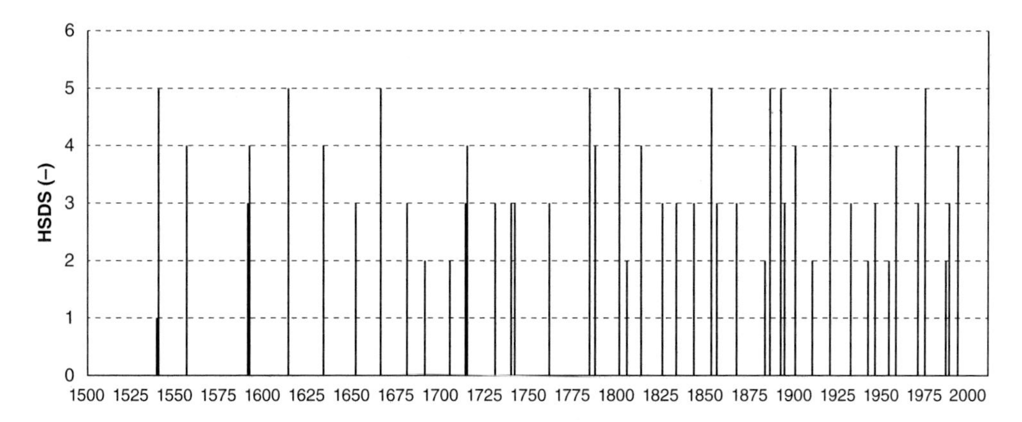

Figure 1.3.3 Chronology and severity of droughts in the United Kingdom between 1500 and 2014 according to HSDS (total 42 droughts).

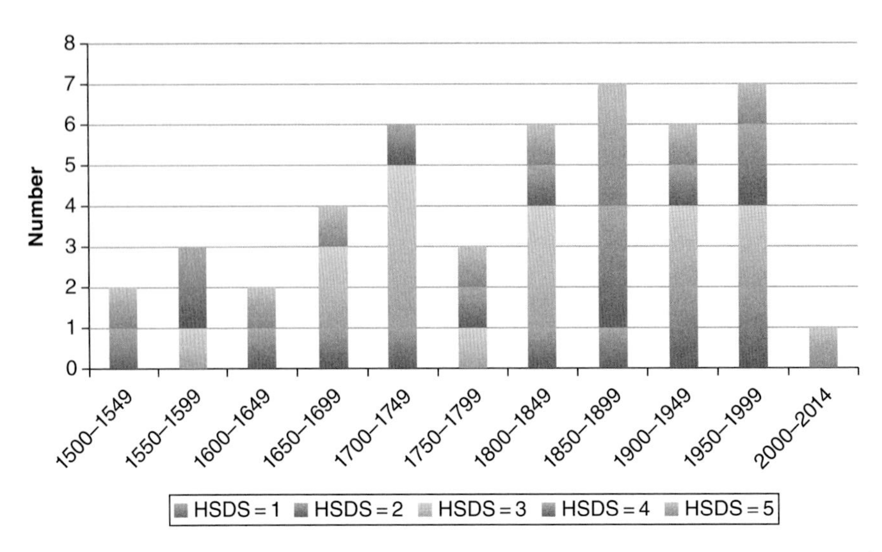

Figure 1.3.4 Distribution and severity of UK droughts by periods of 50 years between 1500 and 2014 according to HSDS. (*See colour plate section for the colour representation of this figure.*)

On the other hand, the chronology and distribution of the droughts according to HSDS does not confirm the previously evoked observations. Hence, the period prior to 1700, which is characterised by a low number of droughts, presents a majority of average severity indications (HSDS = 3). The most severe episodes (HSDS = 5) are found in the second part of the nineteenth century, while the twentieth century seems more moderate, with some notable exceptions (HSDS = 4). These are more numerous between 1950 and 1999, but they are not associated with exceptional droughts (HSDS = 5).

Finally, the monthly pluviometric data from Oxford since the eighteenth century allowed us to compare this series with the droughts estimated according to HSDS (Figure 1.3.5). We observe good similarities between the historic events listed in archives (i.e. HSDS) and the periods of pluviometric deficits. Nevertheless, the droughts with HSDS=5 do not correspond in all cases with the periods of lowest precipitation. Therefore, the extremely severe droughts of 1785 and 1976 (both HSDS = 5) do not coincide with the lowest rain levels in Oxford. This discrepancy can be explained by the local pluviometric conditions being different from the rest of the country, or by other meteorological factors, such as the wind, which could have aggravated the drought.

1.3.3.2 The most extreme events in a 500-year period

From the outset, it is important to specify that a large number of droughts prior to 1800 were systematically associated with epidemics. Up to the great drought of 1666, most of them correspond to the worst UK epidemics of plague (1544–1546, 1592–1593, 1665–1666). After the Great Fire of London in 1666, the droughts coincided more often with epidemics of fevers and smallpox. It would thus seem plausible that low water levels favoured the pollution of water courses, which in turn caused the contagion.

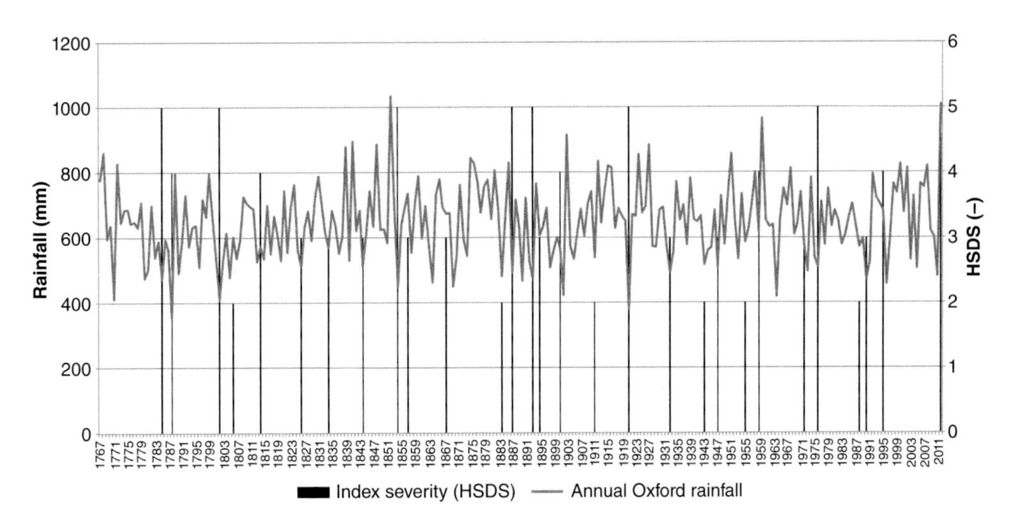

Figure 1.3.5 Comparison of HSDS and the rainfall data from Oxford for UK droughts between 1767 and 2012. *Source*: Meteorological Observations at the Radcliffe Observatory, Oxford. (*See colour plate section for the colour representation of this figure.*)

The 1634 drought In his diary, Samuel Rogers, who was chaplain of Bishop's Stortford, Hertfordshire, says 'it is a sad thing to consider the sad seasons wee have had this year, and this, a most heavy drought, the last yeare in the former part of the summer, which burnt up the grass, scarse raine for a quater together, yet then a michlemasse spring; this yeare in the former part, a singeing, year scorching drought, heavye, gods people mourned, and made the heavens weep; but the lord hath turned raine into a crosse justlye'.

This extract in old English underlines the severity of the drought, but especially reveals the way it was perceived by the population at least till the beginning of the eighteenth century. As often, the extreme event is seen as a sign of god's wrath. Hence, Samuel Rogers declared that 'the Lord is angry with us a most sinful nation'. In reality, the year 1634 was a social and economic disaster. The plague, fevers, and smallpox ravaged the cities of London, Newcastle, Essex, and the ports of the kingdom. At Cambridge, rather than god's wrath, people preferred to explain the drought, which lasted for 3 years, as caused by the draining of the Fens that began at the start of the seventeenth century (Webster and Shipps, 2004; Fuller, 1811).

The disastrous national drought of 1666 According to numerous authors of that time, the drought of 1666 was the cause of the Great Fire of London. Fortunately for the historian, the documentation dealing with this catastrophic climate-related event is particularly rich. These are available in several diaries, of which the best known are those of Samuel Pepys and John Evelyn (Bray, 1901; Fraser, 1905). These reflect exceptional witnesses, because both were direct actors in the event. Samuel Pepys (1633–1703) was the Secretary of the Admiralty and a Member of Parliament, and, later, president of the Royal Society of London, whereas John Evelyn (1620–1706) was a renowned scientist and one of the founders of the Royal Society of London.

Their diaries are indispensable for understanding the history of the 1666 drought and the Great Fire of London. Another remarkable witness was the philosopher John Locke, who studied in Oxford (Christ Church College) at that time. Besides his philosophical works, he also used to record multiple meteorological variables, such as temperature, humidity, atmospheric pressure, wind direction, and wind strength. These historical data are nowadays underestimated, even though they probably constitute the first meteorological reports in the United Kingdom. These data were published by Boyle (1692) (Figure 1.3.6).

In his diary, John Evelyn declared, on 21 October 21 1666, that the season, after a long and extraordinary dryness between July and September, was totally unprecedented. He explains that the dryness began during the plague of 1665 and that it was aggravated by a cold and very dry winter (1666), which was followed by a particularly

(106) *A Register kept by Mr.* Locke, *in* Oxford.

1666 d	*July* h	Th.	Bar.	Hy.	Wind		Weather.
	15	71	29 1		NE	1	Fair.
	17	71	29 1		E	1	Rain.
	21	69	29 2			1	Clouds, Lightning.
24	9	67	29 2		N	1	Clouds.
25	9	67	29 2		N	1	Clouds.
26	6	64	29 3		N	1	Close.
30	23	69	29 3	15			Fair.
31	7	67	29 3	15	NW	0	Fair.
August							
15	10	60	29 2	20			Clouds.
16	9	59	29 2	19	N	1	Fair.
29	21	64	29 2	19			
30	9	59	29 3	19	N	0	Fair.
31	7	61	29 3	20	NE	0	Fair.
	9	61	29 3	20	NE	1	Fair.
September							
1	6	55	29 3	17	NE	1	Fair.
	9	55	29 3	17	E	2	Clouds.
2	9	52	29 3	16	E	3	Clouds.
3	9	49	29 4	16	E	2	Fair.
4	9	47	29 3	15	E	1	Fair.
	13	50	29 3	15	E	2	Dim reddish Sun-shine
	20	51	29 2	14			This unusual Colour of the Air, which without a Cloud appearing, made the Sun-beams of a strange red dim Light, was very remarkable. We had then heard nothing of the Fire of *London*: But it appeared afterwards to be the Smoak of *London* then burning, which driven this way by an *Easterly* Wind, caused this odd Phenomenon.
5	22	49	29 1				
6	9	44	29 2	15	NE	1	Fair.

Figure 1.3.6 Meteorological observations made by the English philosopher John Locke in Oxford during the 1666 drought and the Great Fire of London. *Source*: Bodleian Library, University of Oxford, QC 161 B69 1692.

warm summer. Rain was rare between November 1665 and September 1666. John Locke's meteorological observations (Figure 1.3.6) perfectly reveal the dryness that prevailed then in Oxford. Even if the temperatures did not indicate a particular heat wave, the shortage of precipitation was evident. Between June 24 and September 13 1666, the philosopher recorded only 7 days with precipitation (Figure 1.3.7).

In Scotland, the chronicle of Frasers speaks about a 'hot drought' in May 1666 and indicates that the dryness also became widespread in England (Fraser, 1905). In Oxford, the River Cherwell and River Thames were 'almost dry', said the witnesses of that time, which evoked unemployment of the boatmen, who could not navigate anymore because of the low water. In London, once again, appeals were made to god to put an end to the dryness. The wardens of St. Margaret Church in New Fish Street paid 6 pennies to 'My Lord Mayor officers' for bringing an order to pray for rain. Regrettably, god did not answer their prayers, and, by September 2, London had become a tinderbox ready to ignite.

The socioeconomic impact of the 1785 drought The summer heat wave of 1785 led to a drought of maximal intensity, with disastrous consequences throughout the country. As reported by the English naturalist Gilbert White (1720–1793), it was a severe 'drying exhausting drought', and the kingdom was not more than 'dust'. In another diary, a Norwich mill owner explained in July 1785 that there had not been enough water to turn the wheel of his mill since June 23, and that it had consequently been impossible to grind the wheat. This caused a big increase in bread prices. The drought also affected other industries that were dependent on hydraulic power to function (textile industry, dyeing). The environment did not escape the negative effects of the event either, because there were forest fires in Rothburg Forest (near Newcastle-upon-Tyne), East Hampstead, and Berkshire. The Rothburg Forest fire led to more than 1000 acres (405 ha) of sheep pasture and heath being rendered totally useless.

In the Fens (Cambridgeshire), the summer drought totally dried up the rivers, and a great number of cattle languished and died for want of water (Figure 1.3.8). The situation

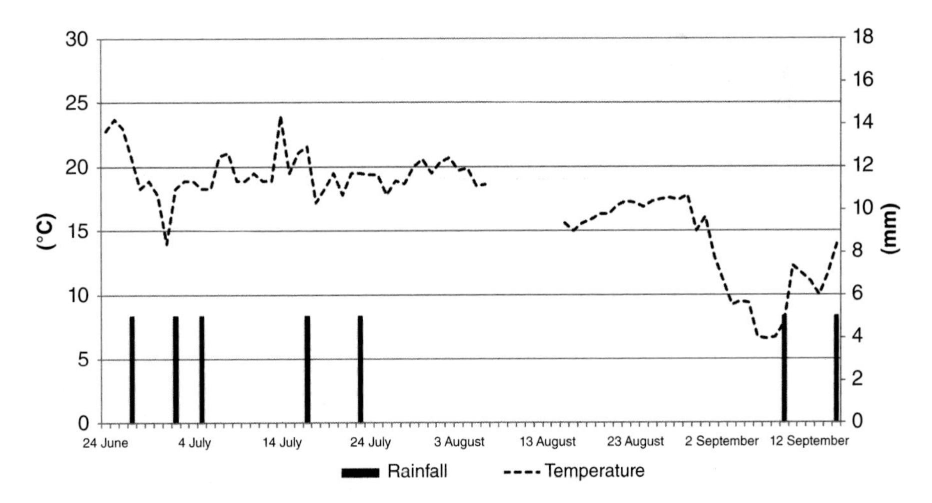

Figure 1.3.7 Temperatures and rainfall recorded by John Locke during the drought of 1666 summer. *Source*: Bodleian Library, University of Oxford, QC 161 B69 1692.

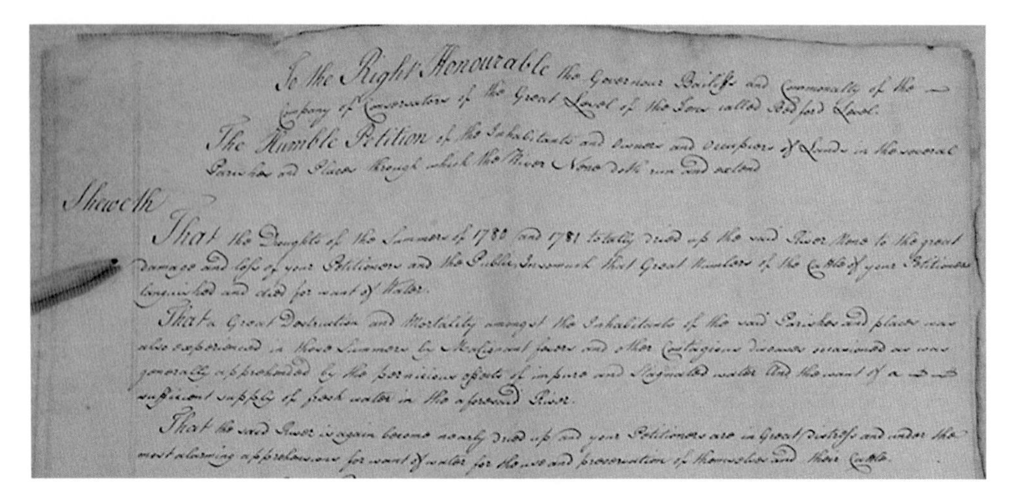

To the Right Honourable the Governour Bailiff and Commonalty of the Company of Conservators of the Great Level of the Fens called Bedford Level.

The Humble Petition of the Inhabitants and Owners and Occupiers of Lands in the several Parishes and Places through which the River Nene doth run and extend

Sheweth

That the Droughts of the Summers of 1780 and 1781 totally dried up the said River Nene to the great damage and loss of your Petitioners and the Public, Insomuch that Great Numbers of the Cattle of your Petitioners languished and died for want of Water.

That a Great Destruction and Mortality amongst the Inhabitants of the said Parishes and places was also experienced in those Summers by Malignant fevers and other Contagious diseases occasioned as was generally apprehended by the pernicious effects of impure and Stagnated water and the want of a sufficient supply of fresh water in the aforesaid River.

That the said River is again become nearly dried up and your Petitioners are in Great distress and under the most alarming apprehensions for want of water for the use and preservation of themselves and their Cattle.

Figure 1.3.8 Petition from the inhabitants and owners of lands to the Governor of the Company of Conservators of the Great Level of the Fens (Bedford Level) about the effects of the 1785 drought. *Source*: Cambridgeshire Archives S/B/SP 742.

was so critical that the inhabitants asked the Governor of the Company of Conservators of the Great Level of the Fens for a reduction in their taxes, which they could not pay anymore.

The 'peril of the drought': The 1921 drought in the United Kingdom The 1921 extreme event extended back to the fairly dry August 1920. High-pressure systems from the Azores remained stuck for almost the entire year. The rainfall of October, November, and December showed considerable deficiencies. With the onset of February, the drought had set in. The whole of Wales, most of England, and large areas in the East Midlands, Scotland, and Ireland had less than a quarter of their average rain, while considerable areas in the north and southwest of England and Wales received less than 10%. There was a temporary break in the dry weather in March, when slightly more than the average rains fell over the country. The subsequent months of 1921 were all markedly dry, except August. The three driest consecutive months of 1921 were April–June, when the rainfall over the United Kingdom generally was 58% of the average rainfall (Brooks and Glasspoole, 1928). According the British Environment Agency, there is no doubt that 1921 was the driest year since 1785 (EA, 2006).

The cinema newsreel producer Pathé News presented the societal consequences of the 1921 drought very well in a programme shown on July 28 1921. Very probably for the first time in Europe, a drought became a national cause and threat, and was presented as such to the entire United Kingdom. Under the suggestive title 'The Peril of the Drought', and with a subtitle devoid of ambiguity ('Burning crops and stacks threatens our winter food supplies'), the Pathé News programme publicly affirmed the danger that this drought presented for the UK population (Figure 1.3.9). Very drained from the catastrophic World War I and still subjected to food rationing, the British were at risk of being deprived of their harvests for the coming winter. Indeed, in the newsreel video, we see fields on fire and trees transformed into torches as firemen try to fight the fire by pumping water from an almost-dried-up river.

Figure 1.3.9 Opening credits of the Pathé News programme about the 1921 drought. *Source*: http://www.britishpathe.com/video/the-peril-of-the-drought/query/menace.

Wrapping-up It is clear from historical records that drought has been a recurring feature of UK climate, with recent drought events by no means exceptional in terms of their intensity or duration. In fact, according to our research, records show that there has been a repeated tendency throughout UK history for dry years to cluster, which have resulted in multi-year droughts that contained shorter, more intense drought periods (e.g. 1700–1740, 1806–1860, 1887–1901).

1.3.4 France: Ile-de-France

Thanks to its long administrative and scientific traditions, France offers exceptional archives to study the history of droughts. The region of Ile-de-France (Paris and the area around it) was chosen as our site of observation because it benefits from rich documentation, in addition to being situated in an intermediate position between the United Kingdom (Section 1.3.3) and the Rhine Valley (Section 1.3.5).

1.3.4.1 Chronological variation and severity of French droughts

The available documentation allowed us to list 68 droughts. They are chronologically distributed very unevenly between 1500 and 2014, and the severity also varies very much (Figure 1.3.10). As in the United Kingdom, the variation is very strong during the five centuries that were studied.

These disparities are even more striking when these events are assigned to periods of 50 years (Figure 1.3.11). Three major trends can be discerned since the sixteenth

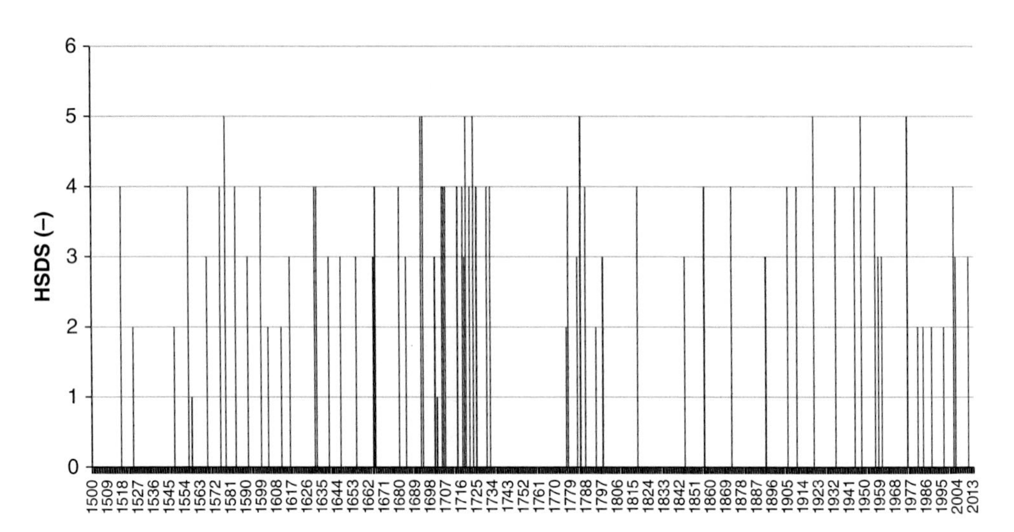

Figure 1.3.10 Chronology and severity of droughts in France between 1500 and 2014 according to HSDS (68 droughts).

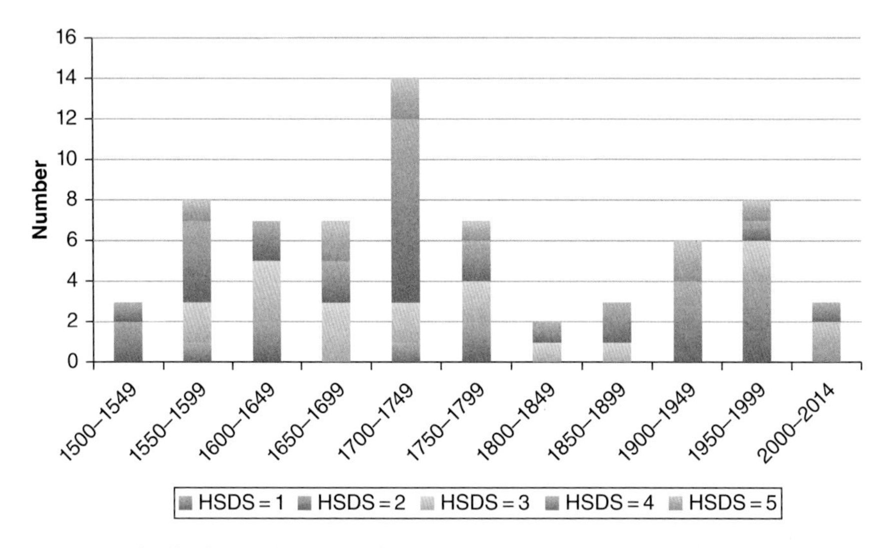

Figure 1.3.11 Distribution and severity of droughts in France by periods of 50 years between 1500 and 2014 according to HSDS. (*See colour plate section for the colour representation of this figure.*)

century. The first one is between 1500 and 1700, which is characterised by a steady frequency of droughts, with on average one drought every 8 years. A distinct change is seen in the eighteenth century, which experiences a very important increase of the number of droughts. There were 21 droughts over the period, with a peak of 14 droughts in the first half of the century, or one drought every 3.5 years. This is quite unparalleled when compared to the rest of the 500 years studied; prior to 1700 and post 1800, there were no comparable frequencies observed.

This series of droughts is particularly well documented in the archives and memoirs of the Royal Academy of Sciences in Paris. The scientists at the Paris Observatory were particularly worried by this long, dry, and very hot episode.

Contrary to the UK drought history, in case of France, the nineteenth and twentieth centuries are not characterised by more frequent droughts, and there was a falling off from the 1850s with the Little Ice Age. Nevertheless, from 1950 onwards, the twentieth-century data show a real increase in the number of droughts, although without any observable turning point. Even then, this increase is remarkable only within the framework of that 200-year period, and remains very clearly lower than the frequency seen in eighteenth-century drought data.

In terms of severity, it is also very difficult to distinguish notable differences or trends in the historical variation (Figure 1.3.11). Until 1700, low- and average-severity droughts (HSDS = 1–3) do not show any noticeable evolution, which also applies to the most extreme droughts (HSDS = 4 and 5). Once again, the period 1700–1750 clearly stands out, with 10 droughts of HSDS = 4 and 5 out of a total of 21 events. Consequently, the first half of the eighteenth century was simultaneously affected by an increase in the number and severity of exceptional droughts.

The eighteenth century is rather homogeneous because the rare droughts of the century divide up between HSDS = 3 and 4. If there is a break in the contemporary period, it is paradoxically situated at the turn of the twentieth century, and not at the midpoint of 1950. The first half of the century adds up to only six extreme events with HSDS = 4–5. In the most recent period, we observe more low-intensity droughts (HSDS = 2), while the strongest-intensity droughts were infrequent.

Thanks to the scientists (Cassini, La Hire, Réaumur) of Paris Observatory and the meteorologists of the Meteorological Station of Montsouris, who regularly recorded Parisian precipitation, the data we have for France is one of the longest meteorological series in Europe, more or less continuous from 1676 up to the present day. The comparison of the annual precipitation with the severity of the droughts (Figure 1.3.12) shows a rather good similarity. The latter confirms the reliability of HSDS to estimate the severity of pre-instrumental droughts.

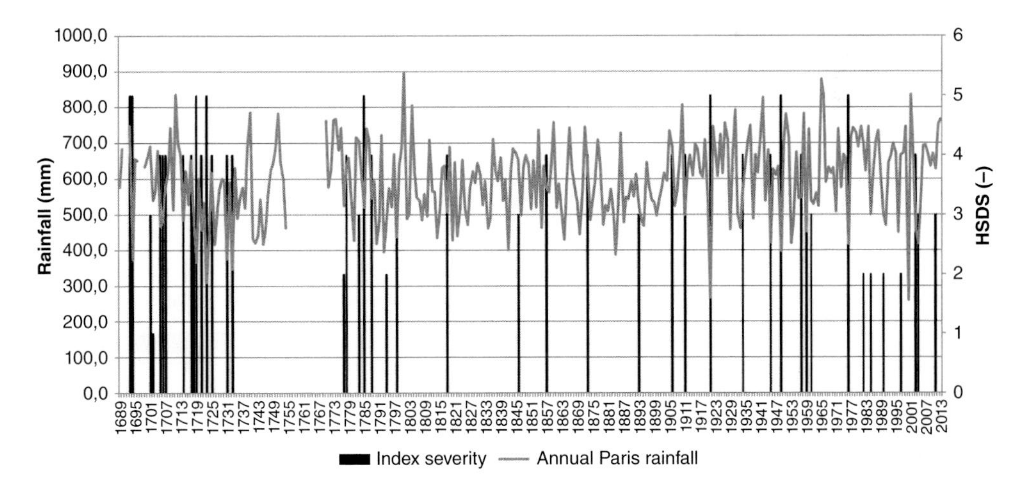

Figure 1.3.12 Comparison of HSDS and the annual Paris rainfall during 1689–2013 for droughts in France. *Source*: Meteo France-OPHELIE ANR Project and Bibliothèque de I'Académie Nationale de Médecine de Paris. (*See colour plate section for the colour representation of this figure.*)

1.3.4.2 Two examples of very severe droughts

Two severe droughts are described in the following text – that is, the 1578 drought, and the droughts and heat waves in the eighteenth century.

The 1578 drought The 1578 drought is very probably one of most extreme in the history of France in the last five centuries. Despite the lack of instrumental data at the time, there are plenty of written testimonies in the archives because of its severity (Figure 1.3.13). It is during the winter of 1577 that archives mention for the first time a 'rather dry' season, followed by a very hot spring. The low precipitation continued, and, in April 1578, archives evoke the general concern about the situation of the harvests. They speak of barley and hemp, which did not grow, and 'pro pluvia' processions were organised in Paris by the municipal authorities.

The environmental crisis reached its climax in the autumn of 1578 in Ile-de-France with the springs and fountains drying up. In the countryside, farmers spoke about the ground being difficult to plough because of it being completely dry to a depth of 1.20 m. The hydrology of the whole Seine basin was influenced. Low water affected the Seine, the Yonne, and the Marne rivers, and, in June, it was even possible to cross the Seine by foot in Paris. The low water levels had catastrophic socioeconomic consequences. Wood and wheat supplies were unable to reach the city by water on the eve of winter, leading to the lay-off of mills. Wheat was rare in the market, and grinding was impossible, so bread prices shot up. Diseases gained ground because of lack of food. The archives report a total of 565 days of dryness in 1577 and 1578. There were only 15 rainy days (Garnier, 2010b).

Droughts and heat waves in the eighteenth century in Ile-de-France As we have already described, eighteenth-century data seems totally atypical, because of the exceptional number of droughts (Figure 1.3.11). Figure 1.3.14 highlights a block of droughts between 1700 and the 1740s, which is significant because the events were both long lasting and very severe (HSDS = 4–5). One such example in this block is the major episode of 1719, which had a national dimension, and simultaneously set a record for dry days (220 days) and low pluviometry.

The temperatures increased from May 1719 onwards and was enhanced throughout the summer. In the village of Varrèdes (Seine-et-Marne, near Paris), the farmers feared the worst when they noticed the low harvest because of a 'very dry and very warm' August. The scientists at the Paris Observatory monitored the event very closely and

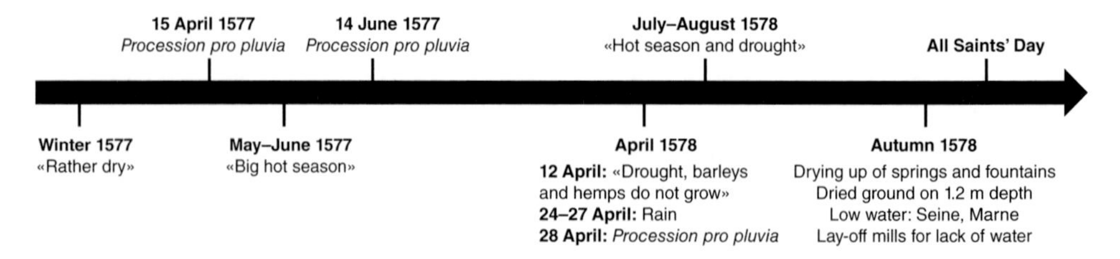

Figure 1.3.13 Progress of the great drought of 1577–1578. *Source*: In Ile-de-France according to written archives.

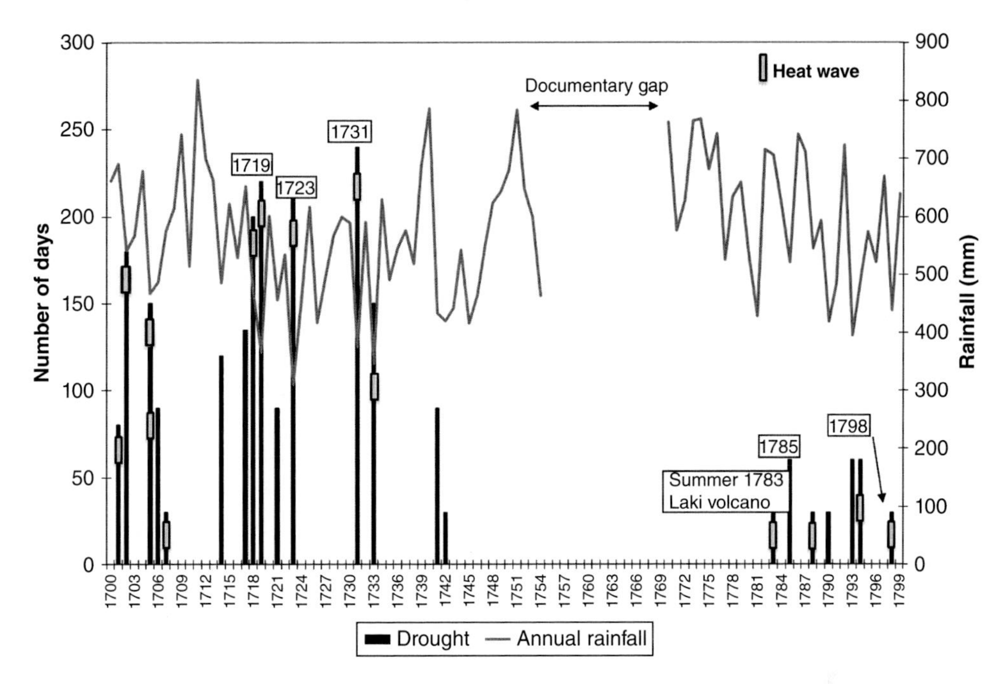

Figure 1.3.14 Droughts, heat waves, and annual rainfall in Ile-de-France in the eighteenth century. (*See colour plate section for the colour representation of this figure.*)

noted that the annual average of precipitation amounted to only 366 mm. This number prompted them to say that 'there is no year so dry … these 3 years (1717–1719) are the driest for more than 30 years' (Anonymous, 1720).

The exceptional heat wave also did not escape their attention, and the academicians followed the fast temperature increase very closely. The climax occurred on July 16 1719. At 3 pm on that day, the scientist named La Hire recorded a temperature of more than Florentine 82°degrees – in other words, at least 36.1°C! The heat and lack of water finally led to an epidemic of dysentery, associated with other infections. It was a national tragedy for the Kingdom of France, which suffered 450 000 deaths out of the total population of 21 million in 1719.

The extreme drought and heat wave of 1719 was slightly less severe than its recent counterpart in 2003, which might provide us information on future conditions due to global warming.

1.3.5 Valley of the Upper Rhine (Germany, Switzerland, France)

The Rhine Valley area (South Rhine, Baden-Württemberg, Switzerland, and Alsace) is particularly well documented by religious chronicles, meteorological journals, and municipal archives (Hegel, 1869; Dietler, 1994; Mercklen, 1864; Dostal, 2005; Glaser, 1991; Muller, 1997; Pfister et al., 2006). This rich corpus enables a reliable chronological reconstruction.

1.3.5.1 Chronological variation and severity of Rhenish droughts

The Rhineland basin offers a far more puzzling scenario than France regarding the outbreak of extreme events since the 1950s. Compared to the English and French scenarios, the Rhine Valley (Alsace, Baden-Württemberg, Basel region) has a lower number of droughts, with 37 events listed between 1500 and 2013 – against 42 in the United Kingdom and 68 in Ile-de-France (Sections 1.3.3.1 and 1.3.4.1). In terms of variation, we observe more or less five dry episodes during the last 500 years (Figure 1.3.15):

1500–1540 (five droughts)
1630–1670 (eight droughts)
1715–1750 (four droughts)
1820–1890 (nine droughts)
1940–1980 (five droughts)

The first intense episode took place in the first half of the sixteenth century, paradoxically called the 'beautiful sixteenth century' by historians because of the favourable weather conditions in general. A break of about 40 years within the Little Ice Age led to an agricultural and demographic boom in this part of Europe. The second episode is clearly more challenging to interpret because climate historians classify it as a cold and wet period attributable to the 'Maunder Minimum' (Le Roy Ladurie, 2004; Luterbacher et al., 2004).

The distribution of these Rhenish droughts by half-centuries turns out to be very surprising, because we observe a regular and important decline in the number of droughts after the peak (six droughts) in the second half of the nineteenth century (Figure 1.3.16). In the twentieth century, a time rich in documentation and easy to exploit, the droughts did not stop decreasing, particularly during the period 1950–2000. Only three droughts were observed in the second part of the twentieth century. On the other hand, the period between 1500 and 1850 gives clear drying trends, particularly in the seventeenth century and between 1850 and 1900.

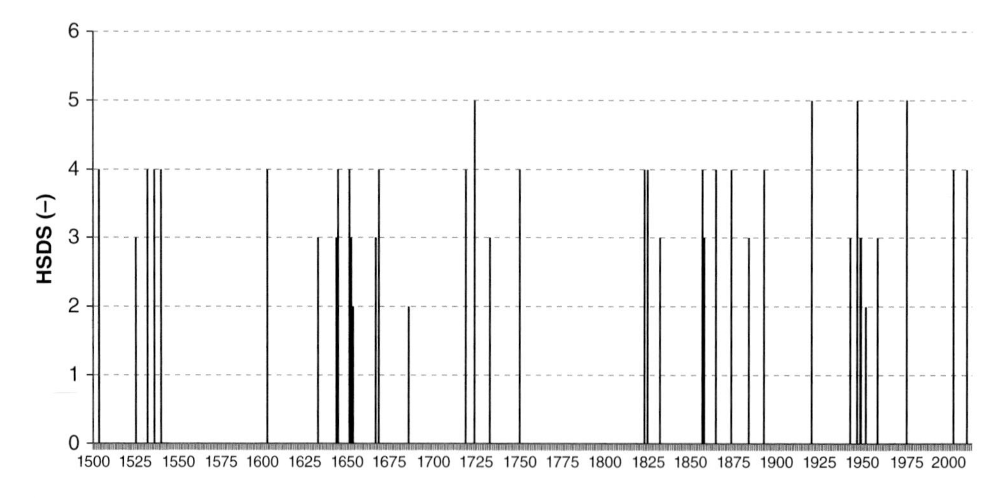

Figure 1.3.15 Chronology and severity of Rhenish droughts between 1500 and 2013 according to HSDS.

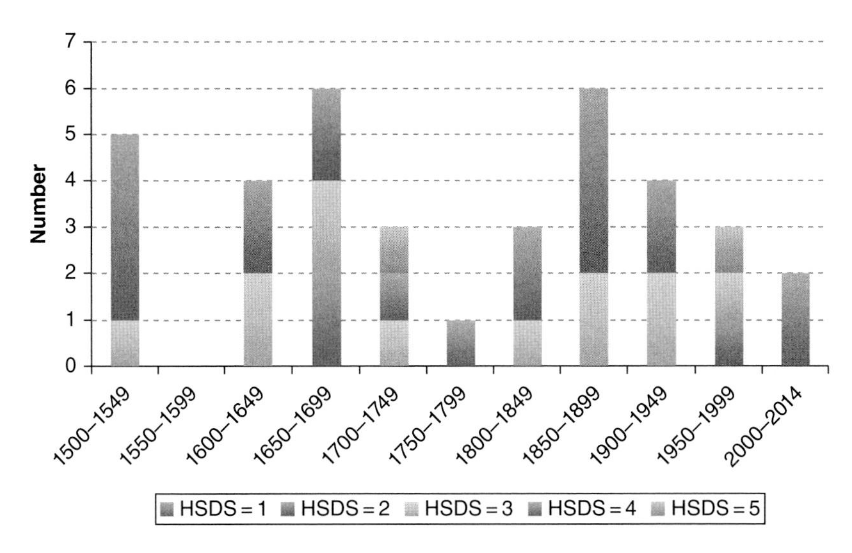

Figure 1.3.16 Distribution and severity of Rhenish droughts by periods of 50 years between 1500 and 2014 according to HSDS. (*See colour plate section for the colour representation of this figure.*)

Another important observation is that, similar to the number of droughts, the severity of the Rhenish droughts too did not stop declining (Figure 1.3.16). While very severe droughts (HSDS = 4 and 5) are omnipresent between 1500 and 1900, with climaxes in 1500–1549 and 1850–1899, they decrease strongly from 1900 onwards. Of the three droughts listed between 1950 and 2000, only one was exceptional (HSDS = 5), and the two others were less severe (HSDS = 2 and 3).

Because of the lack of precipitation data series, the flows of the Rhine at Basel for the period 1808–1944 were compared with the severities of the droughts according to HSDS. Figure 1.3.17 shows that the flows also constitute relevant instrumental information. Most low flows correlate rather well with extreme droughts, except 1830, when low river flow only had small socioeconomic impact. It should be noted that HSDS is based on the economic and social consequences engendered by droughts rather than being a statistical metric derived from physical data (Section 1.3.2.2).

1.3.5.2 The drought, a subject of collective memory, the Laufenstein

The drought history carved in the Laufenstein is described in the following text, as well as the 1540–1541 drought.

The memory engraved in stone: The 'Laufenstein' The droughts in the Rhine Valley gave birth to a very early cultural memory, which was maintained and passed on from one generation to the next. The archetypal symbol of this memory was the stone named 'Laufenstein', literally meaning the 'stone which sinks' in German. More exactly, it consisted of a rock situated 100 meters to the west of the old bridge of Laufenburg (Figure 1.3.18). This small German town in the current state of Baden-Württemberg is located approximately 40 km east of Basel, just downstream of the confluence of the

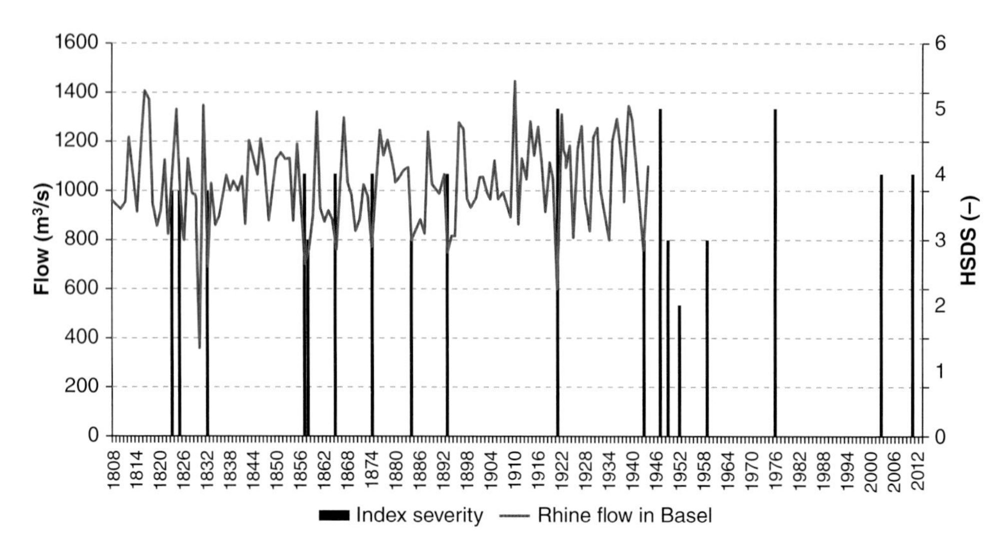

Figure 1.3.17 Comparison of HSDS and the annual flow of the Rhine in Basel 1808–2013 for Rhenish droughts. *Source*: Oesterhaus (1947). (*See colour plate section for the colour representation of this figure.*)

Vierteljahrsschrift d. naturf. Ges. Zürich. 46. Jahrg. 1901.

Laufenstein und grosse „rote Fluh"

Januar 1891

:::::: Granitgänge. — — — — Ausserord. Hochwasser vom 18. VI. 1876.

Figure 1.3.18 Photo from January 1891 showing the mark of low water on the 'Laufenstein' (white circle), situated near the old bridge of Laufenburg, Baden-Württemberg. *Source*: Walter (1901).

Rhine and Aare rivers (Pfister et al., 2006). The stone would be revealed in the river when the water level went low, serving as an indication of low water to the local population for centuries. The locals commemorated such occasions by chiselling the current year on the stone. This mark unfortunately disappeared in 1909 when the authorities decided to blow up the stone to facilitate the construction of a hydroelectric power plant nearby.

Fortunately, a German hydrology engineer had managed to salvage the chiselled data from the stone before its destruction (p. 27, table IV in Walter, 1901). On 18 February 1891, he took advantage of the low river level to study it closely, having beforehand questioned the inhabitants about its function. He noticed that the dates of nine droughts since the sixteenth century were engraved in the stone. Among them, some were perfectly recognisable – that is, the extreme events in 1541, 1750, 1823, 1858 and 1891 (Figure 1.3.19). Walter (1901) was careful, and proposed only the years 1692, 1764, 1797 and 1848. In this last case (i.e. 1848), neither the archives nor the annual river flow corresponding to the year (884 m^3/sec) indicate a particularly severe drought. It is the same for the lowest water level, carved in the stone in 1823 – the year does not correspond to a low flow at Basel station (1098 m^3/sec).

The drought of 1540–1541 Among the nine drought years that were engraved on the Laufenstein (See the subsection titled 'The drought, a subject of collective memory, the Laufenstein' under Section 1.3.5.2), the year 1541 merits particular attention, because the drought really covered the whole region of the Upper Rhine. Consequently, it has been perfectly preserved in the written memory of the time. For climatologists, the summer of 1540 was one of the warmest and driest in the history of Central Europe

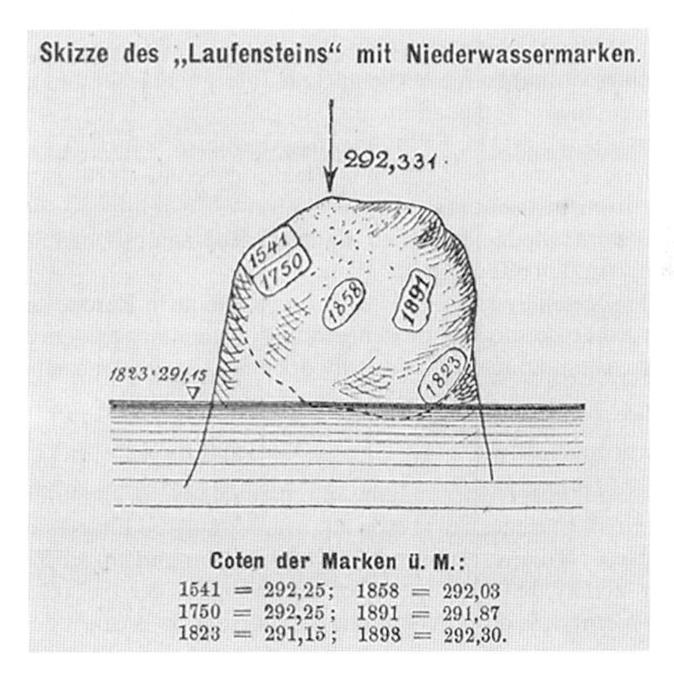

Figure 1.3.19 Sketch by Walter (1901) before the destruction of the Laufenstein in 1909. The historic marks of the droughts of 1541, 1750, 1823, 1858 and 1891 are clearly visible. *Source*: Walter (1901).

(Luterbacher et al., 2002). From the middle of March till the end of September 1540, an anticyclone would have blocked atmospheric moisture flow to the centre of Europe. During the winter of 1540–1541, the level of Lake Constance was so low that the city of Constance was able to begin the construction of new fortifications.

In Alsace and in the Canton of Basel, June 1540 was dry, and July even drier, to the point that water shortages became widespread both in town and in the countryside. In their chronicles and diaries, the bourgeois of Basel and Strasbourg describe how jars of water were being sold for the same price as jars of wine. Hunger was added to thirst in August, because mills on the Rhine and its tributaries could not grind wheat anymore because of the lack of water.

In the countryside around Thann and around Guebwiller (Haut-Rhin, Vosges), fruits dried on trees. The religious chronicles of Colmar, Guebwiller, and Thann all speak of apples, pears, and walnuts falling from the trees. The grape was not spared either, because vines were dry and did not give fruits. Even more serious in this densely wooded region, forest fires multiplied, and the archives even speak of forests that ignited 'spontaneously'. As the ultimate curse connected indirectly to the drought, plague reappeared and quickly spread in the cities of Basel, Mulhouse, and Colmar (Vischer and Stern, 1872; Bibliothèque municipale de Colmar; Wunderbuch aus Colmar, T.CH. 60 Mss; Jahrbücher der Dominker zu Colmar, I.CH. 60Mss; Chronik von Sigmund Billing, I.CH. 64 Mss).

1.3.6 Conclusions

Long-term historical reconstructions of droughts are very useful for climatologists and hydrologists. History can be an excellent tool to assess and improve the reliability of climate models by testing these against long historic time series that include hydrological extremes such as droughts. While the various Intergovernmental Panel on Climate Change (IPCC) scenarios predict an increase in the number of extreme events in multiple regions around the world, including the Mediterranean (IPCC, 2014; see also Chapter 1.4), it is necessary to place them in a historical context. The same holds for the trends obtained from instrumental records that cover a relatively short period (see Chapter 1.2). The knowledge provided by historical analysis can help distinguish between climate variability and climate change.

This study has successfully identified key drought events in Europe since 1500, based on a range of hydrometeorological evidence, and documentary evidence of the impacts for each event. It examined the rather patchy documentary evidence and found several European clusters of dry years in these five centuries. Finally, it has highlighted the challenge of examining European drought and its impact in the historical context. This historical-data-based research proves the importance of long data series, because it is clear that contemporary datasets (post-1950) may be unrepresentative of the full historical series, and thus drought risk (particularly protracted events) may be underestimated. The historical records reveal regular variations in England, France, and Upper Rhine, and do not demonstrate any increase in the frequency of dry episodes since the sixteenth century.

The historical method, as described in this chapter, deserves careful interpretation as well. A region's vulnerability to drought in the eighteenth century can be lower than in the two succeeding centuries, because it is frequently changing, depending on, for instance, the supply infrastructure, the demands placed on the water supply system by users, and various other factors (Brooks et al., 2005; Cardona et al., 2012; Pulwarty and Verdin, 2013; Blauhut et al., 2016; Carrão et al., 2016; González-Tánago et al., 2016). Still more important, in the sixteenth–nineteenth centuries, droughts directly affected peoples' livelihoods and water supplies, while today the impact in Europe is mainly economic or environmental (e.g. Stahl et al., 2016).

Acknowledgements

The funding for this study was provided by the EU project DROUGHT-R&SPI (Fostering European Drought Research and Science–Policy Interfacing), contract 282769.

References

Anonymous (1720). Histoire de l'Académie royale des sciences. Avec les mémoires de Mathematiques et de Physique, Imprimerie royale, Paris.

Blauhut, V., Stahl, K., Stagge, J.H., Tallaksen, L.M., De Stefano, L., and Vogt, J. (2016). Estimating drought risk across Europe from reported drought impacts, drought indices, and vulnerability factors. *Hydrology and Earth System Sciences* 20: 2779–2800.

Boyle, R. (1692). *The General History of the Air*. London: Awnsham & John Churchill, pp. 1–259.

Bray, W. (1901). *The Diary of John Evelyn*. Vol. 2. Washington, D.C.: Walter Dunne, pp. 1–436.

Brooks, C.E.P. and Glasspoole, J. (1928). *British Floods and Droughts*. London: Ernest Benn Limited, pp. 1–199.

Brooks, N., Adger, W.N., and Kelly, P.M. (2005). The determinants of vulnerability and adaptive capacity at the national level and the implications for adaptation. *Global Environmental Change* 15: 151–163.

Cardona, O., Van Aalst, M., Birkmann, J., Fordham, M., Mcgregor, G., Perez, R., Pulwarty, R., Schipper, E., and Sinh, B. (2012). Determinants of risk: exposure and vulnerability. In: *Managing the Risks of Extreme Events and Disasters to Advance Climate Change Adaptation* (ed. C. Field, V. Barros, T. Stocker, D. Qin, D.J. Dokken, K.L. Ebi, M.D. Mastrandrea, K.J. Mach, G.-K. Plattner, S.K. Allen, M. Tignor, and P.M. Midgley), A Special Report of Working Groups I and II of the Intergovernmental Panel on Climate Change (IPCC). Cambridge, UK, and New York, NY, USA: Cambridge University Press, pp. 65–108.

Carrão, H., Naumann, G., and Barbosa, P. (2016). Mapping global patterns of drought risk: An empirical framework based on sub-national estimates of hazard, exposure and vulnerability. *Global Environmental Change* 39: 108–124.

Cooper, C.H. (1827). *Annals of Cambridge*, Vol. III. Cambridge: Warwick, pp. 1–504.

Dietler, S. (1994). *Chronique des Dominicains de Guebwiller*. Société d'Histoire et du Musée du Florival, Guebwiller, pp. 1–359.

Dostal, P. (2005). *Klimarekonstruktion der Regio TriRhena mit Hilfe von direkten und indirekten Daten vor der Instrumentenbeobachtung*. Berichte des Meteorologischen Institutes der Universität Freiburg, Freiburg.

EA (Environmental Agency) (2006). The Impact of Climate Change on Severe Droughts. Major Droughts in England and Wales from 1800 and Evidence of Impact. *Environment Agency Science Report SC40068/SR1*.

Fraser, J. (1905). *Chronicles of the Frasers*. Edinburgh: T. and A. Constable, pp. 1–626.

French National Agency of Research (2009). Available from: http://www.gisclimat.fr/en/project/renasec; http://www.agence-nationale-recherche.fr/fileadmin/user_upload/documents/aap/2009/finance/cep-financement-2009.pdf (last accessed 8 March 2017).

Fuller, T. (1811). *The History of the University of Cambridge*. Cambridge: Cambridge University Press, pp. 1–743.

Garnier, E. (2010a). *Climat et Histoire, XVIe-XIXe siècles, numéro thématique 57-3*. Revue d'Histoire Moderne et Contemporaine, Belin, Paris, pp. 1–159.

Garnier, E. (2010b). *Les dérangements du temps, 500 ans de chaud et froids en Europe*. Plon, Paris, pp. 1–244. https://www.asmp.fr/prix_fondations/fiches_prix/gustave_chaix.htm.

Garnier, E. (2015). A historic experience for a strengthened resilience. European societies in front of hydro-meteors 16th–20th centuries. In: *Prevention of Hydrometeorological Extreme Events – Interfacing Sciences and Policies* (ed. P. Quevauviller), 3–26. Chichester: John Wiley & Sons.

Glaser, R. (1991). *Klimarekonstruktion für Mainfranken, Bauland und Odenwald anhand direkter und indirekter Witterungsdaten*. Paläoklimaforschung 5, Publisher, Stuttgart, New York, pp. 1–138.

González-Tánago, I.G., Urquijo, J., Blauhut, V., Villarroya, F., and De Stefano, L. (2016). Learning from experience: a systematic review of assessments of vulnerability to drought. *Natural Hazards* 80: 951–973.

Hegel (1869). *Die Chronik der Stadt Straßburg*. Leipzig: Verlag Hirzel, pp. 1–1169.

IPCC (Intergovernmental Panel on Climate Change) (2014). *Climate Change 2014: Synthesis Report. Contribution of Working Groups I, II and III to the Fifth Assessment Report of the Intergovernmental Panel on Climate Change* (Core Writing Team, ed. R.K. Pachauri and L.A. Meyer), pp. 1–151. Geneva, Switzerland: IPCC .

Le Roy Ladurie, E. (1972). *Times of Feast, Times of Famine: A History of Climate Since the Year 1000*. London: Georges Allen & Unwin, pp. 1–428.

Le Roy Ladurie, E. (2004). *Histoire humaine et comparée du climat*, Vol. 1. Paris: Fayard, pp. 1–748.

Luterbacher, J., Xoplaki, E., Dietrich, D., Rickli, R., Jacobeit, J., Beck, C., Gyalistras, D., Schmutz, C., and Wanner, H. (2002). Reconstruction of sea level pressure fields over the Eastern North Atlantic and Europe back to 1500. *Climate Dynamics* 18: 545–561.

Luterbacher, J., Dietrich, D., Xoplaki, E., Grosjean., M., and Wanner, H. (2004). European seasonal and annual temperature variability, trends since 1500. *Science* 303(5663): 1499–1503.

Mercklen, F.J. (1864). *Annales oder Jahrs-Gesachichten der Baarfüseren oder Minderen Brüdern S. Franc. Ord.* Insgeneim Conventualen gennant, zu Than durch Malachias Tschamser, Colmar, pp. 1–148.

Muller, C. (1997). Chronique de la viticulture alsacienne au XVIIe siècle, J.D. Reber, Riquewihr, pp. 1–254.

Oesterhaus, M. (1947). *Mehrjährige periodische Schwankungen der Abflussmengen des Rheins bei Basel*. Thesis, University of Basel.

Pfister, C., Weingartner, R., and Luterbacher, J. (2006). Hydrological winter droughts over the last 450 years in the Upper Rhine basin: a methodological approach. *Hydrological Sciences-Journal-des Sciences Hydrologiques* 51(5): 966–985.

Pulwarty, R. and Verdin, J. (2013). *Crafting early warning systems*. In: *Measuring Vulnerability to Natural Hazards: Towards Disaster Resilient Societies* (ed. J. Birkmann), pp. 14–21. Tokyo: UNU Press.

Smith, J. (1825). *Memoirs of Samuel Pepys, Esq. F.R.S., Secretary to the Admiralty in the Reigns of Charles II and James II, Comprising His Diary from 1659 to 1669*. London: Henry Colburn, pp. 1–618.

Stahl, K., Kohn, I., Blauhut, V., Urquijo, J., De Stefano, L., Acacio, V., Dias, S., Stagge, J. H., Tallaksen, L.M., Kampragou, E., Van Loon, A.F., Barker, L.J., Melsen, L.A., Bifulco, C., Musolino, D., De Carli, A., Massarutto, A., Assimacopoulos, D., and Van Lanen, H.A.J. (2016). Impacts of European drought events: insights from an international database of text-based reports. *Natural Hazards and Earth System Sciences* 16: 801–819. doi:10.5194/nhess-16-801-2016.

Vischer, W. and Stern, A. (1872). Basel Chroniken herausgeberen von der historischen gesellschaft in Basel. Leipzig, Verlag von S. Hirzel.

Walter, H. (1901). *Uber die Stromschnelle von Laufenburg*. Diss. Phil. II, University of Zürich, pp. 1–31.

Webster, T. and Shipps, K. (2004). *The Diary of Samuel Rogers*, 1634–1638. Woodbridge: The Boydell Press, pp. 1–287.

1.4
Future Drought

Henny A.J. Van Lanen[1], Christel Prudhomme[2], Niko Wanders[3], and Marjolein H.J. Van Huijgevoort[4]

[1]*Wageningen University, Wageningen, The Netherlands*
[2]*European Centre for Medium-Range Weather Forecasts, Reading, United Kingdom*
[3]*University of Utrecht, Utrecht, The Netherlands*
[4]*KWR Watercycle Research Institute, Nieuwegein, The Netherlands*

1.4.1 Introduction

Drought is one of the most severe natural hazards with large environmental and socio-economic impacts, and it needs to receive attention to safeguard future water, food, and energy security (Tallaksen and Van Lanen, 2004; Sheffield and Wood, 2011). Seneviratne et al. (2012) report that there is medium confidence that, since the 1950s, some regions of the world have experienced longer and more severe droughts (e.g. southern Europe), and that droughts will intensify in the twenty-first century in some seasons and areas, as a result of climate change. Several large-scale trend studies have shown that drought has become more severe in many European regions. For example, Spinoni et al. (2016; 2017) present a comprehensive overview of trends in meteorological drought in Europe. They present trends for three different periods: 1951–1970, 1971–1990 and 1991–2010. For the first period, drying trends (drought severity, drought duration) are mainly found in eastern Europe, the Balkans, and central Scandinavia, whereas the drying trends for the last period largely occurred in southern Europe, including the Balkans. Meteorological droughts affect drought in groundwater and streamflow (see Chapter 1.1), which results, through propagation, in similar patterns for trends in hydrological drought. Generally, trends in catchments that are not disturbed by human measures show lower annual streamflow in southern Europe and the opposite in northern Europe (see Chapter 1.2). An urgent question is whether these trends will continue in the future. Spinoni et al. (2016; 2017) also explore how meteorological drought will develop. They report that the frequency of drought in precipitation (SPI-12, SRES, A1B scenario) is projected to become higher in the Mediterranean, and central and eastern Europe. Larger areas are anticipated to face higher drought frequencies, including more northern European regions (e.g. northern France, Benelux, Germany) when the climate–water balance is considered (SPEI-12). Stagge et al. (2015), who used more

Drought: Science and Policy, First Edition.
Edited by Ana Iglesias, Dionysis Assimacopoulos, and Henny A.J. Van Lanen.

recent Intergovernmental Panel on Climate Change (IPCC) emission scenarios (Representative Concentration Pathways, RCPs, 2.6–8.5), confirmed for the most part their findings (South Europe, France, Benelux, except for Central Europe, where no increase of drought frequency is projected to occur).

This chapter elaborates on the impact of global change on hydrological drought, that is, how changes in meteorological drought affects streamflow. Knowledge on projected changes in river flow (hydrological drought) is extremely important for water resources management (water security). The chapter starts with an overview of selected large-scale studies that examine the impact of climate change on hydrological drought at the global or pan-European scale. Then the impact of human influences (dams, water use, gradually changing hydrological regime) on hydrological drought is explained. The chapter concludes with a description of the uncertainties that are inherent in the assessment of future hydrological drought.

1.4.2 Overview of studies

The impact of global warming on annual runoff and how it will likely change has been the topic of many studies (e.g. Milly et al., 2005). However, for water, food, and energy security, the influence on hydrological extremes is equally important. One of the first large-scale studies that used early-twenty-first-century emission scenarios (IPPC's SRES) to investigate future drought characteristics was done by Arnell (2003). Previous work was based on climate scenarios from the 1990s (e.g. Nijssen et al., 2001) and largely focussed on annual flow. Arnell (2003) applied a single hydrological model that was driven by the output of several climate models and emission scenarios. An important finding was that the coefficient of variation of annual runoff is projected to increase, which will lead to higher frequency of drought runoff in 2050, particularly in parts of Europe and southern Africa.

Several years later, multi-global hydrological models were introduced in the WaterMIP project to explore the impact of global warming on hydrological extremes (Harding et al., 2011; Corzo Perez et al., 2011), using the output from multi-climate models and multi-emission scenarios (SRES). In the framework of WaterMIP, uncertainty in future hydrological drought was explored by including only those models that performed reasonably in the past (Van Huijgevoort et al., 2014; Section 1.4.5.1). Single large-scale models were still used to investigate specific drought aspects. For example, Van Lanen et al. (2013) and Wanders and Van Lanen (2015) introduced the similarity index, based on the bivariate probability distributions of the drought duration and deficit volume. Future drought events based on this approach is presented in Section 1.4.3.1. A single hydrological model was also used to illustrate how the European river network will be affected by future drought (Forzieri et al., 2014; Section 1.4.3.2). This model was also applied to explore the impact of some human influences on future drought (Section 1.4.4.2).

The Inter-Sectoral Impact Model Intercomparison Project (ISI-MIP) was introduced as a follow-up to WaterMIP. In ISI-MIP, the most recent emission scenarios, RCPs (Meinshausen et al., 2011), were used, and, in addition to the natural hazard, inter-sectoral impacts were assessed (Warszawski et al., 2014). Prudhomme et al. (2014) provide a comprehensive overview of future drought (multi-RCPs, multi-climate models, multi-hydrological models), including an uncertainty measure, and the influence of large-scale models and climate models (Sections 1.4.3.3 and 1.4.5). Single hydrological models were applied to study human impacts on future drought – for example, the influence of reservoirs (Wanders and Wada, 2015), or the adaptation to a gradually changing hydrological regime (Wanders et al., 2015).

The timeline in the preceding text shows that a robust assessment of the signal of future change in hydrological drought depends on multiple considerations: (i) plausible pathways for future atmosphere emissions of greenhouse gases, as these affect how radiation from the sun reaches the earth's surface; (ii) understanding and representation of the complex coupled ocean–atmosphere response to greenhouse gas emissions, and how this will influence climatic patterns; and (iii) understanding and representation of land–atmosphere feedback in different geo-climatic settings. The scientific community has addressed these challenges by introducing a set of time evolutions of the atmospheric concentration of greenhouse gases (RCPs), each based upon the evolution of contrasting but realistic politicoeconomic philosophies (Meinshausen et al., 2011). In addition, multi-model ensembles (MMEs) have been established to cover the complexity of the physical processes of the land–atmosphere–ocean system (Haddeland et al., 2011; Warszawski et al., 2014).

1.4.3 Assessment of future hydrological drought

1.4.3.1 Future drought across climate regions

Several studies have tried to quantify the global drought hazard (Section 1.4.2). Intercomparison of the drought hazard obtained in the different studies is hampered by the way the interplay between catchment characteristics and climate type (Van Lanen et al., 2013) is modelled. In a detailed attribution study, Wanders and Van Lanen (2015) show that catchment and climate characteristics have a significant impact on future drought. This work was a follow-up of earlier work by Van Lanen et al. (2013), where the authors tried to quantify the impact of climate and catchment properties in the period 1958–2010. They used a synthetic Global Hydrological Model (GHM) to quantify the impact of changing climate on drought characteristics across the globe. The synthetic model assumes identical land–surface characteristics (i.e. catchment properties) for every location for which the model simulates the hydrological response to future precipitation, temperature, and evaporation. The meteorological forcing was obtained from three General Circulation Models (GCMs) – ECHAM, CNRM, and IPSL – that were downscaled and bias-corrected within the EU-WATCH project.[1] Droughts were identified by using the variable Q_{80} threshold method[2] (see Chapter 1.1) to determine the average drought duration and deficit volume. To allow comparison across different hydroclimates (Köppen–Geiger climate types, Peel et al., 2007), they used the standardised deficit volumes that normalise the deficit volume by the mean discharge. By varying both the catchment properties and climate scenarios, they were able to quantify and attribute their impact to changes in future drought characteristics for the SRES A2 emission scenario (Nakicenovic and Swart, 2000), which is the most extreme emission scenario.

[1] WATer and global CHange: http://www.eu-watch.org.

[2] The threshold is the so-called Q_{80}, which is the flow that is equalled or exceeded in 80% of the time in the control period.

One of the major findings of Wanders and Van Lanen (2015) is that, although drought frequency is expected to decrease on a global scale, an increase in their severity and duration is to be expected (Table 1.4.1). On average, drought duration is projected to increase by 180% and 230% for the near and far future, respectively. The regions that can expect the most severe impact of these changes are to be found in the dry desert climate (B) and the cold, snow-dominated climates (D and E). The changes in future climate have the strongest impact on the water availability in these regions. For desert climates, a strong increase in drought severity due to reduction in the annual rainfall can be seen, leading to increasing water scarcity. In the cold, snow-dominated climates, they anticipate that the discharge peak will shift from early summer snow melt to spring snow melt, leading to a strong reduction in autumn water availability. These strong shifts indicate that human water management and nature are forced to adjust to the seasonal changes in water availability.

The most important conclusion from Wanders and Van Lanen (2015) is that, in general, the number of drought events will become lower – that is, the frequency will decrease due to the pooling of events. However, individual drought events are expected to become more prolonged with an increase in drought severity. This will lead to higher-impact drought events that require more proactive measures to ensure that water managers can deal with these future drought events, especially in the tropical and desert regions.

1.4.3.2 Future streamflow drought in Europe

Long time series of streamflow (1961–2100) were simulated for the European domain with a hydrological model that was forced with an ensemble of bias-corrected climate simulations (IPPC's SRES) (Forzieri et al., 2014). The climate simulations came from 12 combinations of nested GCM-RCMs, with daily output and a spatial resolution of 25 km. The SRES A1B scenario was selected, which reflects, among others, fast economic growth and rapid introduction of new and more efficient technologies (Nakicenovic and Swart, 2000). Daily runoff of 140 years was simulated with the hydrological model LISFLOOD (Van der Knijff et al., 2010), using the 12 different climate scenarios. LISFLOOD is a GIS-based, distributed, physically based model, developed for large-scale studies (e.g. a GHM applied to Europe). It simulates the spatial and temporal patterns of water infiltration into soils, actual evapotranspiration, soil water storage, snowmelt, groundwater storage, and surface water runoff at a 5-km grid level. The gridded runoff is routed using a digital terrain model to obtain time series of daily discharge at any 5-km pixel in the European river network. This allowed calibrating LISFLOOD using historical river flow from over 250 catchments spread over Europe.

The deficit volume (*Def*) was derived from the discharge time series (cumulative deficit in periods when the discharge stays below a threshold flow) for time slices of 30 years – that is, control period (1961–1990), 2020s (2011–2040) and 2080s (2071–2100). Eventually, an extreme value analysis was applied to get *Def* values for the selected 30-year periods for different return periods ranging between 2 and 100 years in each river pixel.

Reduction of the minimum flow (10–20% lower 7-day minimum flow) due to climate change becomes apparent by the 2020s, particularly in southwest and southeast Europe (Forzieri et al., 2014). Reduction is progressing further in time, and, by the end of

Table 1.4.1 Changes in median of drought characteristics (% relative to control period, 1971–2000, including standard deviation) in the near future (2021–2050) and far future (2071–2100) for climate types: equatorial (A), arid (B), warm temperate (C), snow (D), and polar climates (E).

Drought characteristic	Climate	2021–2050			2071–2100		
		ECHAM	CNRM	IPSL	ECHAM	CNRM	IPSL
Duration (day)	A	142 ± 4	138 ± 4	131 ± 12	175 ± 7	169 ± 5	181 ± 15
	B	142 ± 4	133 ± 3	144 ± 6	175 ± 6	160 ± 4	181 ± 12
	C	133 ± 4	123 ± 3	115 ± 5	150 ± 7	162 ± 6	162 ± 7
	D	107 ± 7	93 ± 15	100 ± 11	129 ± 8	121 ± 29	114 ± 12
	E	100 ± 4	108 ± 16	108 ± 8	123 ± 6	138 ± 23	131 ± 7
	All	115 ± 3	114 ± 6	121 ± 5	146 ± 3	143 ± 9	157 ± 6
Standardised Deficit Volume (day)	A	193 ± 7	194 ± 7	182 ± 25	301 ± 18	317 ± 13	327 ± 40
	B	206 ± 9	179 ± 7	218 ± 16	305 ± 15	268 ± 12	310 ± 64
	C	164 ± 10	145 ± 7	134 ± 9	217 ± 21	220 ± 20	247 ± 22
	D	131 ± 8	103 ± 18	117 ± 11	144 ± 12	152 ± 36	126 ± 25
	E	115 ± 7	128 ± 20	115 ± 8	147 ± 14	170 ± 35	167 ± 18
	All	155 ± 4	139 ± 7	146 ± 8	206 ± 7	214 ± 12	222 ± 22

the twenty-first century, most of southern and western Europe, including the United Kingdom and the Benelux, is projected to experience lower minimum flows. Lower precipitation combined with a higher potential evapotranspiration causes reduced minimum flows, although subsurface water storage determines if actual evapotranspiration equals the potential. Minimum flow in areas with low subsurface storage will not be so much affected by increases in potential evapotranspiration rate (see Chapter 1.1).

Changes in discharge deficits (*Def*) for a 20-year return period (20-year deficits) are presented in Figure 1.4.1 for river pixels with an upstream catchment area above 1000 km^2. Red colours indicate an increase of the deficit volume (i.e. less water in the river) relative to the control period, which means lower flows below the threshold[3] than during the control period and/or longer periods below the threshold. The maps show substantial impact of climate change on river flow deficits that is projected to clearly progress over time in the twenty-first century. By the end of the century, over half of Europe will experience more water shortage than the previous era. Already in the 2020s, substantial parts of southern Europe are expected to suffer from larger flow deficits. In some rivers in the Iberian Peninsula and the Balkans, the 20-year deficits increase by 20%. By the end of the century, the increase in these deficits is almost 80% (relative to the twentieth-century control period). The simulation by Forzieri et al. (2014) also shows that west and northwest Europe, including vast parts of France, the Netherlands, Belgium, United Kingdom, and the Alpine regions, are projected to suffer from larger deficits in river flow in the 2080s (increases of up to 50%).

The majority of the rivers in northeast Europe, including in the Baltic countries, are expected to have lower flow deficits in the nonfrost season, meaning that there is more water in the river (Figure 1.4.1). The deficits volumes can be 50% lower than in the

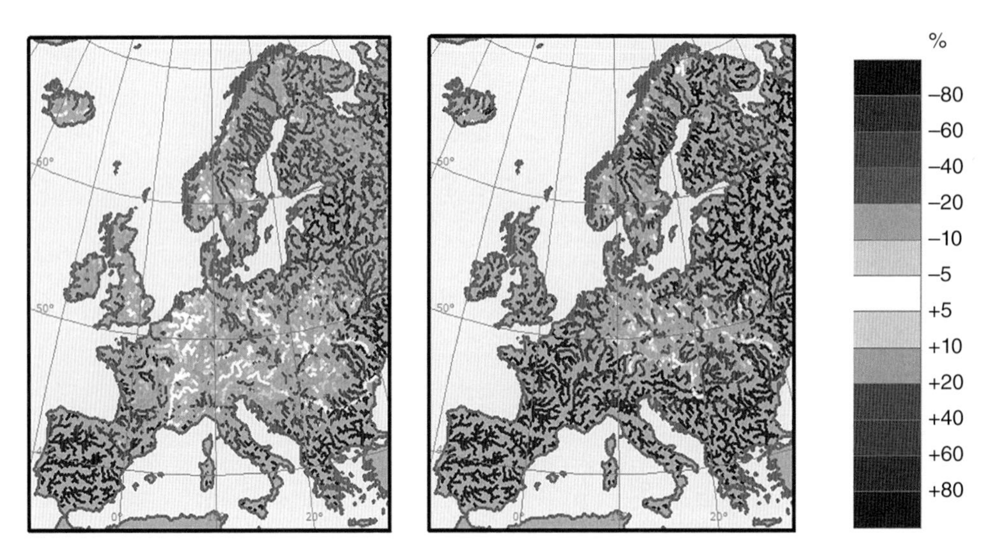

Figure 1.4.1 Change in the ensemble average in the 20-year return period deficit volumes due to climate change, left: 2020s and right: 2080s relative to the control period. *Source*: Derived from Forzieri et al. (2014). (*See colour plate section for the colour representation of this figure.*)

[3] Q_{80} was selected as threshold, which was derived from the control period (1961–1990).

control period, or even more. Spatiotemporal variability of the change in 20-year deficits in this region is higher than in most other parts of Europe, generally due to an increase in precipitation. Moreover, a reduced snowmelt contribution affects the timing and associated identification of drought events (e.g. Van Huijgevoort et al., 2014). Depending on the relative magnitude of these processes in a future climate, deficits in river flow may have a different severity in space and time.

1.4.3.3 Hotspots of future drought

Following evidence that Global Impact Models (GIMs[4]) had different skills in representing the spatiotemporal evolution of hydrological extremes (e.g. Haddeland et al., 2011; Prudhomme et al., 2011), future drought hazard and global hotspots were investigated from an MME of GIMs forced by outputs of Global Climate Models (GCMs) from the ISI-MIP (Warszawski et al., 2014) dataset. The GIMs ensemble used includes seven hydrological models: H08 (Hanasaki et al., 2008); Mac-PDM0.9 (Gosling et al., 2011); MATSIRO (Takata et al., 2003); MPI-HM (Stacke and Hagemann, 2012); PCR-GLOBWB (Wada et al., 2010); VIC (Liang et al., 1994); WaterGAP (Döll et al., 2003), one land–surface model (LSM) JULES (Best et al., 2011) and one dynamic vegetation model LPJmL (Bondeau et al., 2007) to provide a range of possible representations of land–surface interactions (Schewe et al., 2014). Forcing datasets were derived from the climate projections of five GCMs from the Climate Model Intercomparison Project Phase 5 (CMIP5, Taylor et al., 2011) to span the space of global mean temperature change and relative precipitation changes as best as possible (Warszawski et al., 2014). Prior to use as input of GIMs, precipitation and temperature daily time series were bias-corrected towards an observation-based dataset (Weedon et al., 2011) using a trend-preservation algorithm (Hempel et al., 2013). The ensemble comprised runs for four RCPs (Meinshausen et al., 2011), including the lowest (RCP 2.6) and highest (RCP 8.5) concentration scenarios. The signal of change was evaluated for the future simulation of the end of the century (2071–2100).

Drought spots were identified based on the threshold method (see Chapter 1.1), introduced in the late 1960s (Yevjevich, 1967) and now considered as a standard method (Tallaksen and Van Lanen, 2004). Following Prudhomme et al. (2011), a time-varying threshold T_d defined as the 90th-percentile streamflow (Q_{90}) estimated over a 30-days window was used, so that the seasonality of the streamflow regime can be accounted for. The method was applied to simulated runoff time series at the grid scale, using thresholds defined using the control period (1976–2005) for each [GCM;GIM] combination, to maintain multi-model consistency and avoid possible biases in the simulation to influence the results. Two measures were used to identify changes in drought hazard – that is, *drought occurrence*, the fraction of days with runoff below T_d; and *drought severity* (also called global drought index), the percentage of land area with runoff below T_d at the same time. Changes in both metrics were calculated as the percentage difference between 30-year control and future periods for each MME member, and summarised as the MME mean.

The largest mean changes are seen under RCP 8.5, with a widespread increase in MME drought occurrence covering most of the land masses during the northern

[4] Global Impact Models (GIMs) are large-scale models and include both Global Hydrological Models (GHMs) and Land Surface Models (LSMs).

summer (June–August). In December–February, the increase is concentrated in the Southern Hemisphere, and in low to mid-latitudes in the Northern Hemisphere (Figure 1.4.2). At the annual time scale, drought occurrence increases everywhere except in northern Canada, northeast Russia, the Horn of Africa, and parts of Indonesia (Prudhomme et al., 2014). Increases are systematically greater for RCPs associated with stronger radiative forcings, suggesting that mitigation measures aimed at curbing greenhouse gas emissions might be efficient in damping the drought hazard increase. In hotspots around the Mediterranean basin, Central America, Mexico, north of Venezuela, Guyana, Suriname and French Guiana, southwest Argentina, and Chile, drought days are expected to occur 30% more often than now in all seasons under RCP 8.5.

At the global level under RCP 8.5, MME mean drought severity is projected to increase by 13%, with increases reaching 17% in June–August. Even under the most optimistic RCP 2.6, global drought severity is expected to increase by just under 4%, which suggests that food and water security might be at risk in the future.

Using the climate change signal from a 28-member GCM ensemble from CMIP5 and the concept of 'emergence of signal' from Wilby (2006), and considering simulations regrouping four different RCPs, Orlowsky and Seneviratne (2013) found no emerging signal in precipitation and temperature. However, they observed that an increase in drought frequency based on soil moisture anomaly could be detected for the Mediterranean, South Africa, and Central America/Mexico region – regions also identified as hydrological drought hotspots by Prudhomme et al. (2014).

1.4.3.4 Future low flows

The effect of climate change on drought and low flows was studied by Van Huijgevoort et al. (2014) in 41 river basins with contrasting climates and catchment characteristics across the globe. A multi-model analysis was done using results from five global hydrological models (JULES, LPJml, MPI-HM, WaterGAP, and Orchidee) forced with data from three GCMs for the SRES A2 scenario (ECHAM5/MPIOM, CNRM-CM3, and

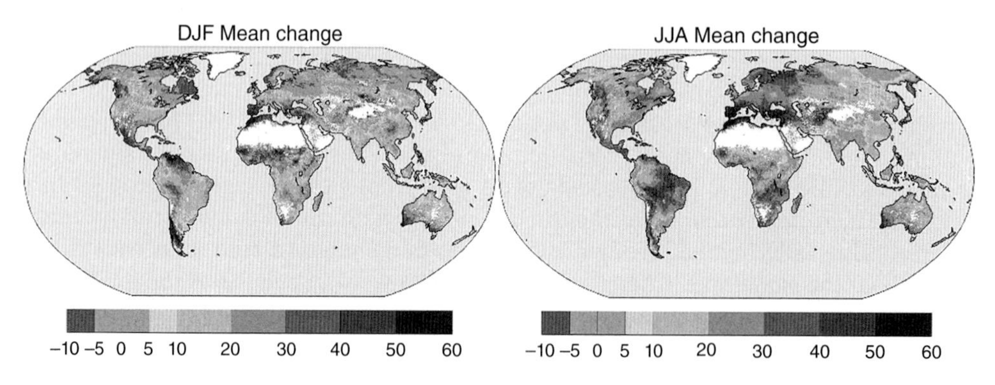

Figure 1.4.2 MME mean change in drought occurrence (in %) between control (1976–2005) and future (2070–2099) simulations under RCP 8.5 forcing for northern winter (December–February, left) and summer (June–August, right). *Source*: Prudhomme et al. (2014) Supplementary Material. Reproduced with permission from authors. (*See colour plate section for the colour representation of this figure.*)

IPSL-CM4) over two periods – the control period (1971–2000) and future period (2071–2100). Most of these GHMs were also used by Prudhomme et al. (2014, Section 1.4.3.3), and the GCMs by Wanders and Van Lanen (2015, Section 1.4.3.1). Low flows and drought were determined from simulated daily discharge (sum of surface and subsurface runoff) series for the selected river basins. The monthly 80th percentile (Q_{80}) was used as a measure of low flows. To focus on anomalies rather than absolute values, all monthly Q_{80} values were normalised with the yearly mean. Drought characteristics were determined with the variable threshold level method (Hisdal et al., 2004; Yevjevich, 1967; see Chapter 1.1), using Q_{80} as the threshold. The threshold was based on the control period, and the same threshold was applied in the future period to identify changes in drought characteristics.

Climate change did not lead to an overall consistent drying or wetting trend in low flows (Q_{80}) for all selected river basins. However, similar changes in Q_{80} were found for river basins within the same major climate zone. The changes in Q_{80} are shown in Figure 1.4.3 for representative river basins in the major climate zones (Köppen–Geiger classification; Peel et al., 2007). Rivers in arid climates and on the border of such climates (B climate) were projected to become drier in the future (i.e. negative mean change in Q_{80}; Figure 1.4.3a). In tropical climates (A climate), changes in Q_{80} were both negative and positive, depending on the model (Figure 1.4.3b–c), although patterns also depend on the geographical location. Projections of changes in low flows for rivers in tropical regions (A climate) in Africa and in the Amazon region were uncertain (Figure 1.4.3b).

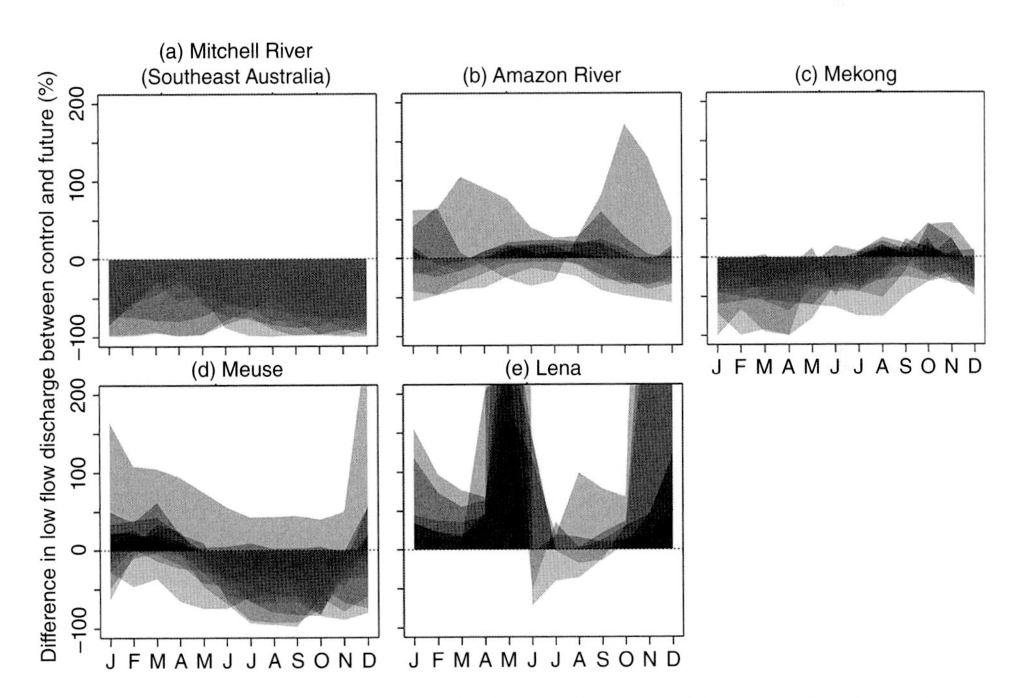

Figure 1.4.3 Relative change in low flows (Q_{80}) between future period (2071–2100) and control period (1971–2000), given as percentage of low discharge in the control period for representative river basins. Red indicates a decrease in Q_{80} in the future period, blue indicates an increase in Q_{80}. Transparency of colours indicates model agreement. (*See colour plate section for the colour representation of this figure.*)

In rivers in other humid regions (e.g. parts of Asia), low flows decreased towards the future (Figure 1.4.3c). In temperate regions (C climate), river basins were under-represented; however, the Meuse river showed an increase in the seasonal cycle, leading to a decrease in low flows in summer (Figure 1.4.3d). Low flows in rivers in cold climates (D and E climates) mainly increased due to a shift in the hydrological regime (Figure 1.4.3e). This shift was characterised by an earlier snow melt peak and less snowfall. A change in snow melt peak was not projected in all rivers, but in most cases the peak occurred 1 month earlier.

The changes found in low flows in the selected river basins were compared with other studies that mainly focused on changes in mean discharge, because other global low flow studies were lacking. The drying trend found in arid river basins corresponds well to other studies (Milly et al., 2005; Nohara et al., 2006; Sperna Weiland et al., 2012; Tang and Lettenmaier, 2012). Changes in low flows in rivers in tropical regions were uncertain; however, other studies also did not show clear trends. Decreases in discharge or runoff in the Amazon regions were reported by Arnell (2003), Arnell and Gosling (2013) and Arora and Boer (2001), whereas increases in mean annual discharge were found by Manabe et al. (2004), Nijssen et al. (2001) and Nohara et al. (2006). For the Meuse, similar changes in discharge were given by De Wit et al. (2007). In cold climates, several studies agree on the shift of the snowmelt peak and increase in low discharge – for example, Arnell and Gosling (2013), Milly et al. (2005), Nohara et al. (2006) and Sperna Weiland et al. (2012).

The changes in low flows were not reflected in the changes in drought characteristics (duration and severity) in all river basins. Low flows decreased in 51% of the river basins, whereas drought characteristics increased in 65%. Some river basins, mainly in tropical regions, did not show a clear signal regarding change in low flows, but drought characteristics increased. In other regions, the changes in seasonality led to increased drought in summer, even though the mean Q_{80} did not decrease. The drought identification method also partly causes differences, because shifts in the hydrological regime can lead to drought events when the threshold of the control period is used.

In general, both low flows and drought characteristics need to be taken into account in order to quantify water availability in the future.

1.4.4 Human influences on future drought

The impact of humans on drought through global warming (emission of greenhouse gasses) is described in the previous section. In this section, the impact of other important human influences on future drought – that is, reservoirs and water use – is explained.

1.4.4.1 Impact of reservoirs on future drought across the globe

Human water use and reservoir management can have a substantial impact on future drought characteristics (Van Dijk et al., 2013; Wanders and Wada, 2015; Van Loon et al., 2016a; 2016b). Generally, water use for agriculture, households, and industry has a negative impact on downstream water availability (Wada et al., 2013). On the other hand, reservoir management can alter drought conditions by alleviating or aggravating drought impact, both locally and downstream (Wanders and Wada, 2015; Rangecroft et al., 2016). The impact of reservoirs is highly dependent on their management, location, and policy.

Wanders and Wada (2015) quantify the impact of reservoirs on changes in drought severity for the twenty-first century. They used a single global hydrological model (PCR-GLOBWB; Van Beek et al., 2011) and forcing datasets from five GCMs that have been downscaled and bias-corrected within the ISI–MIP (Warszawski et al., 2014). Four RCPs (Van Vuuren et al., 2011; Meinshausen et al., 2011) were analysed to quantify the impact of climate change (Section 1.4.3.3).

Wanders and Wada (2015) report that the impact of human water use and reservoir management would be significant in many regions of the world by the end of the twenty-first century (Figure 1.4.4). They found that the increased water holding capacities of basins result in increased coping capacity to severe drought events, resulting in decreased drought severity. This negatively impacts water availability in the high flow season, when most reservoirs increase their storage with excessive runoff or snowmelt. This alleviating impact of reservoirs is even capable of mitigating the impacts of multi-year droughts, due to the relatively high year-round water availability in most regions. This study also found that the alleviating effect of reservoirs is severely hampered in regions with low precipitation excess (e.g. deserts and arid regions, Figure 1.4.4). This is confirmed by Rangecroft et al. (2016), who show that, after multiple drought years, reservoir storage declines significantly, effectively increasing drought impact.

The reservoir type has a major impact on future drought, in that reservoirs that are used for hydropower generally offer drought relief, and reservoirs used for irrigation aggravate downstream drought impacts. The latter type aims at storing water for use in the dry season, thereby depleting downstream water availability. Wanders and Wada (2015) clearly show that regions with high irrigation water use show strong increases in

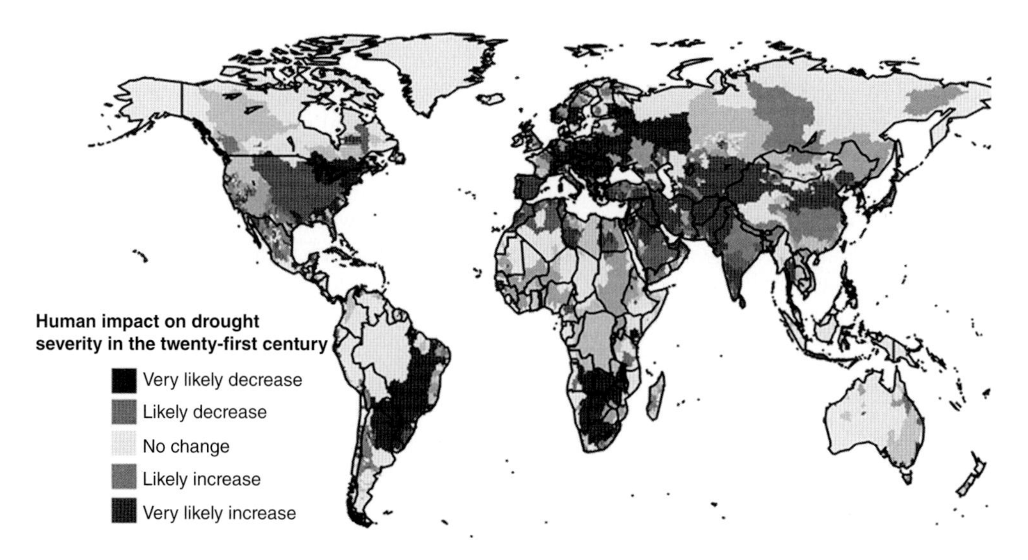

Figure 1.4.4 Impact of human water use and reservoir management on drought severity for the twenty-first century (2070–2099 relative to 1971–2000). (*See colour plate section for the colour representation of this figure.*)

the projected drought severity. Not only do reservoirs impact local drought severity, but the increased impact can be felt throughout the whole catchment.

In most catchments, human alterations to the catchment are more dominant and impactful than climatic changes in the twenty-first century, as has also been demonstrated for the past (Van Loon and Van Lanen, 2013; Wada et al., 2013). This clearly indicates that, for global hydrological drought assessments, people should be aware of the changing nature of the current hydroclimate, on top of a substantial influence of human on the water cycle.

1.4.4.2 Impact of water use on streamflow drought in Europe

The previous section described the impact of reservoirs and water use on future drought across the world. In this section, the impact of water use on streamflow drought in Europe is explained using the model setup introduced by Forzieri et al. (2014). They used a single hydrological model (LISFLOOD) that simulates gridded daily discharge across Europe for the period 1961–2100, using daily output from 12 different climate models. The simulated discharge was used as input for the so-called *threshold approach* (see Chapter 1.1) to obtain streamflow droughts for the past and future – that is, discharge deficit volumes with a 20-year return level for 30-year periods. The outcome was used to investigate the impact of climate change as defined by the IPCC SRES A1B emission scenario (Section 1.4.3.2). Forzieri et al. (2014) also investigated the impact of a certain water use scenario (Economy First, EcF). The EcF scenario was defined on the basis of consistent projections of drivers, such as total population, gross domestic product, agricultural production, thermal electricity production, and technological changes. Under the EcF scenario, use of all available energy sources is encouraged, together with a marked agricultural intensification (Kok et al., 2011). This scenario was selected to drive the WaterGAP3 model (Flörke et al., 2013), which calculates water withdrawal and consumption in different sectors (domestic, tourism, energy, manufacturing, irrigation, and livestock). Forzieri et al. (2014) used the simulated spatially distributed transient water use from this model as input for the hydrological model LISFLOOD.

The simulated change water use in the 2050s with the WaterGAP3 model for the EcF scenario is presented in Figure 1.4.5 (left), as change relative to the twentieth-century control period. In most European river basins, projected water use will increase under the Economy First scenario. In northern, western, and eastern Europe, water use is anticipated to increase owing to the rising need of water for cooling power plants and in the manufacturing sector. In the northern Mediterranean (e.g. northern Iberian Peninsula), water use will likely increase because of the increased irrigation demand, which is the result of developments in the agricultural sector and higher temperatures. Only in the southernmost part of Europe (e.g. southern Spain, major parts of Greece) is water use expected to decrease. In contrast to the northern part of the Mediterranean, water use for irrigation is projected to shrink in southernmost Europe (Flörke et al., 2013).

Change, both in climate and water use (the latter partly reflects on global warming), affects future streamflow drought. The 20-year deficit volume in the 2080s across the European river network has been calculated for two conditions: with climate change only (Figure 1.4.1), and with changes in climate and water use. The difference between these two conditions, that is, the change in 20-year deficit volume due to the

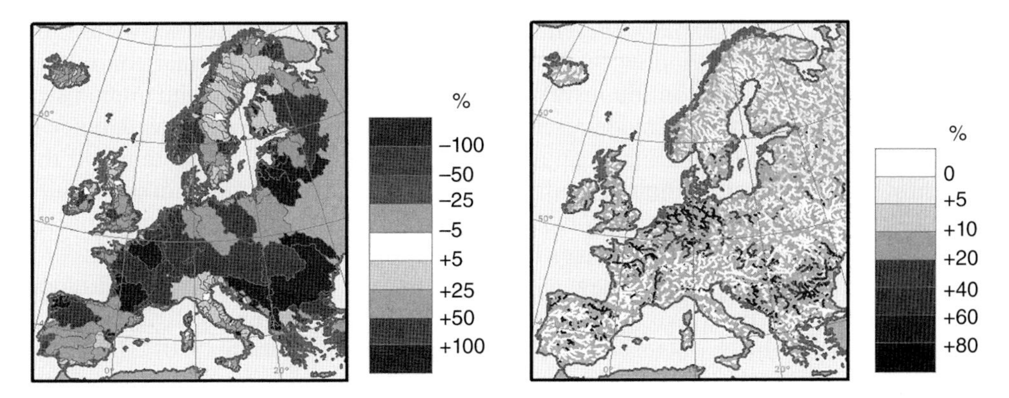

Figure 1.4.5 Future water use and impact on streamflow drought in the 2080s[5] (relative to the control period, 1961–1990). Left: change of total annual water use aggregated to river basin scale for the EcF scenario; right: change in the ensemble average in the 20-year return period deficit volumes due to human water use. *Source*: Derived from Forzieri et al. (2014).

development in water use, is shown in Figure 1.4.5 (right). The higher water use in many European regions leads to an enhancement of the streamflow drought on top of the drought caused by climate change. In particular, rivers in the southern United Kingdom, northern France, Benelux, and northern Germany are expected to suffer from the higher water use. The same applies to some parts of the Balkans and eastern Europe (e.g. Romania). The streamflow droughts under a future climate can be 50% or more severe because of water use changes.

1.4.4.3 Impact of gradual change of hydrological regime on future drought

A changing hydroclimate has significant impact on the 'normal' water availability, especially under human-induced climate change. In standard drought analysis, drought is defined as a deviation from the normal conditions (Tallaksen and Van Lanen, 2004). However, in an ever-changing climate, this 'normal' is changing, begging the question of how to define 'normal' conditions to benchmark our drought severity. This question is especially relevant when we study future projections of drought conditions.

To facilitate the calculation of drought characteristics in the twenty-first century, Wanders et al. (2015) defined the *transient threshold approach*. This method calculates the normal conditions from the previous 30-year period, using a running temporal window. A clear advantage of this method is that it can be applied under fast-changing conditions and still provide a useful drought signal. A downside of this method is that the reference frame is constantly changing, which makes it more difficult to quantitatively compare drought trends.

In Wanders et al. (2015), the authors suggest that, for proper drought quantification, researchers should analyse the changes in both the threshold and drought characteristics. The changes in the threshold indicate the impact of changes in the 'normal' conditions and help to benchmark drought quantification, while the changes in

[5] Water use scenarios run until 2050. Forzieri et al. (2014) assumed that water consumption remains unchanged for the remaining period (2051–2100).

drought characteristics explain the changes in drought impact. They show that, by using a 30-year moving window to determine the climatological normal conditions, they better represent the continuously ongoing adaptation in natural and human systems. As an example, they provide a scenario where the natural water availability is continuously reducing, resulting in an adaption from the natural ecosystem and the human system in adjusting their agricultural water demand to the new 'normal' conditions. Another cited example involves glaciered regions that suffer from glacier retreat, resulting in significant changes in the climatological 'normal' conditions.

In their work, they apply this new methodology to an ensemble of future drought projections, derived from five GCMs and four RCPs (Van Vuuren et al., 2011). They simulate the drought characteristics and thresholds for the period 1960–2100, using both the conventional and newly developed transient threshold approach on the daily discharge values.

They show that most regions in the world are expected to suffer from reduced water availability due to increasing temperature and evapotranspiration. Exceptions are found in the boreal and polar regions in the world. In these regions, increased melt from glaciers and increasing precipitation totals will cause an increase in the annual average discharge, resulting in an increase in the total water availability.

Wanders et al. (2015) show that the impact of the transient approach is highest in regions that show strong changes in future water availability (e.g. Mediterranean, Africa, South America, and the United States). These regions see strong changes in annual water availability and the resulting drought characteristics, leading to a high impact of the transient threshold approach. On the global scale, they find that drought adaptation can mitigate around 50% of the drought impact by reducing the average drought impact area by a factor of two.

These results indicate that research focused on future drought impacts should include the ever-changing nature of both the natural and human systems to facilitate accurate drought impact assessments. They suggest using the transient drought threshold approach in combination with a conventional constant threshold to facilitate benchmarking of the changes in the hydrological regime. Finally, they conclude that using only the conventional approach results in an underestimation of the drought adaptation and, therefore, an overestimation of how the drought impact will be perceived in many regions of the world.

1.4.5 Uncertainties in future drought

Projections of future drought are uncertain by definition, because the pathways of atmosphere emissions of greenhouse gases and physical processes of the land–atmosphere–ocean system are only partly known (Section 1.4.2). Different approaches are used to define, model, and identify the uncertainties connected to these unknowns. A few recent approaches are described in the following text.

1.4.5.1 Uncertainty assessments

Some uncertainty approaches intercompare simulated streamflow drought events against those derived from observed discharge. It is hypothesised that, if a modelling approach adequately performs on observed data, it is also able to satisfactorily predict future data. Forzieri et al. (2014) validated the river deficit volumes (*Def*, Sections

1.4.3.2 and 1.4.4.2) with a set of about 450 gauging stations across Europe for the period 1961–1990. It appeared that LISFLOOD, driven by a model ensemble of bias-corrected regional climate simulations, is able to reproduce *Def* values reasonably well for the non-frost season across a wide range of geo-climate settings in Europe. Simulations were less reliable for the frost season due to conceptualisation of frost and snow processes, and uncertainties in the input winter precipitation and temperatures. In addition, Forzieri et al. (2014) investigated the uncertainty of future drought by studying the consistency in projections among the 12 ensemble members. They counted the number of ensemble members (out of 12) for each river pixel that agreed on an increase or decrease. Consistency increases with time. By the end of the twenty-first century (2080s), simulated future drought will become consistent in large areas of Europe. Most models agree on an increase of streamflow drought in the Mediterranean, the Balkans, France, Benelux, and the United Kingdom, and a decrease in Scandinavia, the Baltic countries, and northeast Poland. Models disagree in sign (low consistency) in a belt stretching from northern Germany, over the Alpine countries, and into Ukraine, which approximately coincides with the area with no or small changes in the 20-year return period of the discharge deficit (Figure 1.4.1).

Van Huijgevoort et al. (2014) compared the low flows (Q_{80}; see Section 1.4.3.4) simulated by the models with observed discharge data from GRDC to estimate the uncertainty in the model results. To reduce the uncertainty in future projections, a selection of model combinations (GHM and GCM) was made for the analysis of future changes based on this comparison. Since only the selected models were used for the low flow analysis, the range in model estimates was reduced. From the comparison with observed data, it was found that, in most river basins, the hydrological model had a larger influence on the results than the forcing datasets from different GCMs. This corresponds with the findings of Hagemann et al. (2013). Information might get lost by selecting only a few hydrological models based on their performance in the past (e.g. Gosling et al., 2011; Reifen and Toumi, 2009); however, uncertainty can be reduced when models are constrained by observations (Hall and Qu, 2006; Stegehuis et al., 2013).

In addition to the mean signal of changes, the MME contains information on the uncertainty in the projection. MME mean changes in drought occurrence and severity, discussed in Section 1.4.3.3 (Signal), were divided by the spread of the ensemble measured by its inter-quartile range (Noise). For annual drought occurrence, strong Signal-to-Noise ratios (S2N > 1) were found in southern Europe, Middle East, southeast United States, Chile, and southwest Australia, which can be considered as possible hotspots for future water security issues. The variability in the signal due to different GIMs was found to be larger than that of the different GCMs. The increase in the occurrence of severe droughts was found to be statistically significant, with nearly half of the ensemble projecting drought severity affecting more than 40% of the land masses under the RCP 8.5 by the end of the twenty-first century. This drought increase signal was not found, however, for the unique model that accounts for the dynamic response of plants to CO_2 and climate (JULES), which considers more effective plant respiration and lower transpiration reduced under enriched CO_2 atmosphere, despite the associated warming. Considering a diverse range of GIMs is hence critical for a more robust assessment of the impact of climate change on hydrology.

In the context of the DROUGHT–R&SPI project, Van Lanen et al. (2015), Alderlieste and Van Lanen (2013) and Alderlieste et al. (2014) also investigated the S2N ratios for a number of case study areas across Europe. The data from this case study were obtained from the EU-WATCH project that covered the whole of Europe (Harding et al., 2011). In total, data from 244 grid cells was retrieved to cover the cases. Time series of runoff simulated with six GHMs (almost identical to Van Huijgevoort et al., 2014; Section 1.4.3.4) were used for the uncertainty analysis. These time series were available for each grid cell, each GHM was forced with daily meteorological variables from a re-analysis data-set (1971–2000) (WFD, Weedon et al., 2011) and three GCMs (control period: 1971–2100, intermediate future: 2021–2050, far future: 2071–2100). Similar to Van Huijgevoort (2014, Section 1.4.3.4), GCMs such as CNRM, ECHAM5 and IPSL for two emission scenarios (A2 and B1) were used. First, they calculated the MME mean runoff (MME-R) leading per grid cell to: (i) four MME-R time series for the control period (WFD, three GCMs), (ii) three MME-R time series (three GCMs) for each of two emission scenarios for the intermediate future, and (iii) three MME-R time series (three GCMs) for each of two emission scenarios for the far future. Second, the following flow characteristics were determined for each MME-R time series and grid cell: (i) mean annual flow, (ii) mean monthly flow, and (iii) May–November MAM7 – that is, the mean annual 7-day minimum runoff, over May–November. The same was done to obtain the drought characteristics (duration, deficit volume) using the variable threshold method. The smoothed monthly thresholds were different for each combination [GCM;GHM]; and, to assess future drought, the threshold of the control period was used.

The analysis for the control period showed that the multi-model annual flow was clearly overestimated when using GCM forcing instead of WFD forcing (i.e. the noise). The overestimation varies between 37% and 55% (Figure 1.4.6, top left, most left bar of each of the GCMs) (i.e. uncertainty in flow characteristics). The noise in the MAM7 is substantially smaller for two GCMs (<25%), but substantial for CNRM (>100%, Figure 1.4.6, top right). The average drought duration is slightly underestimated with the GCM forcing relative to the WFD forcing (10–16%) (Figure 1.4.6, bottom left, most left bar of each of the GCMs) (i.e. noise in drought characteristic). The difference in drought deficit volume between the GCM and WFD forcing is not mono-directional; the deficit is underestimated by 3% when the IPSL forcing was used, whereas the overestimation varies between 22% and 32% in case of CNRM and ECHAM5 forcing (Figure 1.4.6, bottom right). The climate change signal in the annual runoff is opposite to the direction of the noise, and, for most projections, smaller than the noise (except [IPSL;B1]) (Figure 1.4.6, top left). In most cases, the signal for the MAM7 is larger than the noise (except CNRM). Of the projections, 50% have S2N $\geq$ 2. The climate change signal in the drought characteristics (both duration and deficit volume) is, for all projections, larger than the noise. For instance, the drought duration is expected to increase by at least 50% for the A2 scenario (Figure 1.4.6, bottom left) with S2N > 3. The percentage change in the deficit volume can be very large (Figure 1.4.6, bottom right), which is predominantly caused by the rather low number of the deficit volume.

1.4.5.2 Impact sources of uncertainty

A formal comparison of different sources of uncertainty can be made using the Analysis Of VAriance (ANOVA) technique (Hawkins and Sutton, 2011), where the uncertainty associated with different sources can be compared with each other and with that from

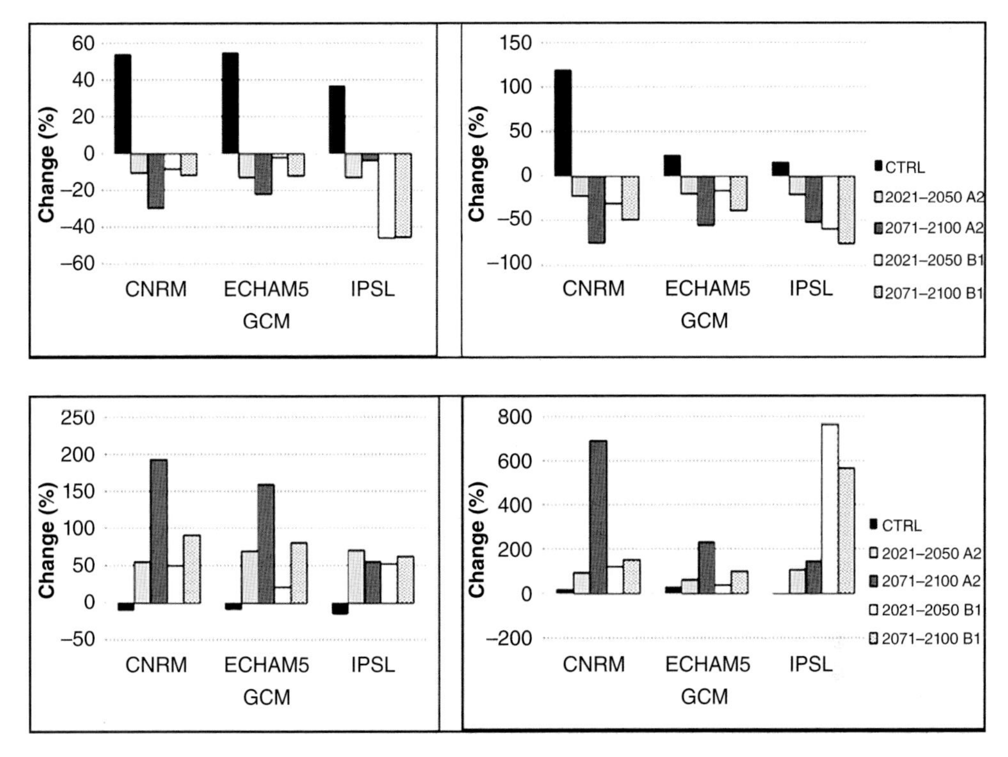

Figure 1.4.6 Projected change (signal) in mean annual runoff (top left); mean annual 7-day minimum runoff, MAM7 (top right); average drought duration in runoff (bottom left); deficit volume in runoff (right) for three GCMs, two emission scenarios (A2 and B1), and the intermediate and far future. 'CTRL' specifies the difference between runoff (top)/drought characteristics (bottom) simulated with the GCM and the runoff/drought characteristics simulated with re-analysis data (WFD), that is, the noise. Results are shown for the average change of all 244 grid cells. *Source*: Derived from Alderlieste et al. (2014).

internal variability. Excluding the JULES model from the ensemble used by Prudhomme et al. (2014), Giuntoli et al. (2015) quantified the uncertainty in drought change signal by the end of the twenty-first century, which is attributable to GIMs and GCMs, using a two-way ANOVA method. They found that, globally, uncertainty in GCMs accounts for 43% of the total uncertainty, as compared to 35% for GIMs, but that strong regional variation exists, with GIMs being a dominant source of uncertainty in some northern (e.g. northeast Russia) and southern (e.g. southern Africa, southwestern Australia) regions. A similar pattern was found for July–August, while internal variability globally dominates in northern winter (December–February). When further investigating the contribution of GCMs and GIMs to the overall uncertainty in different climatic zones, they found that GIM-related uncertainty dominates in snow- and ice-dominated regions, and in arid zones. Dominant processes in low flow generation (and, at its extreme, droughts) are influenced by land–surface processes including evaporation, infiltration, and storage. The marked part of uncertainty due to GIMs is likely to reflect the differences in which GIMs conceptualise and parametrise those processes. The spread due to land–surface processes was also identified by Orlowsky and

Seneviratne (2013) on the soil moisture anomaly signal, based on a 28-member GCM ensemble from CMIP5.

1.4.6 Conclusions – Future needs

1.4.6.1 Conclusions

In the last 15 years, several different modelling approaches have been introduced to explore future hydrological drought (drought in groundwater and rivers) as a response to global warming. They built upon studies about projections of annual river flow, which appeared to be insufficient for hydrological extremes. Since the beginning, it has been common to use multi-emission scenarios. The IPCC SRES for the twenty-first century have been used until recently, but these are gradually being replaced with RCPs. Recent hydrological drought projections are driven by these RCPs. Most investigations also used outcomes from multiple GCMs, providing the large-scale hydrological models (GHMs) with an ensemble of possible future climates for each emission scenario. In the first phase, single GHMs were applied to study future drought, which were gradually replaced with multiple GHMs that accounted for the differences in the structures of these models. Eventually, the spread in future hydrological drought addresses possible emission scenarios and different structures of GCMs and GHMs. The results of studies on future hydrological drought over the last 15 years reported in this chapter have to be interpreted in the context of this development of approaches. Although the use of an ensemble of multi-models is advocated, it does not mean that single-model studies should be discarded. Single-model applications (e.g. Section 1.4.4) are still valuable for testing specific hypotheses that could guide further multi-model studies.

The studies on future hydrological drought (Section 1.4.3) show that future drought (total duration, severity) is likely to become more extreme in many European regions. There are specific differences between regions, depending on the climate. For example, the Mediterranean is identified as a hotspot in many studies. Regional differences are driven by changes in the dominant hydrological processes associated with global warming. One of the best examples is the impact of changes in snow accumulation and melt on the river flow regime. Plotting of river flow drought on the pan-European river network (Figure 1.4.1) clearly illustrates how future hydrological drought is projected to develop.

Studies that include the influence of human interferences on the assessment of future hydrological drought (Section 1.4.4) clearly illustrate the impact of humans, which could lead to either an alleviation or an enhancement of drought. In many river basins, human alterations play a more dominant role than climate change in the twenty-first century. An important issue is the baseline period that is used to assess the change in hydrological drought (Section 1.4.4.3), whether 1971–2000 is used (static approach), or a gradually forwards-moving time window (e.g. 30 years) into the twenty-first century to account for slow adaptation.

Uncertainty is inherent to the assessment of future hydrological drought. Different approaches are reported (Section 1.4.5). Some methods are based on the analysis of the outcome of the MME ([GCM;GHM] combinations) for the future – for example, count the number of ensemble members (i.e. combinations of) pointing in the same direction (increase or decrease of percentage change relative to the control period).

Others compare the spread in the outcome of the MME (Noise) with the median or mean of the ensemble (Signal). Climate change projections are assumed to be more credible for high Signal-to-Noise ratios. Another group of approaches investigates [GCM;GHM] performance in the control period against observations. It is assumed that performance in the control period is a relevant measure for the credibility of assessment in the future. For instance, one method compares the low flow regime simulated with a [GCM;GHM] combination against a regime simulated with the same GHM driven with observed weather. Based on the performance, some [GCM;GHM] combinations were excluded for the assessment of future drought. All reported uncertainty assessments provide relevant, but different, information, which makes them hard to compare. Clearly, a distinction needs to be made between the projection of flow characteristics and drought characteristics, which are a derivate of the flows. It seems that the noise in drought characteristics is lower than in flow characteristics.

Model intercomparison projects, such as WaterMIP and ISI-MIP, bringing together expertise from tens of different hydroclimatic modelling groups, is good way to progress on knowledge and skills to explore future hydrological extremes, including drought. These consortia enable the creation of large global databases with consistent forcing data (e.g. weather, human interferences) to run the comprehensive chain of multi-emission scenarios, multi GCMs, and multi GHMs; to efficiently store the output data; and to analyse these using an a-priori-agreed consistent procedure.

1.4.6.2 Future needs

The work on the influence of human water needs and associated interferences on future hydrological drought, as reported in Section 1.4.4, needs to strengthened – for instance, by integrating human interferences, instead of separate analysis. Possible feedbacks between the human and natural systems also warrant more attention (Van Loon et al., 2016a; 2016b), which is also strongly encouraged by programmes such as 'Panta Rhei', under the umbrella of the International Association of Hydrological Sciences' programme on socio-hydrology (Montanari et al., 2013). In this context, pragmatic approaches to assess future drought that gradually assume changes in the baseline (Section 1.4.4.3) should be investigated to check whether tipping points are passed in the environmental and economic systems (e.g. Scheffer et al., 2012).

Model intercomparison projects, such as WaterMIP and ISI-MIP, should be regularly repeated if new knowledge becomes available on emission scenarios, GCMs, or GHMs. These projects should be put on the international research agenda, and funding agencies should reserve budget to make them possible. As illustrated in this chapter, the different uncertainty approaches provide various aspects of uncertainty in the assessment of future drought. A concerted action is required to bring the different approaches together in one conceptual framework, which clearly states the multi-model chain and data needed, the outputs to be generated, and which removes possible inconsistencies.

Acknowledgements

Funding was provided by the EU WATCH project (WATer and global CHange), contract 036946, and the EU project DROUGHT–R&SPI (Fostering European Drought

Research and Science–Policy Interfacing), contract 282769. In the last phase, support came from the EU H2020 ANYWHERE project (contract no. 700099). Niko Wanders was supported by the Netherlands Organisation for Scientific Research (NWO), Rubicon Fellowship 825.15.003. The study is also a contribution to UNESCO IHP–VII programme (Euro FRIEND-Water project). The authors thank all data providers and national authorities for the observed streamflow records.

References

Alderlieste, M.A.A. and Van Lanen, H.A.J. (2013). Change in Future Low Flow and Drought in Selected European Areas Derived from WATCH GCM Forcing Dataset and Simulated Multi-Model Runoff. *DROUGHT–R&SPI Technical Report No. 5*, 316 p. Available from: http://www.eu-drought.org/technicalreports.

Alderlieste, M.A.A., Van Lanen, H.A.J., and Wanders, N. (2014). Future low flows and hydrological drought: how certain are these for Europe? In: *Hydrology in a Changing World: Environmental and Human Dimensions* (ed. T.M. Daniell, H.A.J. Van Lanen, S. Demuth, G. Laaha, E. Servat, G. Mahe, J.-F. Boyer, J.-E. Paturel, A. Dezetter, and D. Ruelland), 60–65. Wallingford: IAHS Publ. No. 363.

Arnell, N.W. (2003). Effects of IPCC SRES* emissions scenarios on river runoff: a global perspective. *Hydrology and Earth System Sciences* 7: 619–641. doi:10.5194/hess-7-619-2003.

Arnell, N.W. and Gosling, S.N. (2013). The impacts of climate change on river flow regimes at the global scale. *Journal of Hydrology* 486: 351–364. doi:10.1016/j.jhydrol.2013.02.010.

Arora, V.K. and Boer, G.J. (2001). Effects of simulated climate change on the hydrology of major river basins. *Journal of Geophysical Research* 106: 3335–3348. doi:10.1029/2000JD900620.

Best, M.J., Pryor, M., Clark, D.B., Rooney, G.G., Essery, R.L.H., Ménard, C.B., Edwards, J.M., Hendry, M.A., Porson, A., Gedney, N., Mercado, L.M., Sitch, S., Blyth, E., Boucher, O., Cox, P.M., Grimmond, C.S.B., and Harding, R.J. (2011). The joint UK land environment simulator (JULES), model description – part 1: energy and water fluxes. *Geoscientific Model Development* 4: 677–699.

Bondeau, A., Smith, P., Zaehle, S., Schaphoff, S., Lucht, W., Cramer, W., Gerten, D., Lotze-Campen, H., Müller, C., Reichstein, M., and Smith, B. (2007). Modelling the role of agriculture for the 20th century global terrestrial carbon balance. *Global Change Biology* 13: 679–706.

Burke, E.J. and Brown, S.J. (2008). Evaluating uncertainties in the projection of future drought. *Journal of Hydrometeorology* 9: 292–299. doi: 10.1175/2007JHM929.1.

Corzo Perez, G.A., Van Lanen, H.A.J., Bertrand, N., Chen, C., Clark, D., Folwell, S., Gosling, S., Hanasaki, N., Heinke, J., and Voss, F. (2011). Drought at the Global Scale in the 21st Century. *WATCH Technical Report No. 43*, 117 pg. Available at: http://www.eu-watch.org/publications/technical-reports/2.

De Wit, M.J.M., Van Den Hurk, B., Warmerdam, P.M.M., Torfs, P.J.J.F., Roulin, E., and Van Deursen, W.P.A. (2007). Impact of climate change on low-flows in the river Meuse. *Climate Change* 82: 351–372.

Döll, P., Kaspar, F., and Lehner, B. (2003). A global hydrological model for deriving water availability indicators: model tuning and validation. *Journal of Hydrology* 270: 105–134.

Flörke, M., Kynast, E., Bärlund, I., Eisner, S., Wimmer, F., and Alcamo, J. (2013). Domestic and industrial water uses of the past 60 years as a mirror of socio-economic development: a global simulation study. *Global Environmental Change* 23: 144–156.

Forzieri, G., Feyen, L., Rojas, R., Flörke, M., Wimmer, F., and Bianchi, A. (2014). Ensemble projections of future streamflow droughts in Europe. *Hydrology and Earth System Sciences* 18: 85–108. doi:10.5194/hess-18-85-201418-85-2014.

Giuntoli, I., Vidal, J.-P., Prudhomme, C., and Hannah, D.M. (2015). Future hydrological extremes: the uncertainty from multiple global climate and global hydrological models. *Earth System Dynamics* 6: 267–285. doi:10.5194/esd-6-267-2015.

Gosling, S.N., Taylor, R.G., Arnell, N.W., and Todd, M.C. (2011). A comparative analysis of projected impacts of climate change on river runoff from global and catchment-scale hydrological models. *Hydrology and Earth System Sciences* 15: 279–294. doi:10.5194/hess-15-279-2011.

Haddeland, I., Clark, D.B., Franssen, W., Ludwig, F., Voß, F., Arnell, N.W., Bertrand, N., Best, M., Folwell, S., Gerten, D., Gomes, S., Gosling, S.N., Hagemann, S., Hanasaki, N., Harding, R., Heinke, J., Kabat, P., Koirala, S., Oki, T., Polcher, J., Stacke, T., Viterbo, P., Weedon, G.P., and Yeh, P. (2011). Multimodel estimate of the global terrestrial water balance: setup and first results. *Journal of Hydrometeorology* 12: 869–884.

Hagemann, S., Chen, C., Clark, D.B., Folwell, S., Gosling, S.N., Haddeland, I., Hanasaki, N., Heinke, J., Ludwig, F., Voss, F., and Wiltshire, A.J. (2013). Climate change impact on available water resources obtained using multiple global climate and hydrology models. *Earth System Dynamics* 4: 129–144. doi:10.5194/esd-4-129-2013.

Hall, A. and Qu, X. (2006). Using the current seasonal cycle to constrain snow albedo feedback in future climate change. *Geophysical Research Letters* 33: L03502. doi:10.1029/2005GL025127.

Hanasaki, N., Kanae, S., Oki, T., Masuda, K., Motoya, K., Shirakawa, N., Shen, Y., and Tanaka, K. (2008). An integrated model for the assessment of global water resources – part 1: model description and input meteorological forcing. *Hydrology and Earth System Sciences* 12: 1007–1025.

Harding, R., Best, M., Blyth, E., Hagemann, S., Kabat., P., Tallaksen, L.M., Warnaars, T., Wiberg, D., Weedon, G.P., van Lanen, H.A.J., Ludwig, F., and Haddeland, I. (2011). Water and Global Change (WATCH) special collection: current knowledge of the terrestrial Global Water Cycle. *Journal of Hydrometeorology* 12(6): 1149–1156. doi: 10.1175/JHM-D-11-024.1.

Hawkins, E. and Sutton, R. (2011). The potential to narrow uncertainty in projections of regional precipitation change. *Climate Dynamics* 37: 407–418.

Hempel, S., Frieler, K., Warszawski, L., Schewe, J., and Piontek, F. (2013). A trend-preserving bias correction; the ISI-MIP approach. *Earth System Dynamics* 4: 219–236. doi.org/10.5194/esd-4-219-2013.

Hisdal, H., Tallaksen, L.M., Clausen, B., Peters, E., and Gustard, A. (2004). Drought characteristics. In: *Hydrological Drought: Processes and Estimation Methods for Streamflow and Groundwater. Developments in Water Science. 48* (ed. L.M. Tallaksen and H. Van Lanen), 139–198. Amsterdam: Elsevier.

Kok, K., Van Vliet, M., Bärlund, I., Dubel, A., and Sendzimir, J. (2011). Combining participative backcasting and explorative scenario development: experiences from the SCENES project. *Technological Forecasting and Social Change* 78: 835–851.

Liang, X., Lettenmaier, D.P., Wood, E.F., and Burges, S.J. (1994). A simple hydrologically based model of land surface water and energy fluxes for general circulation models. *Journal of Geophysical Research: Atmospheres* 99: 14415–14428.

Manabe, S., Milly, P.C.D., and Wetherald, R. (2004). Simulated long-term changes in river discharge and soil moisture due to global warming. *Hydrological Sciences Journal* 49: 642. doi:10.1623/hysj.49.4.625.54429.

Meinshausen, M., Smith, S.J., Calvin, K., Daniel, J.S., Kainuma, M.L.T., Lamarque, J.F., Matsumoto, K., Montzka, S.A., Raper, S.C.B., Riahi, K., Thomson, A., Velders, G.J.M., and Vuuren, D.P.P. (2011). The RCP greenhouse gas concentrations and their extensions from 1765 to 2300. *Climatic Change* 109: 213–241.

Milly, P.C.D., Dunne, K.A., and Vecchia, A.V. (2005). Global pattern of trends in streamflow and water availability in a changing climate. *Nature* 438: 347–350. doi:10.1038/nature04312, 2005.

Montanari A., Young, G., Savenije, H.H.G., Hughes, D., Wagener, T., Ren, L.L., Koutsoyiannis, D., Cudennec, C., Toth, E., Grimaldi, S., Blöschl, G., Sivapalan, M., Beven, K., Gupta, H., Hipsey, M., Schaefli, B., Arheimer, B., Boegh, E., Schymanski, S.J., Di Baldassarre, G., Yu, B., Hubert, P., Huang, Y., Schumann, A., Post, D., Srinivasan, V., Harman, C., Thompson, S., Rogger, M., Viglione, A., McMillan, H., Characklis, G., Pang, Z., and Belyaev, V. (2013). Panta Rhei – Everything flows: change in hydrology and society – The IAHS Scientific

Decade 2013–2022. *Hydrological Sciences Journal* 58(6): 1256–1275. doi:10.1080/02626667.2013.809088.

Nakicenovic, N. and Swart, R. (Eds.) (2000). *IPCC Special Report on Emission Scenarios*. Cambridge, UK: Cambridge University Press.

Nijssen, B., O'Donnell, G.M., Lettenmaier, D.P., Lohmann, D., and Wood, E.F. (2001). Predicting the discharge of global rivers. *Journal of Climate* 14: 3307–3323. doi:10.1175/1520-0442(2001)014<3307:PTDOGR>2.0.CO;2.

Nohara, D., Kitoh, A., Hosaka, M., and Oki, T. (2006). Impact of climate change on river discharge projected by multimodel ensemble. *Journal of Hydrometeorology* 7: 1076–1089. doi:10.1175/JHM531.1.

Orlowsky, B. and Seneviratne, S.I. (2013). Elusive drought: uncertainty in observed trends and short- and long-term CMIP5 projections. *Hydrology and Earth System Sciences* 17: 1765–1781. doi:10.5194/hess-17-1765-2013.

Peel, M.C., Finlayson, B.L., and Mcmahon, T.A. (2007). Updated world Köppen–Geiger climate classification map. *Hydrology and Earth System Sciences* 11: 1633–1644. doi:10.5194/hess-11-1633-2007.

Prudhomme, C., Parry, S., Hannaford, J., Clark, D.B., Hagemann, S., and Voss, F. (2011). How well do large-scale models reproduce regional hydrological extremes in Europe? *Journal of Hydrometeorology* 12: 1181–1204.

Prudhomme, C., Giuntoli, I., Robinson, E.L., Clark, D.B., Arnell, N.W., Dankers, R., Fekete, B.M., Franssen, W., Gerten, D., Gosling, S.N., Hagemann, S., Hannah, D.M., Kim, H., Masaki, Y., Satoh, Y., Stacke, T., Wada, Y., and Wisser, D. (2014). Hydrological droughts in the 21st century, hotspots and uncertainties from a global multimodel ensemble experiment. *Proceedings of the National Academy of Sciences* 111(9): 3262–3267. doi:10.1073/pnas.1222473110.

Rangecroft, S., Van Loon, A.F., Maureira, H., Verbist, K., and Hannah, D.M. (2016). Multi-method assessment of reservoir effects on hydrological droughts in an arid region. *Earth System Dynamics Discussions*. doi:10.5194/esd-2016-57.

Reifen, C. and Toumi, R. (2009). Climate projections: past performance no guarantee of future skill? *Geophysical Research Letters* 36 (13): 13704. doi:10.1029/2009GL038082.

Scheffer, M., Carpenter, S.R., Lenton, T.M., Bascompte, J., Brock, W., Dakos, V., Van de Koppel, J., Van de Leemput, I.A., Levin, S.A., Van Nes, E.H., Pascual, M., and Vandermeer, J. (2012). Anticipating critical transitions. *Science* 338: 344–348. doi: 10.1126/science.

Schewe, J., Heinke, J., Gerten, D., Haddeland, I., Arnell, N.W., Clark, D.B., Dankers, R., Eisner, S., Fekete, B.M., Colón-González, F.J., Gosling, S.N., Kim, H., Liu, X., Masaki, Y., Portmann, F.T., Satoh, Y., Stacke, T., Tang, Q., Wada, Y., Wisser, D., Albrecht, T., Frieler, K., Piontek, F., Warszawski, L., and Kabat, P. (2014). Multimodel assessment of water scarcity under climate change. *Proceedings of the National Academy of Sciences* 111(9): 3245–3250. doi/10.1073/pnas.1222460110.

Seneviratne, S.I., Nicholls, N., Easterling, D., Goodess, C.M., Kanae, S., Kossin, J., Luo, Y., Marengo, J., McInnes, K., Rahimi, M., Reichstein, M., Sorteberg, A., Vera, C., and Zhang, X. (2012). Changes in climate extremes and their impacts on the natural physical environment. In: *Managing the Risks of Extreme Events and Disasters to Advance Climate Change Adaptation. A Special Report of Working Groups I and II of the Intergovernmental Panel on Climate Change (IPCC)* (ed. C.B. Field, V. Barros, T.F. Stocker, D. Qin, D.J. Dokken, K.L. Ebi, M.D. Mastrandrea, K.J. Mach, G.K. Plattner, S.K. Allen, M. Tignor, and P.M. Midgley), 190–230. Cambridge, UK and New York, NY, USA: Cambridge University Press.

Sheffield, J. and Wood, E.F. (2011). *Drought: Past Problems and Future Scenarios*. London: Earthscan.

Sperna Weiland, F.C., Van Beek, L.P.H., Kwadijk, J.C.J., and Bierkens, M.F.P. (2012). Global patterns of change in discharge regimes for 2100. *Hydrology and Earth System Sciences* 16: 1047–1062. doi:10.5194/hess-16-1047-2012.

Spinoni, J., Naumann, G. Vogt, J., and Barbosa, P. (2016). *Meteorological Droughts in Europe. Events and Impacts Past Trends and Future Projections.* Luxembourg, EUR 27748 EN: Publications Office of the European Union. doi:10.2788/450449.

Spinoni, J., Naumann, G., and Vogt, J.V. (2017). Pan-European seasonal trends and recent changes of drought frequency and severity. *Global and Planetary Change* 148: 113–130. http://dx.doi.org/10.1016/j.gloplacha.2016.11.013.

Stacke, T. and Hagemann, S. (2012). Development and evaluation of a global dynamical wetlands extent scheme. *Hydrology and Earth System Sciences* 16: 2915–2933.

Stagge, J.H., Rizzi, J., Tallaksen, L.M., and Stahl, K. (2015). Future meteorological drought: projections of regional climate models for Europe. *DROUGHT-R&SPI Technical Report No. 25*. Oslo. Available from: http://www.eu-drought.org/technicalreports/3, accessed 27 March 2018.

Stegehuis, A.I., Teuling, A.J., Ciais, P., Vautard, R., and Jung, M. (2013). Future European temperature change uncertainties reduced by using land heat flux observations. *Geophysical Research Letters* 40: 2242–2245. doi:10.1002/grl.50404.

Takata, K., Emori, S., and Watanabe, T. (2003). Development of the minimal advanced treatments of surface interaction and runoff. *Global and Planetary Change* 38: 209–222.

Tallaksen, L.M. and Van Lanen, H.A.J. (2004). *Hydrological Drought: Processes and Estimation Methods for Streamflow and Groundwater. No. 48 in Development in Water Science.* Amsterdam: Elsevier Science B.V.

Tang, Q. and Lettenmaier, D.P. (2012). 21st century runoff sensitivities of major global river basins. *Geophysical Research Letters* 39. doi:10.1029/2011GL050834.

Taylor, K.E., Stouffer, R.J., and Meehl, G.A. (2011). An overview of CMIP5 and the experiment design. *Bulletin of the American Meteorological Society* 93: 485–498.

Van Beek, L.P.H., Wada, Y., and Bierkens, M.F.P. (2011). Global monthly water stress: I. Water balance and water availability. *Water Resources Research* 47: W07517. doi: 10.1029/2010WR009791.

Van Dijk, A.I.J.M., Beck, H.E., Crosbie, R.S., de Jeu, R.A.M., Liu, Y.Y., Podger, G.M., Timbal, B., and Viney, N.R. (2013). The Millennium Drought in southeast Australia (2001–2009): natural and human causes and implications for water resources, ecosystems, economy, and society. *Water Resources Research* 49: 1040–1057. doi: 10.1002/wrcr.20123.

Van Huijgevoort, M.H.J., Van Lanen, H.A.J., Teuling, A.J., and Uijlenhoet, R. (2014). Identification of changes in hydrological drought characteristics from a multi-GCM driven ensemble constrained with observed discharge. *Journal of Hydrology* 512: 421–434. doi.org/10.1016/j.jhydrol.2014.02.060.

Van Der Knijff, J., Younis, J., and De Roo, A. (2010). LISFLOOD: a GIS-based distributed model for river basin scale water balance and flood simulation. *International Journal of Geographical Information Science* 24: 189–212. doi:10.1080/13658810802549154.

Van Lanen, H.A.J., Wanders, N., Tallaksen, L.M., and Van Loon, A.F. (2013). Hydrological drought across the world: impact of climate and physical catchment structure. *Hydrology and Earth System Sciences* 17: 1715–1732. doi:10.5194/hess-17-1715-2013.

Van Lanen, H.A.J., Tallaksen, L.M., Stahl, K., Assimacopoulos, D., Wolters, W., Andreu, J., Rego, F., Seneviratne, S.I., De Stefano, L., Massarutto, A., Garnier, E. and Seidl, I. (2015). Fostering Drought Research and Science–Policy Interfacing: achievements of the DROUGHT–R&SPI project. In: *Drought: Research and Science–Policy Interfacing* (ed. J. Andreu, A. Solera, J. Paredes-Arquiola, D. Haro-Monteagudo, and H.A.J. Van Lanen), 3–12. Boca Raton, London, New York, Leiden: CRC Press.

Van Loon, A.F. and Van Lanen, H.A.J. (2013). Making the distinction between water scarcity and drought using an observation-modeling framework. *Water Resources Research* 49: 1483–1502. doi:10.1002/wrcr.20147.

Van Loon, A.F., Gleeson, T., Clark, J., Van Dijk, A., Stahl, K., Hannaford, J., Di Baldassarre, G., Teuling, A., Tallaksen, L.M., Uijlenhoet, R., Hannah, D.M., Sheffield, J., Svoboda, M., Verbeiren,

B., Wagener, T., Rangecroft, S., Wanders, N., and Van Lanen, H.A.J. (2016a). Drought in the Anthropocene. *Nature Geoscience* 9(2): 89–91.

Van Loon, A.F., Stahl, K., Di Baldassarre, G., Clark, J.; Rangecroft, S., Wanders, N., Gleeson, T., Van Dijk, A.I.J.M., Tallaksen, L.M., Hannaford, J., Uijlenhoet, R., Teuling, A.J., Hannah, D.M., Sheffield, J., Svoboda, M., Verbeiren, B., Wagener, T., and Van Lanen, H.A.J. (2016b). Drought in a human-modified world: reframing drought definitions, understanding, and analysis approaches. *Hydrology and Earth System Sciences* 20: 3631–3650. doi: 10.5194/hess-20-3631-2016.

Van Vuuren, P., Edmonds, J., Kainuma, M., Riahi, K., Thomson, A., Hibbard, K., Hurtt, G., Kram, T., Krey, V., Lamarque, J.-F., Masui, T., Meinshausen, M., Nakicenovic, N., Smith, S., and Rose, S. (2011). The representative concentration pathways: an overview. *Climatic Change* 109: 5–31. doi: 10.1007/s10584-011-0148-z.

Wada, Y., Van Beek, L.P.H., Van Kempen, C.M., Reckman, J.W.T.M., Vasak, S., and Bierkens, M.F.P. (2010). Global depletion of groundwater resources. *Geophysical Research Letters* 37: L20402.

Wada, Y., Van Beek, L.P.H., Wanders, N., and Bierkens, M.F.P. (2013). Human water consumption intensifies hydrological drought worldwide. *Environmental Research Letters* 8: 034036. doi:10.1088/1748-9326/8/3/034036.

Wanders, N. and Van Lanen, H.A.J. (2015). Future discharge drought across climate regions around the world modelled with a synthetic hydrological modelling approach forced by three General Circulation Models. *Natural Hazards and Earth System Science* 15: 487–504. doi:10.5194/nhess-15-487-2015.

Wanders, N. and Wada, Y. (2015). Human and climate impacts on the 21st century hydrological drought. *Journal of Hydrology* 526: 208–220. http://dx.doi.org/10.1016/j.jhydrol.2014.10.047.

Wanders, N., Wada, Y., and Van Lanen, H.A.J. (2015). Global hydrological droughts in the 21st century under a changing hydrological regime. *Earth System Dynamics* 6: 1–15. doi:10.5194/esd-6-1-2015.

Warszawski, L., Frieler, K., Huber, V., Piontek, F., Serdeczny, O., and Schewe, J. (2014). The Inter-Sectoral Impact Model Intercomparison Project (ISI-MIP): project framework. *Proceedings of the National Academy of Sciences* 111(9): 3228–3232. doi: 10.1073/pnas.1312330110.

Weedon, G.P., Gomes, S., Viterbo, P., Shuttleworth, W.J., Blyth, E., Österle, H., Adam, J.C., Bellouin, N., Boucher, O., and Best, M. (2011). Creation of the WATCH forcing data and its use to assess global and regional reference crop evaporation over land during the twentieth century. *Journal of Hydrometeorology* 12: 823–848.

Wilby, R.L. (2006). When and where might climate change be detectable in UK river flows? *Geophysical Research Letters* 33: L19407.

Yevjevich, V. (1967). An objective approach to definition and investigations of continental hydrologic droughts. *Hydrology papers 23*, Colorado State University, Fort Collins, USA.

Part Two

Vulnerability, Risk, and Policy

2.5
On the Institutional Framework for Drought Planning and Early Action

Ana Iglesias, Luis Garrote, and Alfredo Granados
Universidad Politécnica de Madrid, Madrid, Spain

2.5.1 Introduction

For decades, droughts have been perceived as another natural hazard. However, the increase in their frequency and intensity over the last 30 years has raised awareness on the issue and has concentrated efforts on the study of their causes, consequences, and potential circumstances, aiming to minimise the impacts of this phenomenon. There are many studies that have conducted scientific analyses and proposed different management alternatives for this natural disaster. The real improvement of management in this field requires an analysis of the current legislation behind the issue, because it compiles the meaning of policies and, at the same time, is a guarantee for citizens towards adopted public commitments.

In the context of natural disaster management, the general trend is a combination of preventive and risk management strategies with emergency responses. This political approach is especially positive in the case of droughts, because, even if drought events are hardly predictable, the application of early warning and prevention mechanisms can considerably reduce the negative consequences of the phenomenon. In addition, effective policies in the context of drought management requires coordinated national and international action due to the extension reached by droughts, which are not dependent on administrative border, and therefore require effort coordination among the affected international stakeholders.

The Mediterranean is one of the regions where the impacts of drought events have shown an exponential increase during the last 20 years. The traditional approach adopted by governments in the basin has been the application of reactive responses in the short term, with little analysis about the consequences, the problem or the effectiveness of the

Drought: Science and Policy, First Edition.
Edited by Ana Iglesias, Dionysis Assimacopoulos, and Henny A.J. Van Lanen.

adopted measures, giving no continuity at all in the management of drought events. This approach has generally been supported by the legislative and institutional frameworks. As shown in the following text, the legislative framework in the countries selected as case studies has developed continuously during the last decades, but there are still important gaps that protect governments from the application of integrated drought management policies. However, there are some international and regional studies that reflect the consensus between countries in the area about the necessity of a policy change and the application of preventive measures (IUCN Centre for Mediterranean Cooperation, 2002).

A general weakness of drought-related legal frameworks is the absence of clear institutional responsibility attributions for management. Most of the analysed texts avoid the specification of bodies in charge of adopting decisions, approving actions, execution, and supervision. In those cases where the body in charge is clearly mentioned, the specific definition of competences is still not clearly determined, keeping an incomplete institutional structure. This unclear institutional situation is a logical consequence of the little attention that has traditionally been paid to this problem in general legislation. The adequate attribution of responsibilities in legislation is essential for efficient drought management due to the generally unpredictable character of drought events.

The absence of complete and adequate legislation reveals the limited political importance attributed to drought events. In the six analysed countries, drought management is equivalent to that adopted for any other natural hazard, applying a reactive, short-term approach for the mitigation of negative impacts. However, due to the demonstrated importance of drought in Mediterranean countries in comparison to other natural disasters, it would seem appropriate to develop more complex and integrated responses for the management of such phenomena, as agreed by the World Meteorological Organization's Working Group on Hydrology Regional Association VI (Europe) in 2005.

2.5.2 Drought planning and water resources planning

Droughts provide a good opportunity to implement water policy. Society recognises that it is necessary to improve drought planning, and additional funds are made available. In political terms, we are 'solving a problem created by others' (Figure 2.5.1).

There is strong interaction between water resources planning and drought contingency planning in areas exposed to water scarcity, as shown in Figure 2.5.1. Droughts always start as a meteorological phenomenon, with persistent precipitation deficiencies over a region. After a while, these deficiencies deplete soil moisture content and produce impacts on natural and rain-fed agricultural systems, which have only limited capacities to store water in the soil. The river catchment has more mechanisms to buffer droughts, mainly through storage of groundwater in aquifers, but if the drought persists, the effects are also seen in hydrological systems: low water tables and reduced river flows, which affect river ecosystems and riparian zones. Water resources planning cannot interfere in these processes when they occur in natural systems; these systems have evolved a variety of methods to cope with droughts, and are usually able to survive

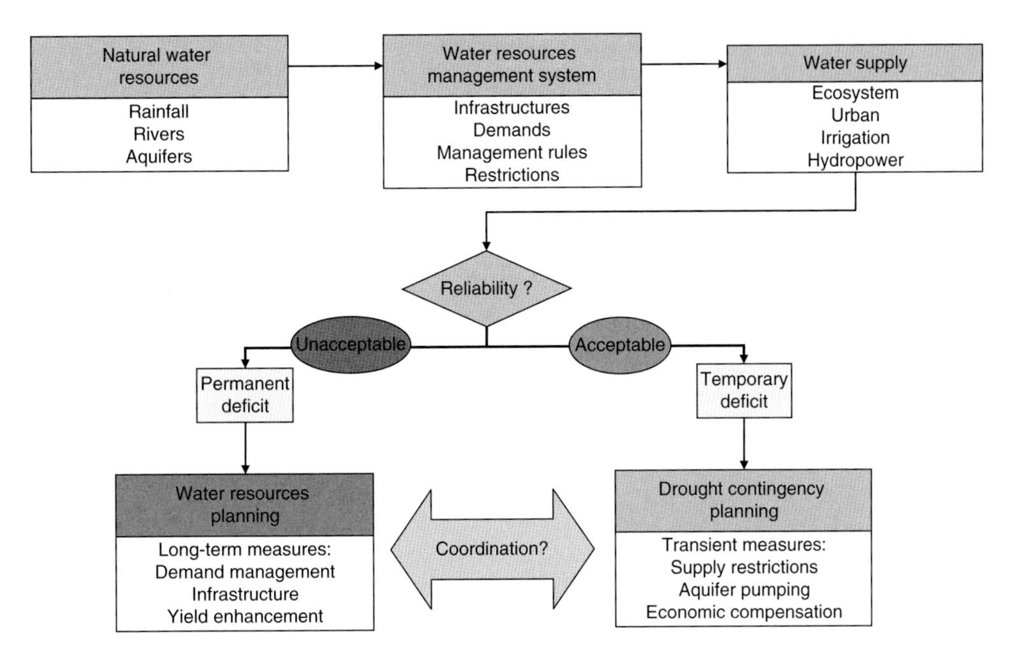

Figure 2.5.1 Overview of the coordination between basin and drought policy to derive legislation.

under strong water shortages and to recover after the drought is over. But, in addition to natural systems, there are also artificial water resource systems that modify the natural conditions of water bodies in order to provide adequate reliability in water supply to users through water abstraction, storage, transportation, and distribution. The alteration of natural systems is more challenging in areas that have large water shortages and when water supply is more irregular. The prolonged absence of precipitation and soil moisture deficits do not necessarily mean water scarcity in these water resource systems, because water can also be supplied from natural or artificial reservoirs – snow packs, aquifers, and regulation dams can sustain water demands during periods of meteorological drought.

Traditionally, water managers have designed these systems to overcome drought situations. The degree of severity at which droughts produce impacts in water resource systems is analysed in water resources planning, and it depends on the relation between the available resources and demands. Water resources planning estimates demand reliability, quantified as the probability that a given demand may suffer water shortages during a given time horizon. This reliability index is normally used for decision-making, identifying demands that do not comply with a pre-specified minimum standard, in order to evaluate the effect of water conservation measures and to define a programme of measures to correct the reliability deficit.

However, it is not economically feasible to design water resource systems for the worst possible drought, and, even if done, these systems may still fail. If drought conditions persist, reservoirs in the system are depleted of their reserves, and there is water scarcity. These situations must also be contemplated in water resources planning. Water scarcity under drought conditions is analysed by identifying demands that are not fully satisfied by the available water resources. Competition for water among urban, agricultural, industrial, and environmental demands is strongest in times of water scarcity. Allocation of the scarce water among multiple demands is a challenging task that requires careful analysis. It is very important for system managers to develop methods, rules, and criteria to evaluate water scarcity and prioritise proactive and reactive measures for drought management, especially in well-developed regions with extensive hydraulic infrastructure and complex socioeconomic interactions. These measures are more effective if they are objectively put together in a drought management plan (DMP). Some countries have already introduced DMPs in order to minimise the negative environmental, economic, and social impacts caused by droughts. DMPs are reference documents and useful, efficient tools to manage water resources under drought episodes. Their action methods and established measures must be applied once they have been previously agreed upon by the interested parties – that is, the society, administration, scientific community, NGOs, etc.

2.5.3 A code for best practices: early action and risk management plans

Both legislation and institutional organisation in Mediterranean countries reveal a clear reactive approach towards the problem. Most countries have developed crisis response policies to face already developing drought events instead of designing risk prevention policies. The common reaction mechanism up to date has been the adoption of emergency planning. In some cases, the application of preventive plans was not foreseen, or the institutional structure did not allow for the application of such instruments – examples are Italy and Tunisia. And in other countries, such as Spain or Cyprus, the designed prevention mechanisms have been adopted very slowly (Iglesias and Moneo, 2005).

For the design of an effective preventive plan, there are some requisites: first, an adequate and objective definition of drought based on indicators that are able to measure the evolution of the problem and determine the level of risk in vulnerability situations; second, the mitigation measures must be defined, together with the necessary actions for their application. It is also essential to define the participating institutions and the responsibilities that can be attributed to each of them. Few of the analysed Mediterranean countries have adopted plans that respond to all these requisites. In the best case, some countries have established legislation for the adoption of preventive actions. Drought management plans have been developed for all basins in Spain (Estrela and Vargas, 2012), but they only cover collective water supply systems. However, the specific normative to apply it has not been adopted, such as in the case of Italy. There is some resistance to design a preventive plan that totally covers the problem. This problem derives from several deficiencies, such as the unclear definition of drought events, lacking a scientific analysis that would provide better options for management in relation to the general water management system. This situation leads to the perception of drought as an unpredictable phenomenon, limiting the adoption of preventive plans for intervention.

From another perspective, approved public policies seek short or medium-term results. However, drought requires a long-term preventive public policy, while reactive policies act in the short term. Reactive policies are sometimes associated with the allocation of subsidies to mitigate the consequences of the event. Even if these subsidies are not effective from the environmental point of view, they make a difference in the electoral field. In order to move away from this approach, the Spanish Ministry of Agriculture and the Environment created, in 2005, a commission with a consultative and assessment character, composed of experts in different areas related to water, aimed at the adoption of public policies related to water management in general and drought events in particular.

2.5.4 Institutions involved in drought planning

2.5.4.1 Main issues and guidelines for the institutional analysis of drought planning

The institutional framework associated with water and drought are all organisations related to the management of water resources. The institutions are classified into policy-level institutions, executive-level institutions, user-level institutions, and NGOs, at the national, regional, district, and local levels. Correctly defining the roles of the different levels of government in planning and co-ordination is a primary requirement in the preparedness and management processes.

This section provides a common methodology for analysing the organisations and institutions relevant to water scarcity and drought management. This common methodology is adequate for providing information that will enable comparison among and across countries and promote cooperation with the existing institutions, organisations, networks, and other stakeholders in the Mediterranean. The methodology proposed and described in this chapter is supported by the experiences of previous researchers such as Iglesias and Moneo (2005), Iglesias et al. (2009). Although the objectives of these guidelines are not directly focused on institutional analysis per se, it is important to understand the conceptual bases, and to identify and map them, to ensure the relevance of subsequent drought management analyses. The approach is intended to cover the following areas:

1. Explicit description of institutions and organisations with competence in water policy and administration, planning, decision-making, operation of water supply systems, drought preparedness, and emergency action, with particular emphasis on municipal and irrigation water supply
2. Explicit description of the linkages and hierarchical relations among organisations and institutions
3. Information on existing drought preparedness and management plans
4. Documentation of the institutional experience on the application of the existing drought preparedness and management plans
5. Description of the data collection systems in each country, specifying the institutions responsible, the types of reporting, and accessibility, and the primary uses of the data

The analysis aims to provide insights that answer the following key questions:

1. Do the set of organisations and institutions interact within a formal or an informal network?

2. Are there networks to provide communication and hierarchical flows of command?
3. Are the stakeholders included in the network?
4. What is the degree of influence and dependence of the stakeholders' decisions on the institutions' core themes?

Most of the strategies for drought management are typically based on ex-post approaches and address only part of the issue of social and environmental sustainability.

Starting from the *institutional* setting of each country, it is possible to define the characteristics of the particular policies that need to be modified in order to promote sustainable drought management plans.

The relations among organisations and institutions are essential for understanding current drought management plans and for improving future actions that mitigate the effects of drought on agriculture, water supply systems, and the economy. Understanding the national institutional regime is a key factor for establishing effective and integrated drought management plans that incorporate monitoring, public participation, and contingency planning (Iglesias and Moneo, 2005). Combating drought risks is viewed in most societies as a public good that justifies government action. Therefore, societies must develop policies that deliver significant drought risk reduction and lesser social vulnerability.

The analytical framework proposed here comprises five main tasks:

1. Elaborate a mental model of organisations and institutions in each country, and describe the institutional and legal frameworks.
2. Collect additional information by interviews and/or other dialog methods. The interview should include 'problem analysis' (i.e. what actions did your institution take during a historical drought in a specific year?) and identification of the stakeholders affected by the decisions of each institution.
3. Validate the model structure. Communicate the results of the previous two tasks back to the organisations and institutions, and complete the analysis.
4. Analyse the strengths and weaknesses of the system organisational processes to take decisions within the institution and within the hierarchical structure in each country.
5. Discuss about the challenges and opportunities for improving drought management.

2.5.4.2 The legal framework and complexity of institutional systems

The absence of a unitary and coherent legal framework for drought management is revealed in the institutional systems in charge of drought management in most countries. Most countries include drought management in the context of general water planning legal frameworks, which are already large and complex enough (Iglesias and Moneo, 2005).

According to Iglesias and Moneo (2005), institutional responses to drought in the Mediterranean countries can be classified into two groups: those that include drought management in the general water management systems, with no special provisions (i.e. Cyprus and Greece), and those that have developed an institutional context different from that of general water management. In this second case, there are also some differences between the analysed countries. In some cases, legislation reflects some specificities of drought events and describes the participation of some institutions that have no competences in the general management of water resources (i.e. Tunisia); in some others, competences of water resources management bodies are modified for the adoption of alternative measures during drought periods (i.e. Spain). In some other cases, drought is considered to be an emergency situation that triggers a response system attributable to any kind of emergency (i.e. Italy) or specific to drought events (i.e. Morocco) that implies, in the latter, the intervention of a different institution (Iglesias and Moneo, 2005). In relation to the current legislation and institutional organisation and coordination schemes, we can conclude that the administrative systems described for drought management suffer importantly from lack of a clear attribution of competences and an excessive number of public participants; these make the system even more complex and exclude the participation of individuals affected by drought. This leads to a general inefficiency in the decision-making process and the execution of alternative actions.

The complexity of institutional systems also affects the decision-making processes, which suffer from a lack of clear definition of the institutions involved and the attributed competences in this process. The absence of integrated drought management plans that include the definition of preventive and reactive measures to be adopted limits the development of decision-making processes to the time of occurrence of drought, restricting the adoption of action to the reactive, short-term approach. The decision-making processes and the adoption of measures under the pressure of a currently developing drought event limit the reaction capacity and efficiency of the adopted mitigation measures.

The limited reaction and planning horizon sums up to the unclear definition of institutional responsibilities and competences for the adoption of drought management measures and the lack of participation mechanisms for those affected by drought events. The example of Spain can be useful in this case. Spanish legislation attributes the management of drought events in a diffuse way to the river basin authorities, but no clear structure is defined to determine the particular competences at each moment. In the case of severe droughts, the competence for adopting drought emergency action, according to article 58 in the revised Water Law, is attributed to the Ministers' Council. This institution takes into account the reports presented by the corresponding river basin authority when adopting decisions in case of a drought event. This mechanism for decision-making is only applied in the case of exceptional drought.

The adequate integrated response to drought events should be based on a good coordination and communication strategy among the Ministers' Council, the river basin authorities, and the other affected public organisations, even if coordinated action is not always possible. The absence of a regulating text that defines the competences and the tasks corresponding to each administration body, or the inexistence of a structured protocol for decision-making processes to overcome drought events, delays the triggering of actions for mitigation, and complicates the participation of affected individuals. In summary, the lack of legislation clarity, in terms of responsibility identification, affects the whole drought management process, including at the decision-making phase.

This situation directly affects the participation assigned to individuals affected by drought. From their perspective, droughts affect farmers, industries, and citizens in general as water consumers. Reactive response and emergency management in response to a drought event excludes the participation of large portions of these groups from the decision-making process. They bear the consequences of the situation with no option for defending their own interests. It is adequate to say, in general terms, that drought management decision processes are not inclusive, and that they exclude the participation of affected individuals, in opposition to the trend defined by the most modernised legislation (Water Framework Directive 2000/60/CE).

The prevalent mechanisms for drought management are in coherence with the situation of institutional systems and the decision-making processes described in the preceding text. Interventions developed in some Mediterranean states to face drought events reflect a fragmented management strategy (Iglesias and Moneo, 2005). This is a direct consequence, on the one hand, of the decentralisation of institutional systems' management and, on the other hand, the lack of a coordination body for drought management – in addition to being the effect of decentralisation on decision-making processes and the lack of clarity in competences distribution in institutional organisation schemes.

A common characteristic of the countries in the region is the weak cooperation among the different institutions related to water management. Another similarity is the fragmented roles of the government, the administrative regions, and the river basin authorities, which result in administrative conflicts that are an impediment for adequate water management. The key issue of transboundary water management is included in drought management plans. Spain shares a large amount of surface water resources among basins in the country, and also some basins that extend to Portugal. The agreements on water transfer amounts between national basins (such as the Tagus–Segura), or between countries sharing a common basin (such as the Spanish and Portuguese portions of the Tagus basin), include strategic regulations in the case of drought. The EU mandates that international basins that include EU member states must approve agreed programmes of measures applicable for the entire demarcation. Other Mediterranean countries, especially in the southern basin, share a significant portion of groundwater, but the regulation during drought needs to be further developed.

No single management action, legislation, or policy can respond to all the aspects and achieve all the goals for effective drought management. Multiple collaborative efforts are needed to integrate the multidimensional effects of drought on society. The United Nations Convention to Combat Desertification (UNCCD, 2000) provides the global framework for implementing drought mitigation strategies. The United Nations International Strategy for Disaster Reduction (UNISDR, 2002) establishes a protocol for drought risk analysis.

Drought management policies should be based on integrated evaluations of all the measures required to implement the objectives of the water policy, together with those measures required under other policies and relevant legislation. This section reviews the extant legal initiatives and present legislation explicitly focussing on drought risks. It provides a description, ordered hierarchically, of all the laws, rules, norms, and statutes that are presently in force in each country, and connected to water use, management, and conservation, as well as land use and the natural environment. The water and drought legal framework includes all acts and regulations related to water resources management, wastewater management, non-conventional water resources, and environment-related issues. The legal framework includes all laws applicable at the national, regional, district, and local levels, including international agreements or regulations in force.

Mediterranean countries have extensive legal provisions (legislation and normative) related to water management focusing on water scarcity. The existing legislation enables governments to develop specific drought mitigation plans, both proactive and reactive in nature. Legislation is an instrument that allows governments to implement drought mitigation plans and drought relief policies. For effectiveness of the legislation, governments need to include adequate budgets for implementation of the measures. In general, the laws focus on drought management strategies adopted under stress situations, providing conditions for emergency actions.

The UNCCD (2000) provides the framework for implementing drought mitigation strategies. The convention is especially relevant to southern Mediterranean countries.

2.5.4.3 Examples of legal provisions in Spain

Table 2.5.1 summarises the legal provisions related to water scarcity and drought contingency plans in Spain. Spain has developed an Agricultural Insurance Law that includes drought hazards, and has established specific drought mitigation plans, both proactive and reactive in nature. In Spain, the Law of the National Hydrological Plans (Law 10/2001 art. 27[1]) explicitly deals with drought and establishes the bases for a proactive and reactive response against hydrological and meteorological drought. The contingency plans include supply reliability and the future development of supply plans for large cities. The reactive responses include emergency works, decision on reservoir management, and user strategies. The legal base to meteorological drought is based on the development of an agricultural insurance law.

Proactive and reactive responses to drought include some action plans in order to prepare for drought and to mitigate its effects. The performance of these action plans could be defined by a legal framework. For example, Spain has defined a specific legal provision for this action plans, and it is currently in force.

Action plans define different proactive responses by means of a programme of measures. Some measures are specifically defined by water resources law – for example, the definition of order of priorities of users during scarcity, or the possibility to implement economic measures, such as the revision of water tariffs. The first measure is to establish that urban use has priority over agricultural, industrial or recreational use. The Spanish water code include a broad range of water allocation mechanisms, such as water pricing and water markets. In this sense, the Spanish Water Law permits the river basin authorities to create water exchanging centres (now called 'water banks'), through which rights-holders can offer or demand usage rights in periods of drought or severe water scarcity situations. In other cases, the agency itself can offer rights-holders compensations for surrendering their rights and allocate the resources to alternative users or to environmental purposes. Perhaps this is because water rights are not well defined.

[1] Although this law was repealed by Law 2/2005, the provisions concerning drought were kept intact. Based on the former article 27, the government is finalising a drought plan and contingency strategy for all Spanish basins.

Table 2.5.1 Legal provisions related to water scarcity and drought in Spain. *Source of data*: Iglesias and Moneo (2005).

Legal provisions	Contingency plans	Institutions / stakeholders	Focus / funding
International Convention (United Nations) 1994 Agreement, 2000 Enforced	Strategy to combat drought and desertification	United Nations and national governments	Strategies to fight desertification and mitigate drought, to be implemented by all nations that signed the convention
Basin Hydrological Plans (2001)		River basin authorities (Ministry of the Environment) Permanent Committee (Drought Management) Ministry of Agriculture, Agricultural Insurance Agency Finance Ministry, Reinsurance Public Agency Permanent Office for Drought (officials of the Ministry of Agriculture)	*Reactive*: new insurance products *Proactive*: taxation abatements or deferrals, drilling wells *Proactive*: water supply reliability, urban priority *Reactive*: emergency works, decisions on reservoir management and user strategies
Civil protection 1983, 1999	Crisis management plans	Civil protection Permanent Office for Drought	Creation of committees that will define the action terms in case of drought Different performance environments, social or agricultural
Civil protection 1995–2000	Emergency measures	Most of them undertaken as mitigation measures after the most severe drought periods	Laws, royal decrees, and orders created to mitigate the impacts of drought Hydraulic supply measures Transfers of water between different river basins Measures for subsectors of agriculture (apiculture, livestock, tree crops)
Definition of the areas where emergency measures are applied; 1993, 2000, 2001	Crisis management		Definition of the criteria used to delimit areas affected by drought Establishment of criteria for aid supply Final criteria used Amount of rainfall Stocking rate
Agricultural Insurance Law 1978, 2001, 2002	Insurance	Agricultural Insurance Agency	Definition of the conditions, application areas, and other characteristics of drought insurance

Table 2.5.1 (Continued)

Legal provisions	Contingency plans	Institutions / stakeholders	Focus / funding
Albufeira Convention, 1998	Transboundary		Albufeira Convention between Spain and Portugal for transboundary basins under the framework of sustainable water resources management and common environmental protection

In general, the advisory authorities have the authority to allocate and reallocate water during drought periods.

EU Water Framework Directive, in the long-term, forces the adoption of full-cost-recovery pricing criteria to ensure that tariffs charged to users cover the full cost of the service. These criteria can be considered as economic instruments to save water by means of demand management. Currently, European countries have not yet implemented this kind of economics instruments, but, in the near future, this will be a clear possibility for developing a proactive response to drought by improving the efficiency of water use.

In Spain, the legislation has evolved as a consequence of severe drought episodes, such as those in 1993–1995 and 2001. An attempt has been made to identify the legal base that enables national and local authorities to develop specific drought mitigation plans, both proactive and reactive in nature. The current legislation offers opportunities to governments to use instruments to develop, and allocate budget to, mitigation plans and drought relief policies. Ultimately, the legislation is the instrument that provides the means to produce drought management plans. Table 2.5.1. summarises the legal provisions related to drought and water scarcity in Spain.

The contingency plans based on legal provisions in Spain include:

1. Specific reactive measures, economic compensations (such as taxation abatement), and emergency measures (such as drilling wells and water transfer)
2. Water reallocations
3. Demand management
4. Insurance schemes
5. Long-term measures: new water infrastructure
6. Reactive measures, depending on the scenario of drought
7. Policy planning process
8. Proactive plans that anticipate costs and effects
9. Drought management operations, depending on the phase of drought: combined methods of physical and socioeconomic data
10. Proactive action plans, based on the most probable scenarios
11. Hydrological national plan: supply reliability and supply plans for cities
12. Drought management operations, depending on the phase and severity of drought (National Drought Plan)

13. National Water Plan: water supply for drinking water and irrigation
14. Modern technologies to monitor the hydrological system
15. Drought insurance system: knowledge of drought risk
16. Good-performance basin agencies: social participation and planning process
17. Drought adapted legal framework: establish priorities, reallocation mechanisms, and legal mandate to develop contingency plans
18. Coordination of the organisations that are relevant for drought planning

2.5.4.4 Key aspects of drought management in mediterranean countries

The drought policy must be flexible to avoid imposition of inappropriate or unnecessarily strict requirements simply for the sake of harmonisation. Such flexibility would also ensure that, where a problem is regionally specific, measures appropriate to that particular area can be taken. The range of environmental conditions in the Mediterranean basin is very diverse, and this must be taken into account.

A cost-effective strategy implies assessment from an economic perspective of advantages and disadvantages of the three basic sets of policy instruments: regulations and standards, new technology, and internalisation of external pollution costs through pricing and market-based incentives. These sets of policy instruments are not mutually exclusive and can be used as complementary or alternative measures, depending on their relative cost-effectiveness to address water pollution as well as water scarcity issues.

Other actions that influence water demand such as awareness campaigns for water conservation, adoption of water-saving measures have a clear framework for implementation in Tunisia, Morocco or Cyprus, where there are some specific contingency plans containing these kinds of mitigation measures. In Spain and Morocco, their national hydrological plans regulated by law include these kinds of measures. Cyprus also has specific contingency plans not regulated by law. In this case, water transfers, emergency schemes, water cuts or water reallocations are examples of the measures considered. Drought management operations settings in Tunisia is not based on a specific law. The Drought National Commission has the role of supervising the execution of the operational actions. Drought Commission establishes drought indicators, which are the triggers of the actions to be taken during drought periods, such as changes in the reservoir operations and management to ensure water for the priority users.

Drought policy is not to be seen in isolation, but as a contributory element in the wider search for balanced and sustainable development. And such a sustainable approach can neither be planned nor implemented in a satisfactory and efficient way without providing for broad consultation and participatory procedures of all the actors concerned.

In general, decisions related to droughts are taken in the context of the formal legal system. There are legal provisions for emergency actions in case of crisis situations, such as extreme drought. Informal customs may evolve into formal decisions. For example, historical users of groundwater without formal rights may be legalised. The legislation does not provide explicit regulations about how to calculate ecological discharge during drought situations; this important question is left to the discretion and responsibility of various institutions. Table 2.5.2 summarises the key aspects of drought management plans in selected Mediterranean countries.

Table 2.5.2 Summary of the drought management actions in selected Mediterranean countries. *Source*: Adapted from Iglesias et al. (2009).

Concept	Cyprus	Greece	Italy	Morocco	Tunisia	Spain
Surface water ownership	Public	Public	Public	Public	Public	Public
Groundwater ownership	Partially private	Public	Public	Partially private	Public	Mixed
Water law	Does not include drought	Includes drought	Includes drought	Includes drought	Includes drought	Includes drought
River basin authorities	Not developed	Developed	Developed	In development	Partially developed	Developed
Drought contingency plan	Not developed	In development	Regional	In development	National	At river basins and urban supply levels
Drought monitoring system	Partially developed	Partially developed	River basin	National	National	River basin
Agricultural insurance	Rainfed agriculture	Not developed	In development	In development	Not developed	Developed for rainfed agriculture
Relation among institutions	Low	Low	Low	NA	High	Medium
Public participation in water management	Low	Medium	High	Low	Low	High

In all cases, there is a clear and constant conflict between water uses during drought periods; however, in some countries, the drought management actions and regulations are generally accepted and perceived as legitimate. And yet, in some others, the related regulations still need development and evaluation. Also, the view on water use rights exchange varies dramatically from one institution to another, making the real application of approved plans and initiatives more difficult. There is also a conflict concerning emergency works. On the one hand, some of these works are necessary for the normal functioning of the basin, and the emergency situation accelerates the approval process; on the other hand, these works result in larger costs and efforts than would normally be the case. The traditional treatment of drought has rarely incorporated environmental issues. The European Water Framework Directive highlights the importance of improving the 'ecological status of the heavily modified water bodies', and mandates that ecological water quality be integrated as an objective of the programmes of measures. However, it foresees derogations of quality targets if severe drought conditions prevail or social costs are high.

2.5.5 Conclusions

Droughts are becoming a constant phenomenon in the Mediterranean area. Countries are aware of this fact and try to adopt policies that help mitigate the impacts of this natural disaster. However, the development of these national policies and the legislation and institutional systems they rely on are commonly slow and unsatisfactory. This limits their capacity to adequately respond to drought control issues.

The development of national policies is important in the field of drought management; however, the development of international initiatives is also essential for dealing with a phenomenon that does not depend on administrative borders for distribution of impacts. These international policies should be based on a well-established foundation of mature legislation that articulates the development of actions to be adopted in the case of drought events.

In coherence with these limitations and the common evolution described previously, the adoption of a regional code on drought management in the Mediterranean would be the most adequate step for the countries in the region for the development of policies for the control of this natural phenomenon.

Current legislation on water and drought management shows different development stages for the Mediterranean countries, which lead to important differences in the way droughts can be faced. While some of the countries have a stable and long tradition of legislative frameworks with functional river basin authorities and clearly defined responsibilities, others are still developing institutions and organisations that take care of water management issues. Drought preparedness requires adequate institutions and agencies with competences to develop and enforce plans. In their absence, governments must necessarily resort to emergency actions and alleviation programmes, but very little can be done to reduce the likelihood and severity of drought risks.

Acknowledgements

Funding for this study was provided by European Commission for the MEDROPLAN and DEWFORA projects. The authors thank all information providers, as well as the national organisations and institutions that participated in focus groups and responded to questionnaires.

References

Estrela, T. and Vargas, E. (2012). Drought management plans in the European Union. The case of Spain. *Water Resources Management* 26: 1537.

Iglesias, A. and Moneo, M. (2005). *Drought Preparedness and Mitigation in the Mediterranean: Analysis of the Organisations and Institutions*. Zaragoza: CIHEAM, 199 pp.

Iglesias, A., Cancelliere, A., Cubillo, F., Garrote, L., and Wilhite, D.A. (2009). *Coping with Drought Risk in Agriculture and Water Supply Systems: Drought Management and Policy Development in the Mediterranean*. The Netherlands: Springer.

IUCN Centre for Mediterranean Cooperation (2002). Strategic Review of the IUCN Centre for Mediterranean Cooperation. https://www.iucn.org/downloads/cmc_synthesis_en.pdf.

UNCCD (2000). United Nations International Strategy for Disaster Reduction.

UNISDR (2002). United Nations Convention to Combat Desertification in Those Countries Experiencing Serious Drought and/or Desertification, Particularly in Africa (UNCCD).

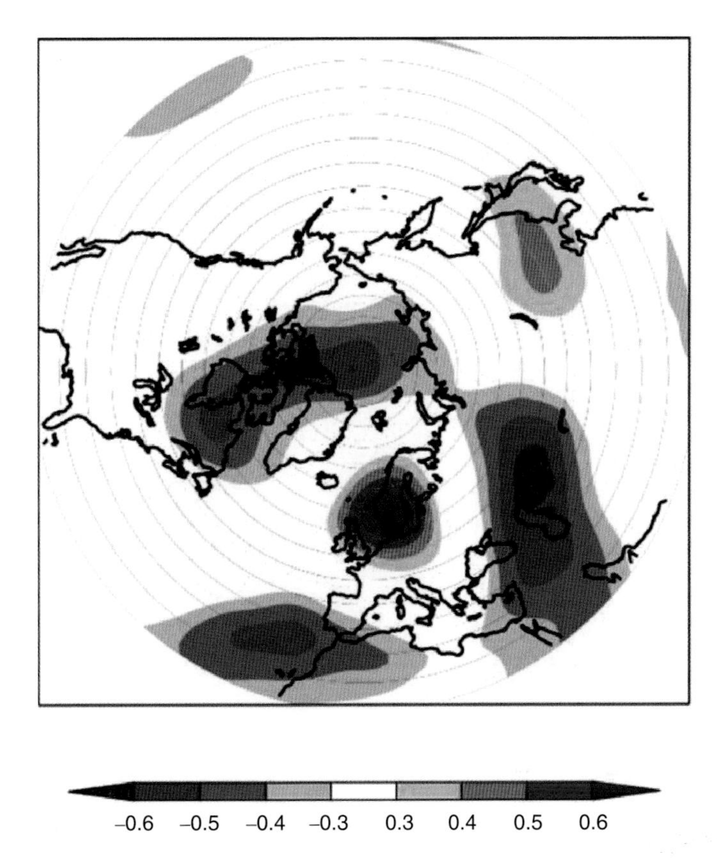

Figure 1.1.2 Correlation of May SPI-6 with December–May mean 500 hPa geopotential height. *Source*: From Kingston et al. (2015).

Drought: Science and Policy, First Edition.
Edited by Ana Iglesias, Dionysis Assimacopoulos, and Henny A.J. Van Lanen.

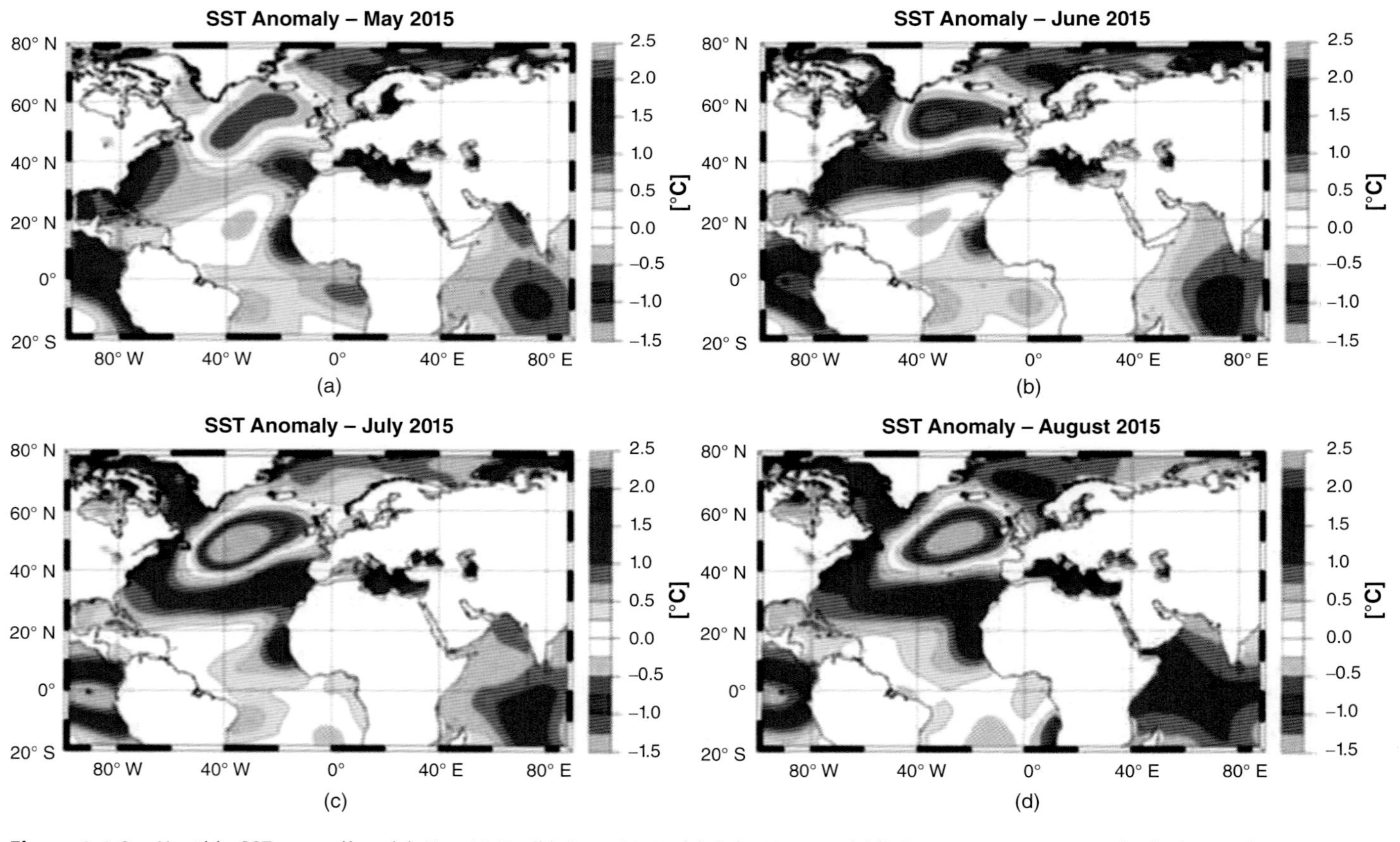

Figure 1.1.3 Monthly SST anomalies: (a) May 2015; (b) June 2015; (c) July 2015; and (d) August 2015; computed relative to the 1971–2000 period. *Source*: From Ionita et al. (2017).

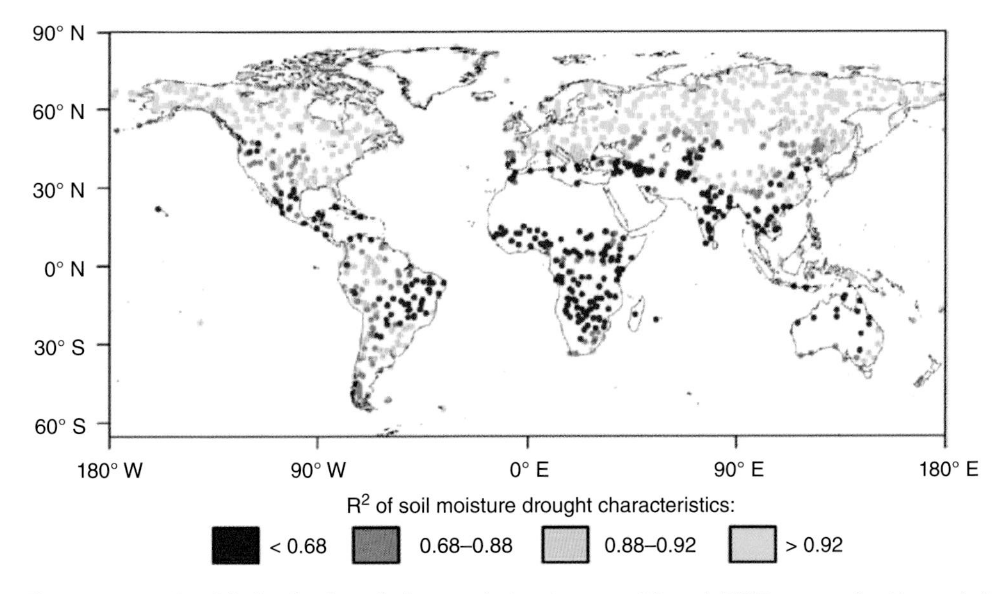

Figure 1.1.4 Spatial distribution of the correlation between DD and DSMD events for the period 1958–2001. *Source*: From Van Loon et al. (2014).

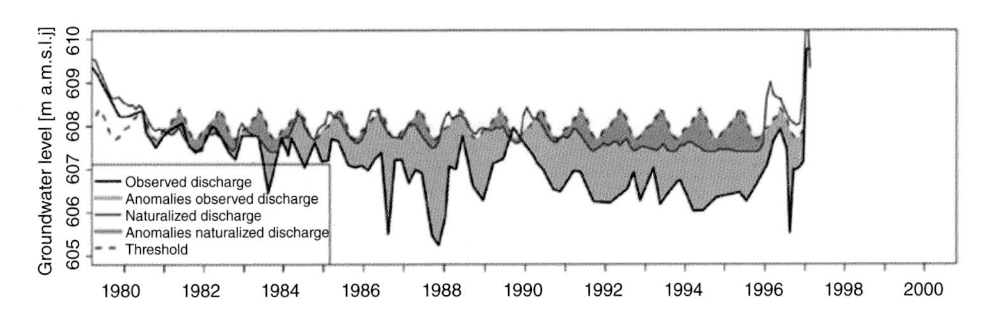

Figure 1.1.6 Anomalies in groundwater level in Upper Guadiana during the period 1980–1997 (disturbed period), using a variable 80% monthly threshold, and derived from the observed and naturalised groundwater levels. Blue areas can be attributed to climate, and grey areas to humans. *Source*: Derived from Van Loon and Van Lanen (2013).

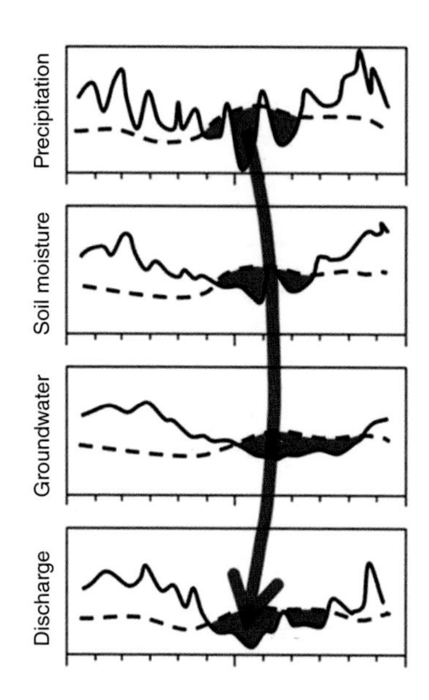

Figure 1.1.7 Propagation of drought through the terrestrial hydrological cycle, from meteorological drought to soil moisture drought to hydrological drought, showing pooling, delay, lengthening, and attenuation.

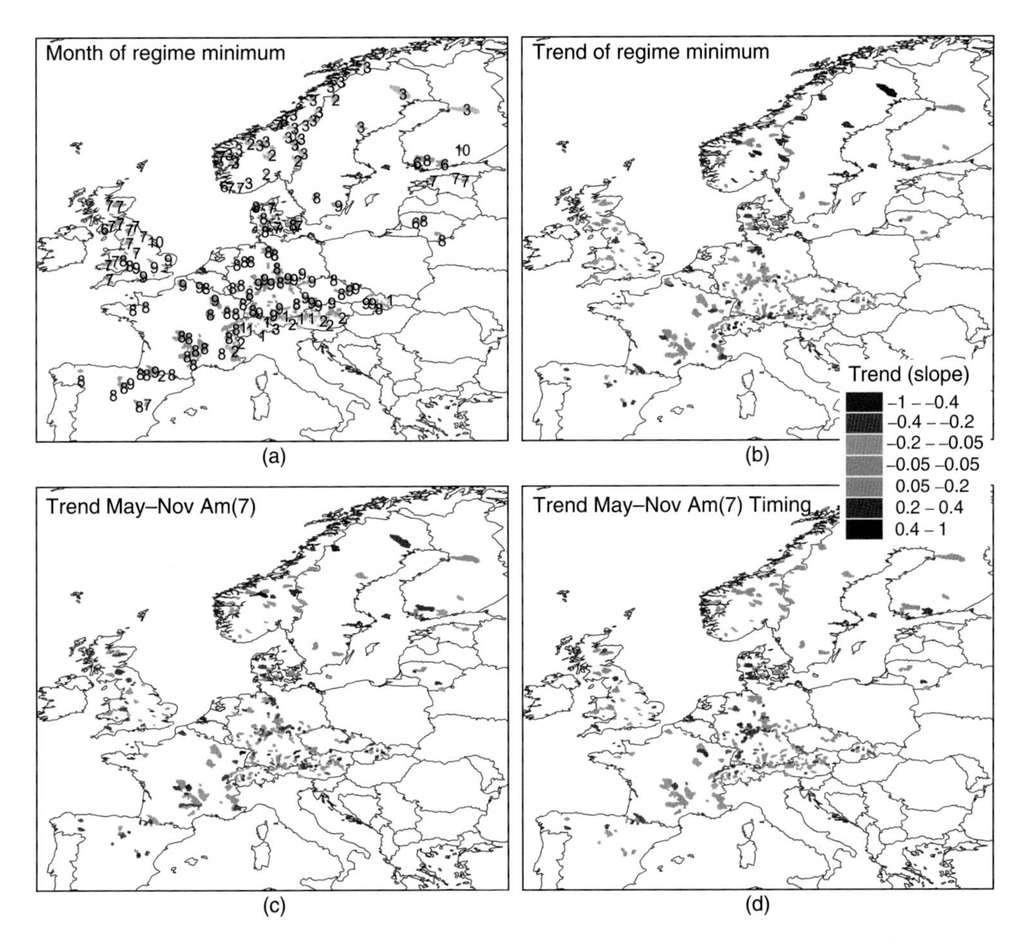

Figure 1.2.1 (a) The month of the regime minimum; (b) trend for mean monthly streamflow for the month of the regime minimum; (c) trend for AM(7) during May–November; (d) trend for the timing of AM(7) during May–November. All trends are for the period 1962–2004 and given in standard deviations per year. *Source*: From Stahl et al. (2010).

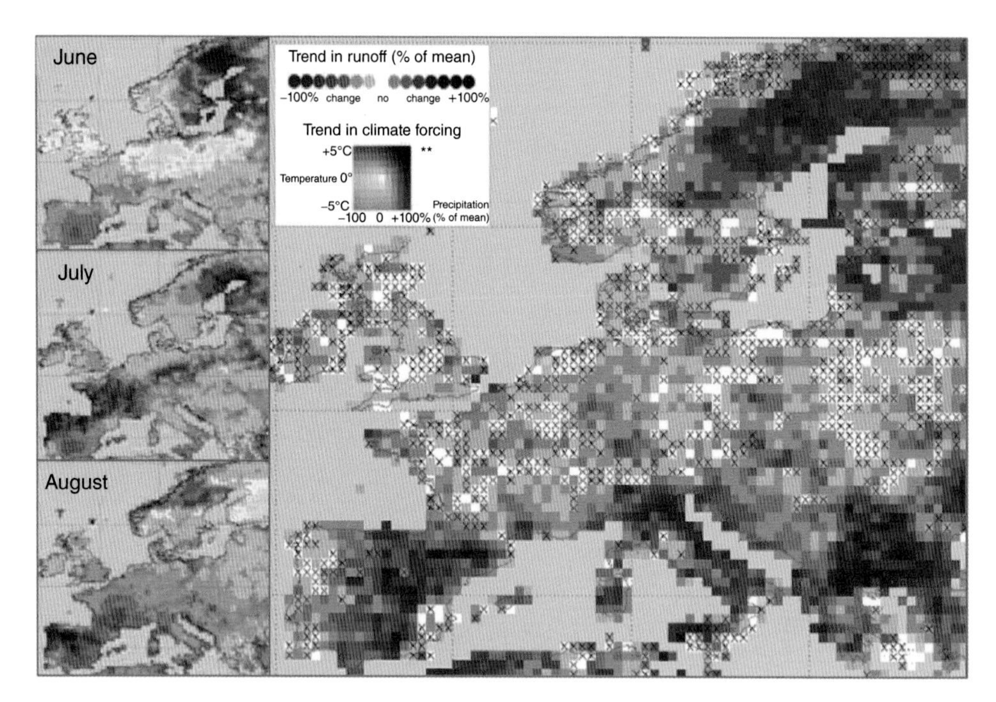

Figure 1.2.2 Trends in summer hydroclimate and modelled low flows in Europe over the period 1962–2000. Left: precipitation and temperature trends from the WATCH Forcing Dataset (www.eu-watch.org/data_availability) for June (upper), July (middle) and August (lower). Right: Low flow (AM(7)) trends for the summer half-year from the WATCH multi-model ensemble (from Stahl et al., 2012). Crosses: Less than three-quarters of the models agree on the sign of the trend. *Source*: Reproduced from Stahl et al. (2014).

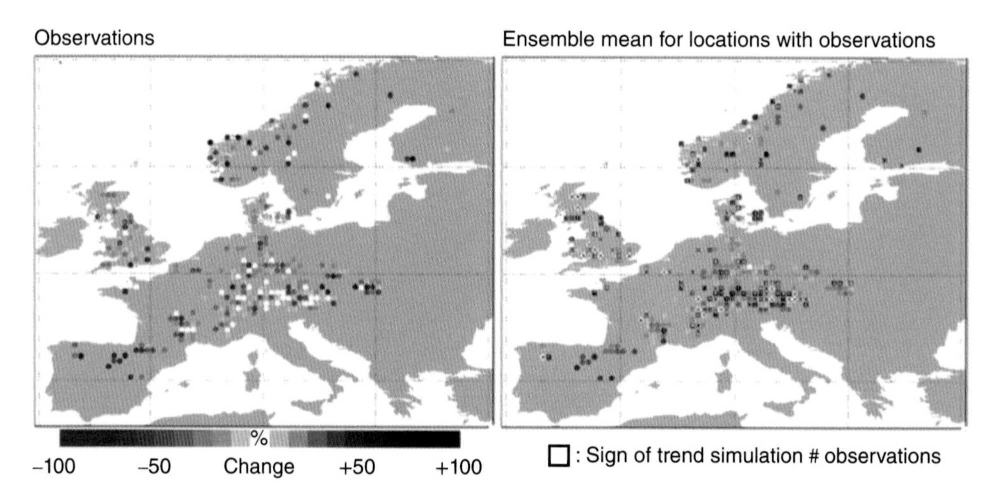

Figure 1.2.3 Spatial distribution of 7-day summer low flow trends from observed streamflow records (left) and from the model ensemble mean (right). As with Figure 1.2.2, crosses denote that less than three-quarters of the models agree on the sign of the trend. *Source*: From Stahl et al. (2012).

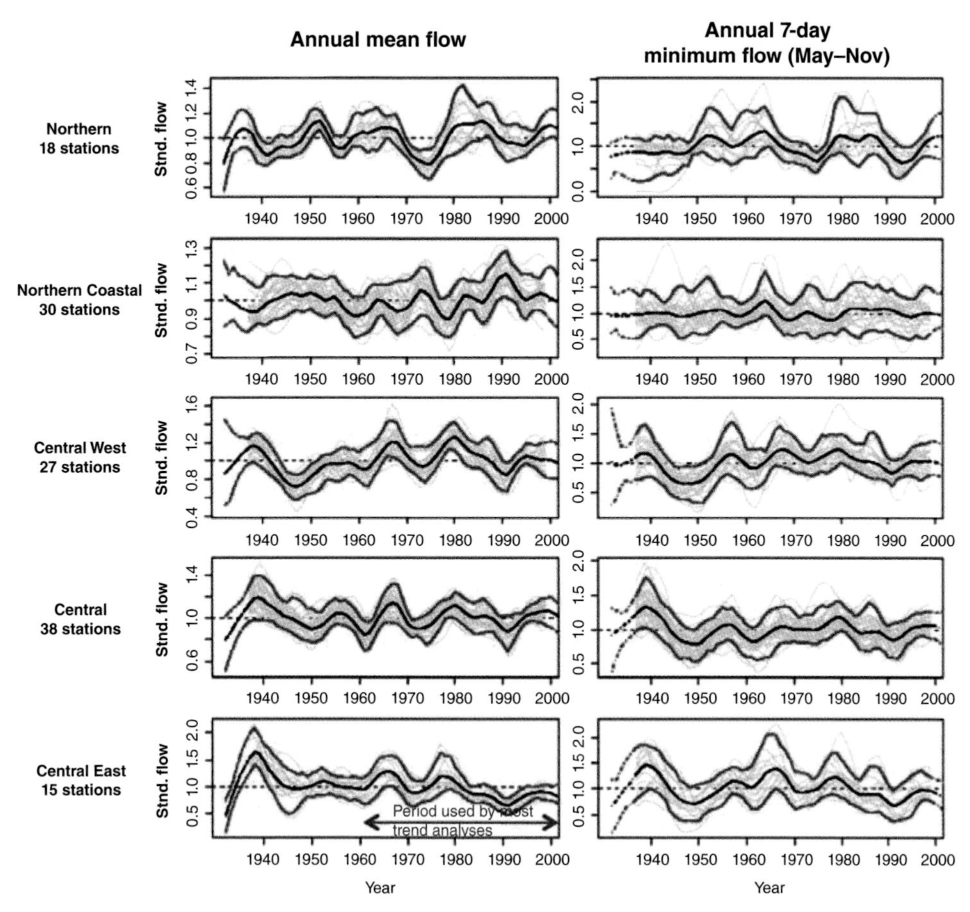

Figure 1.2.4 Interdecadal variability of streamflow records using LOESS smoothing. Grey lines show individual station series; black line shows the cluster average smoothed series for each cluster (region). Red lines show the 5th and 95th percentiles of the combined cluster member series. *Source*: From Hannaford et al. (2013).

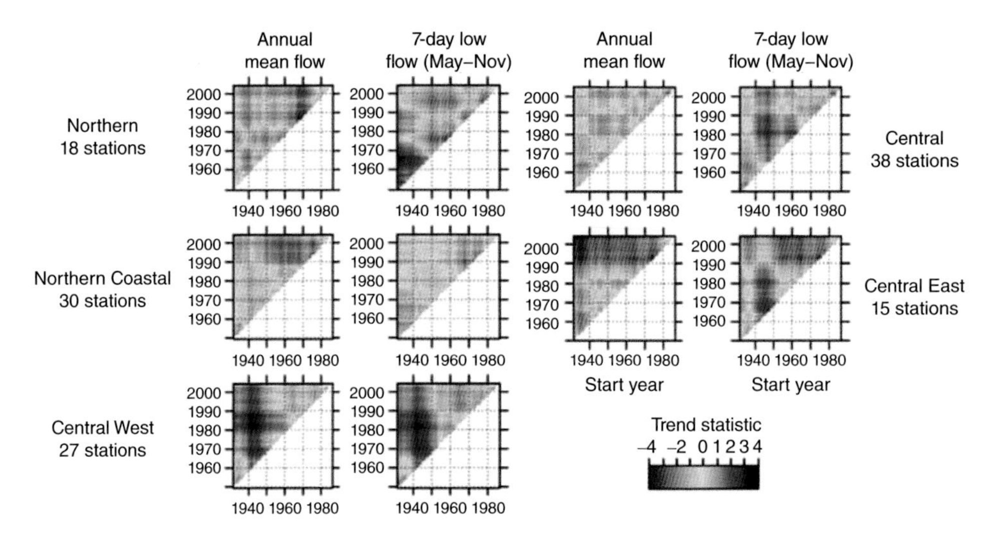

Figure 1.2.5 Multi-temporal trend analysis for annual mean and annual 7-day minimum flow, with x-axis showing start year and y-axis showing end year of the period for which the displayed trend statistic is shown. The MK test applied to all start and end years, and the corresponding pixel is coloured according to the resulting Z statistic, with red representing negative trends and blue representing positive trends. *Source*: Modified from Hannaford et al. (2013).

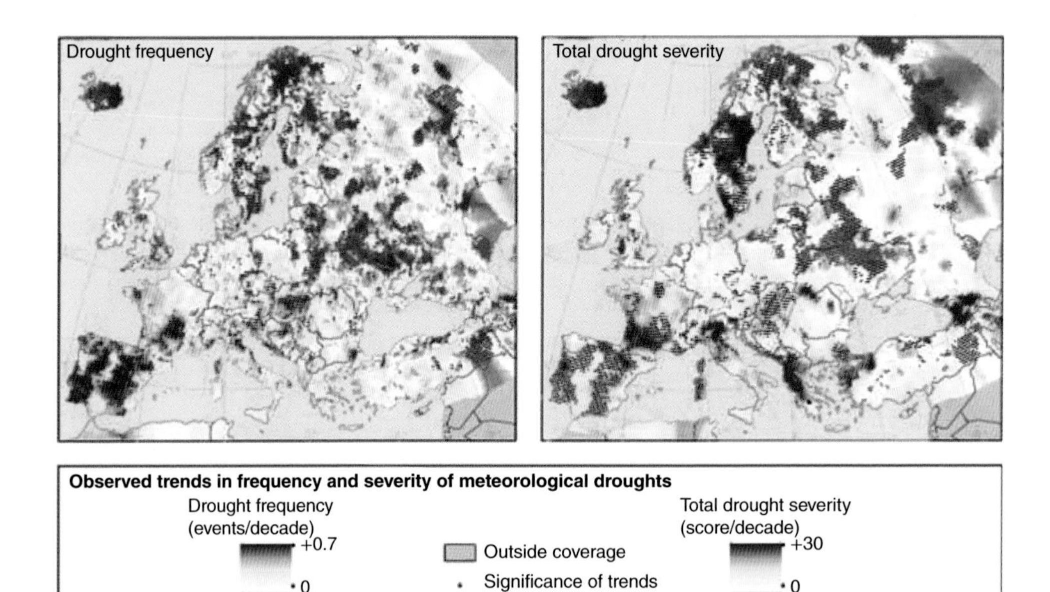

Note: This map shows the trends in drought frequency (number of events per decade; left) and severity (score per decade; right) of meteorological droughts between 1950 and 2012. The severity score is the sum of absolute values of three different drought indices (SPI, SPEI and RDI) accumulated over 12-month periods. Dots show trends significant at the 5% level.

Source: Adapted from Spinoni, Naumann, Vogt et al., 2015.

Figure 1.2.6 Trends in frequency and severity of meteorological droughts. *Source*: From Spinoni et al. (2015). Reproduced from EEA (2017).

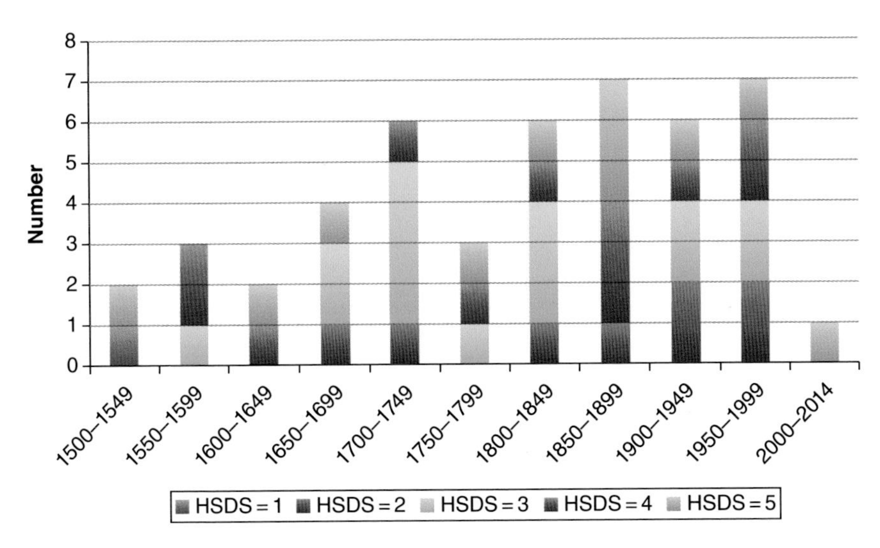

Figure 1.3.4 Distribution and severity of UK droughts by periods of 50 years between 1500 and 2014 according to HSDS.

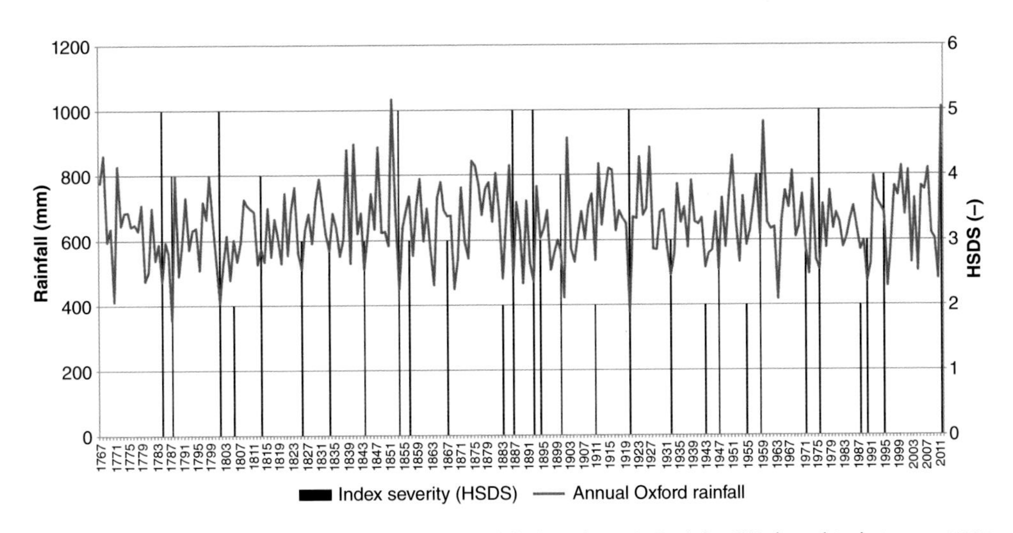

Figure 1.3.5 Comparison of HSDS and the rainfall data from Oxford for UK droughts between 1767 and 2012. *Source*: Meteorological Observations at the Radcliffe Observatory, Oxford.

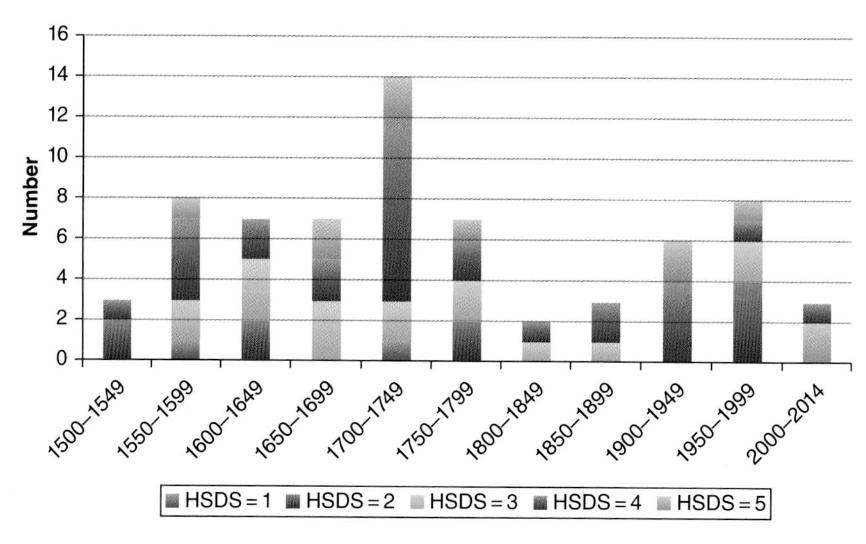

Figure 1.3.11 Distribution and severity of droughts in France by periods of 50 years between 1500 and 2014 according to HSDS.

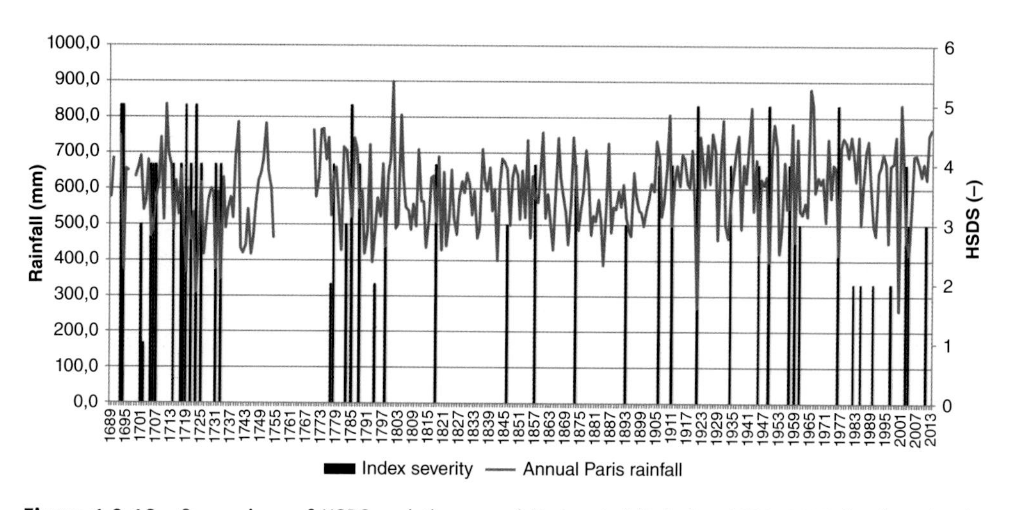

Figure 1.3.12 Comparison of HSDS and the annual Paris rainfall during 1689–2013 for droughts in France. *Source*: Meteo France-OPHELIE ANR Project and Bibliothèque de I'Académie Nationale de Médecine de Paris.

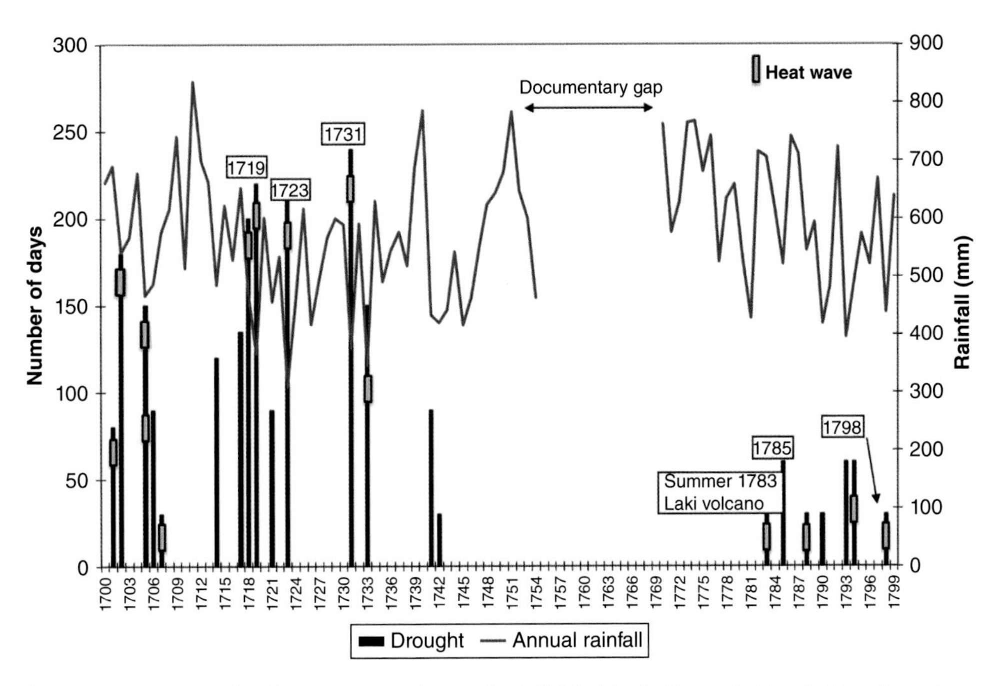

Figure 1.3.14 Droughts, heat waves, and annual rainfall in Ile-de-France in the eighteenth century.

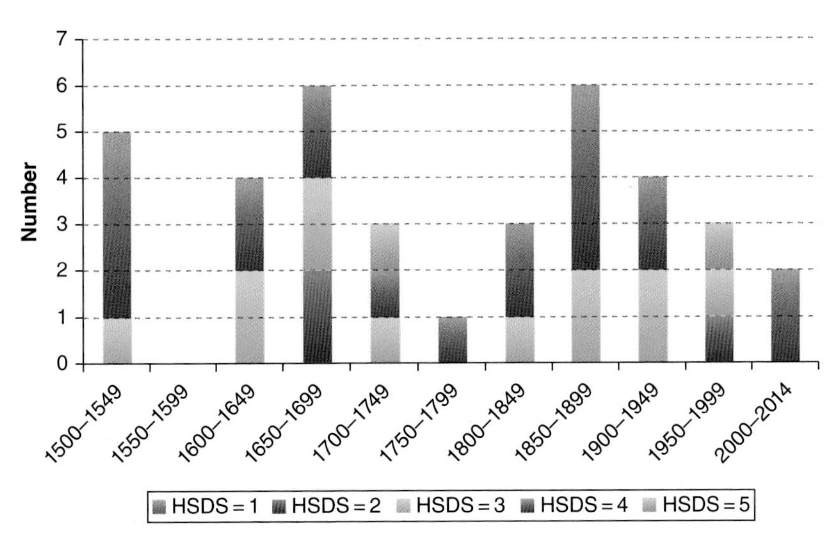

Figure 1.3.16 Distribution and severity of Rhenish droughts by periods of 50 years between 1500 and 2014 according to HSDS.

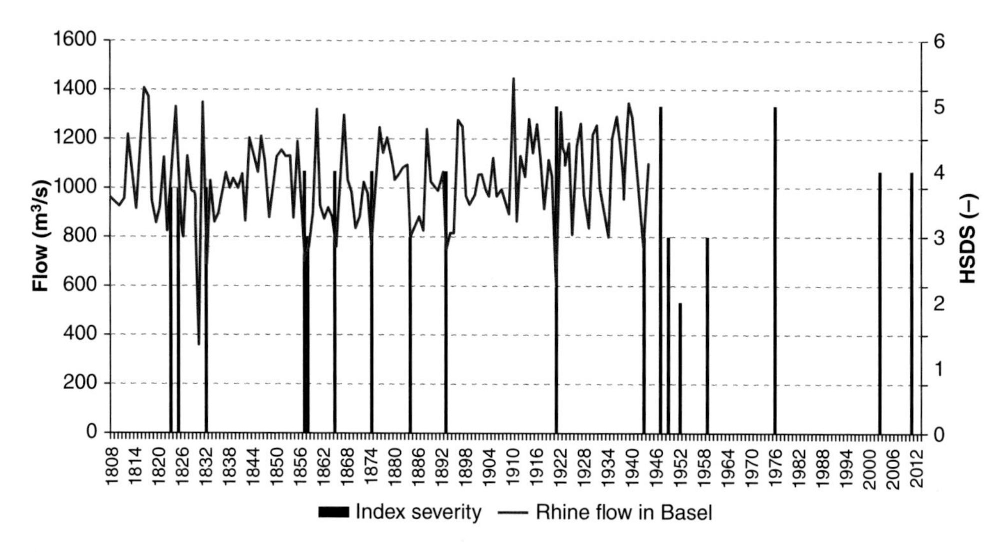

Figure 1.3.17 Comparison of HSDS and the annual flow of the Rhine in Basel 1808–2013 for Rhenish droughts. *Source*: Oesterhaus (1947).

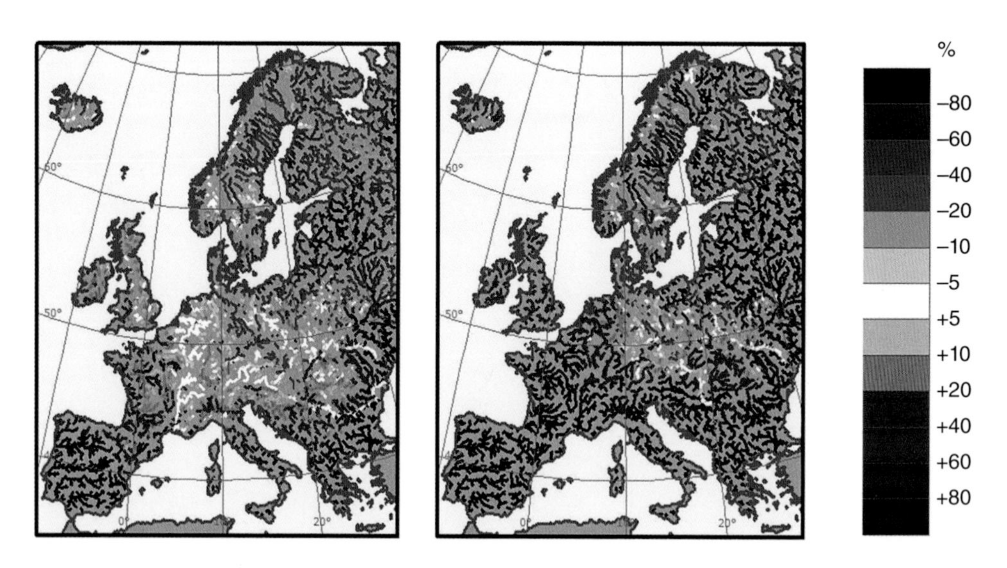

Figure 1.4.1 Change in the ensemble average in the 20-year return period deficit volumes due to climate change, left: 2020s and right: 2080s relative to the control period. *Source*: Derived from Forzieri et al. (2014).

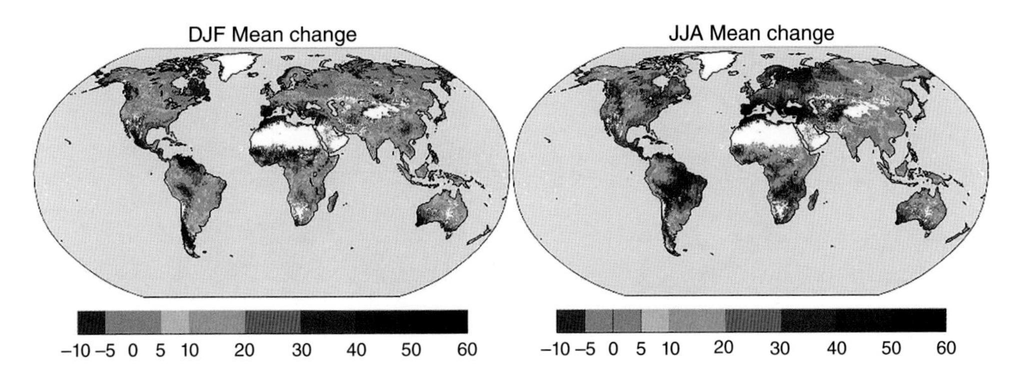

Figure 1.4.2 MME mean change in drought occurrence (in %) between control (1976–2005) and future (2070–2099) simulations under RCP 8.5 forcing for northern winter (December–February, left) and summer (June–August, right). *Source*: Prudhomme et al. (2014) Supplementary Material. Reproduced with permission from authors.

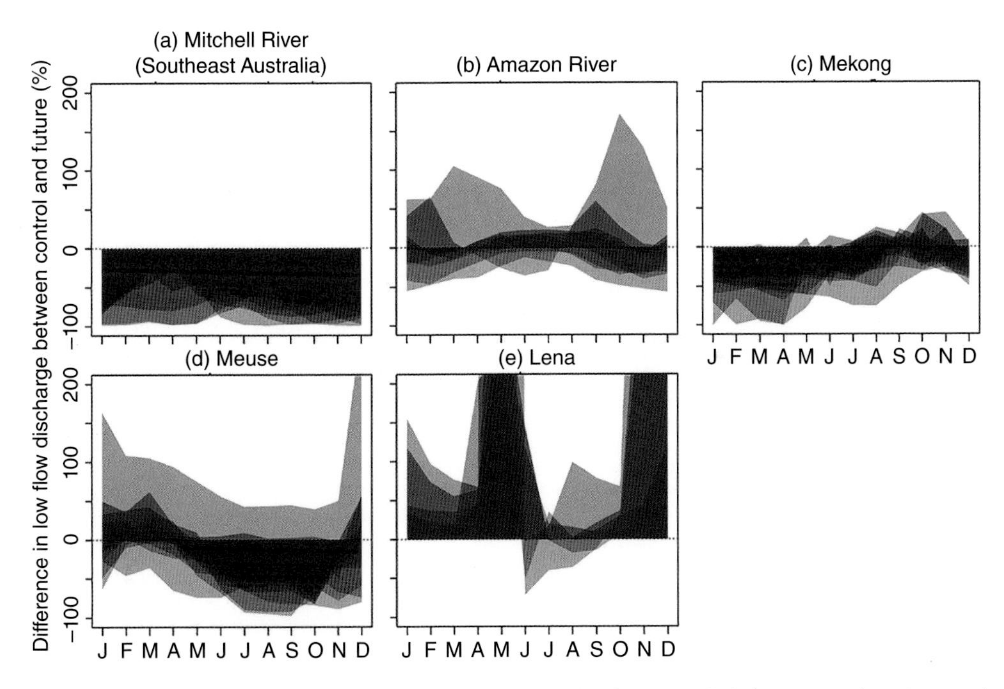

Figure 1.4.3 Relative change in low flows (Q_{80}) between future period (2071–2100) and control period (1971–2000), given as percentage of low discharge in the control period for representative river basins. Red indicates a decrease in Q_{80} in the future period, blue indicates an increase in Q_{80}. Transparency of colours indicates model agreement.

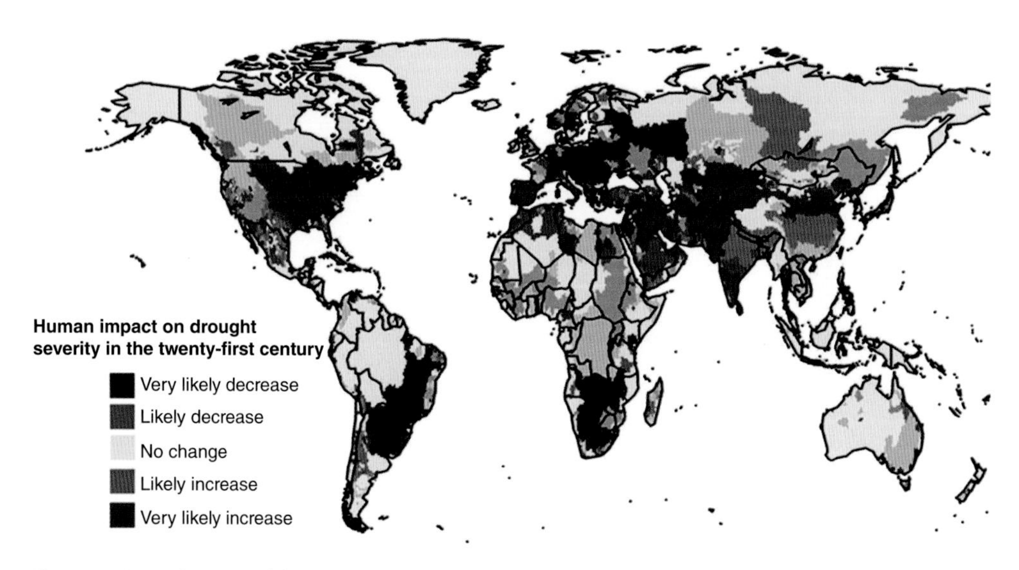

Figure 1.4.4 Impact of human water use and reservoir management on drought severity for the twenty-first century (2070–2099 relative to 1971–2000).

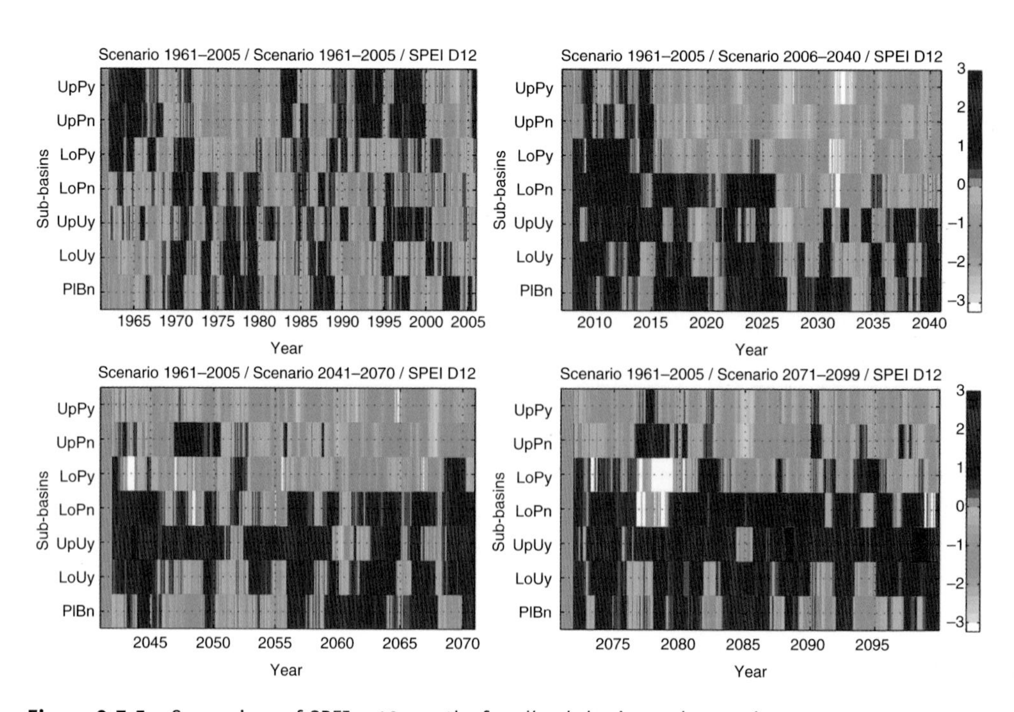

Figure 2.7.5 Comparison of SPEI – 12 months for all sub-basins and scenarios.

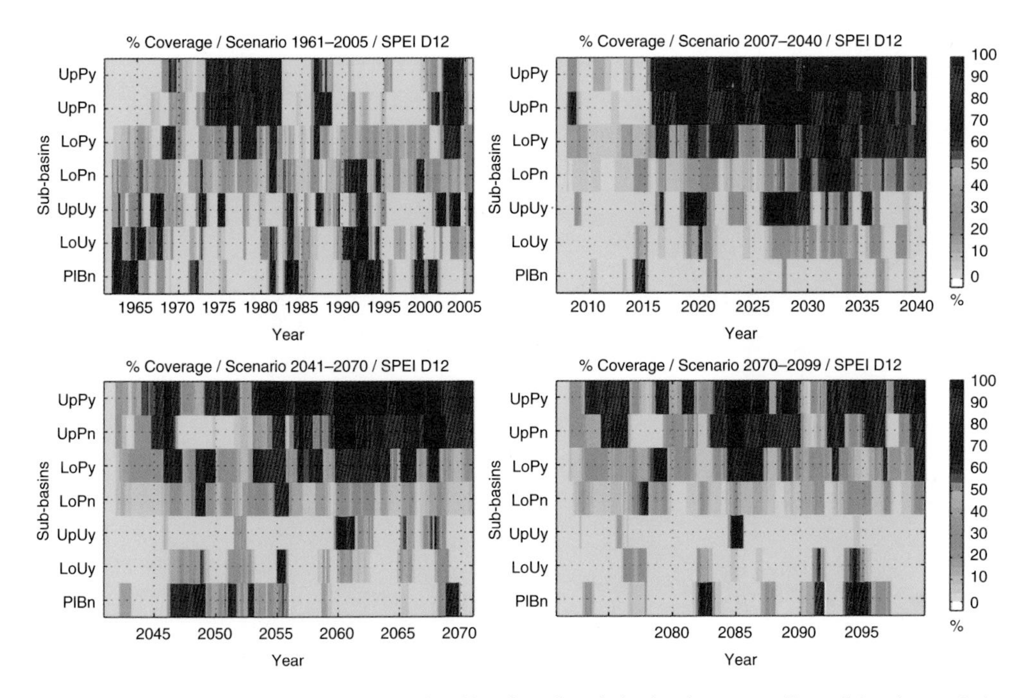

Figure 2.7.6 Percentage from the total cells of each sub-basin that was affected by dry periods (% coverage) per each month of the analysed time series for SPEI – 12 months, for all scenarios.

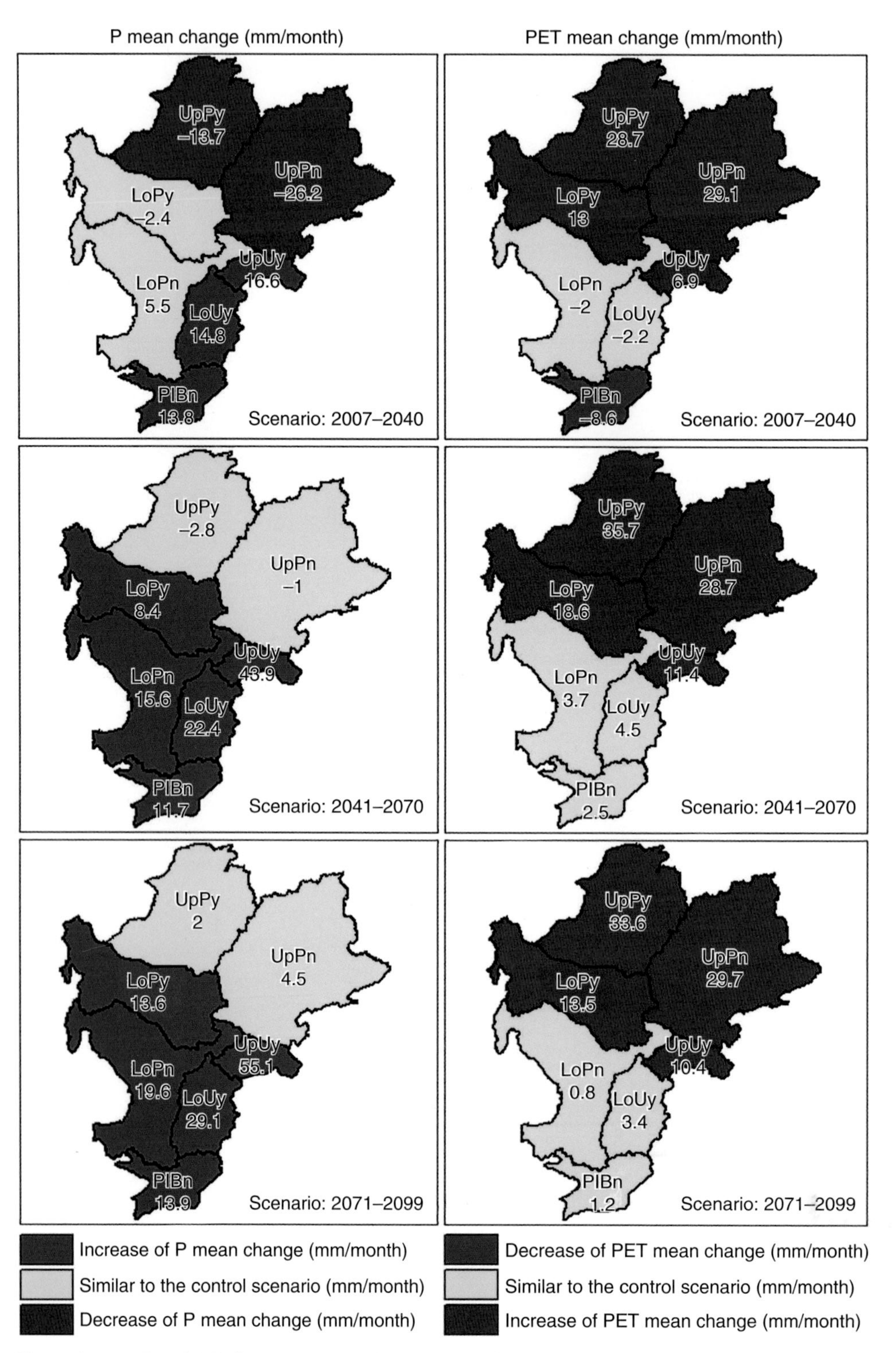

Figure 2.7.7 P and PET changes compared to the control scenario. The values show the mean changes of P and PET in mm for the whole sub-basin and the total time series for each scenario.

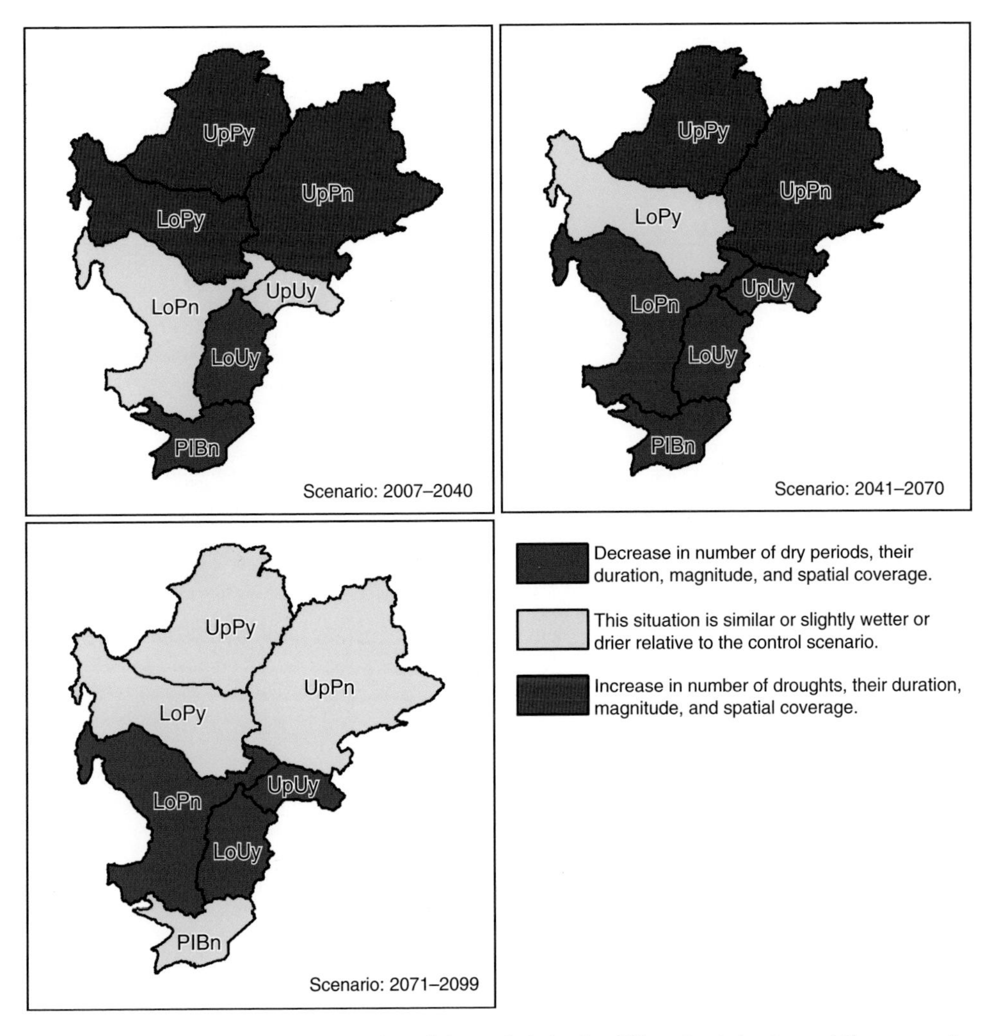

Figure 2.7.8 Variations in the situation of dry periods in the different sub-basins and time scenarios as compared to the control scenario (1961–2005).

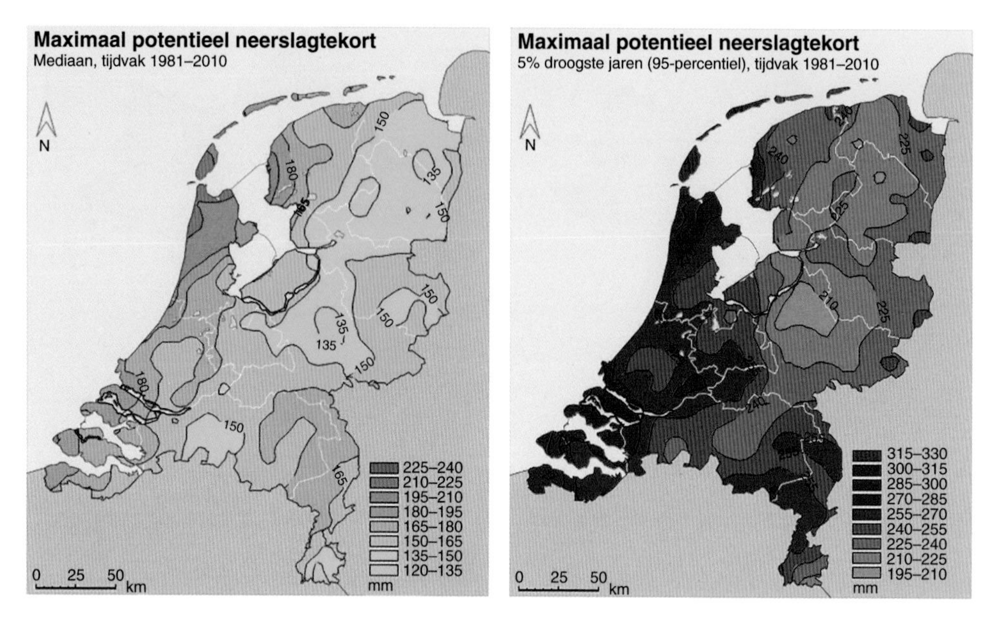

3.9.1 Maximum potential precipitation deficit in summer months over the period 1981–2010. Left: median; right: 95th percentile. *Source*: KNMI (2015a).

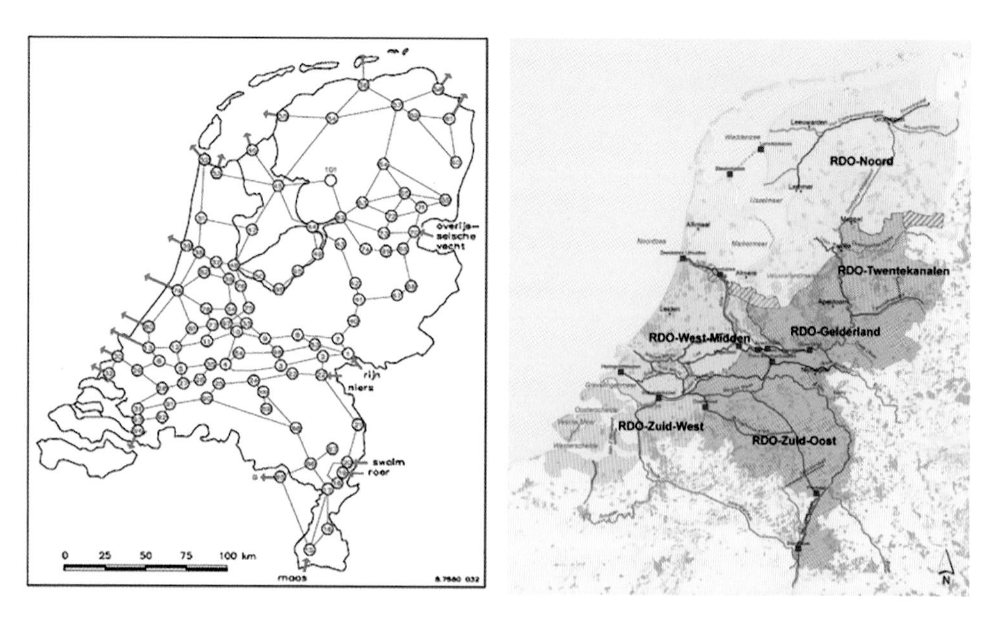

Figure 3.9.4 Surface water system that is used to supply freshwater in the Netherlands (left), and the six regions for regional drought platforms (right). *Source*: Andreu et al. (2015).

Neerslagtekort in Nederland in 2015
Landelijk gemiddelde over 13 stations

5% Droogste jaren
Mediaan
Recordjaar 1976
Jaar 2015: actueel 173 mm
15–Daagse verwachting met midden van verdeling: 199.3 mm

Neerslagtekort (mm)
0
100
200
300
April
Mei
Juni
Juli
Augustus
September

[c] KNMI, bijgewerkt 2015–08–03, 15:21 UT

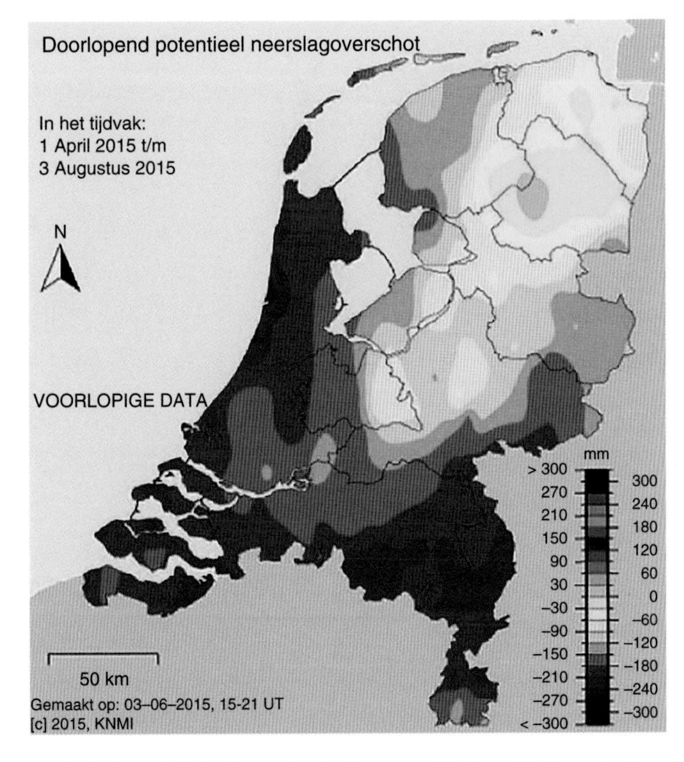

Figure 3.9.7 The precipitation deficit throughout the year 2015 until August. Top: Precipitation deficit in The Netherlands, August 2015; Bottom: Cumulative Potential Rainfall Surplus (1 April 2015–3 August 2015). *Source*: KNMI (2015b).

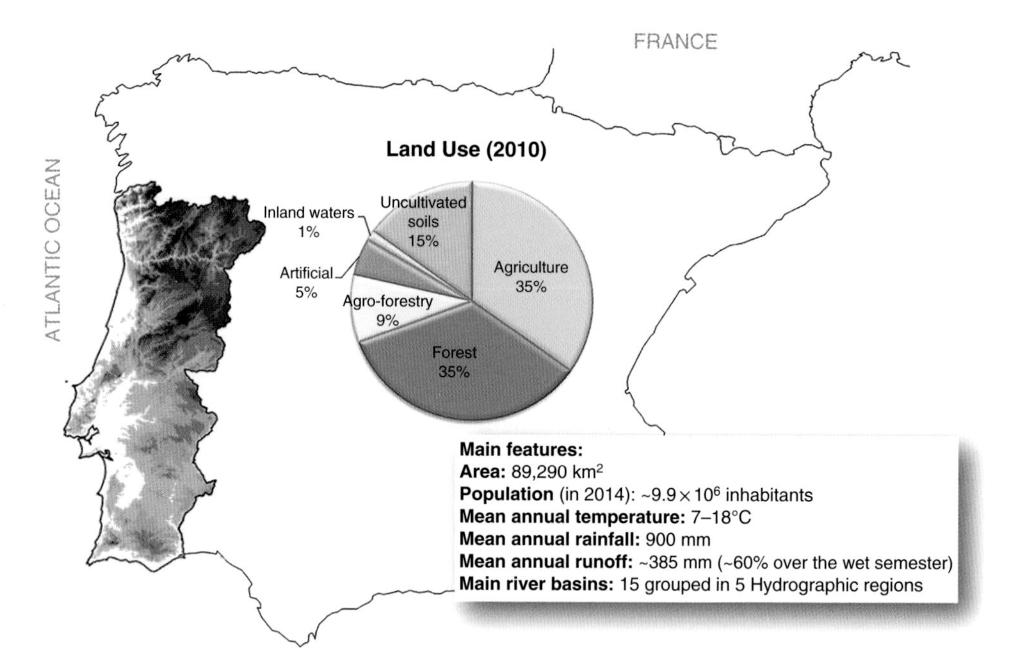

Figure 3.10.1 Elevation map of mainland Portugal; main features and land use composition. *Source*: Pie chart adapted from Vale et al. (2014).

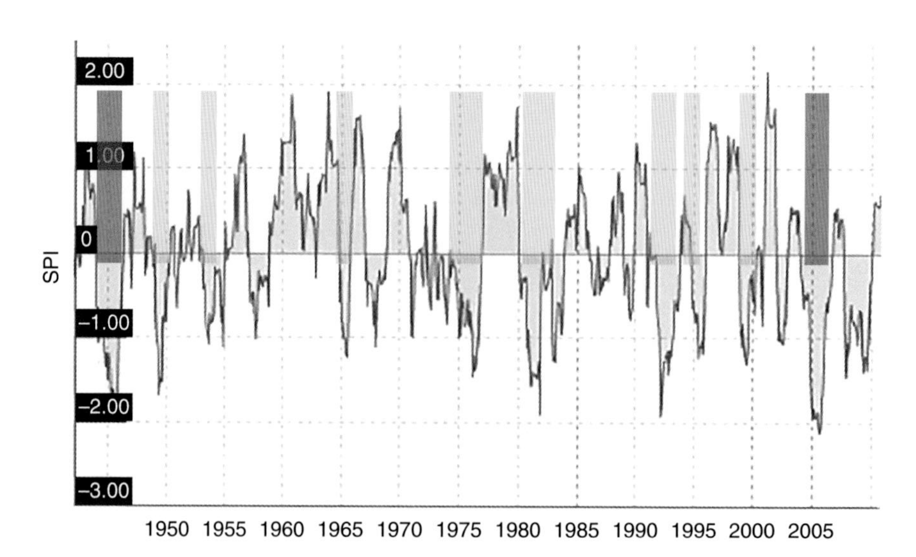

Figure 3.10.3 Six decades of annual precipitation anomalies estimated by the standardised precipitation index (SPI) for mainland Portugal, regarding the reference period of 1970–2000 (IPMA, 2015). Major drought events are highlighted: red – extreme; yellow – moderate or severe.

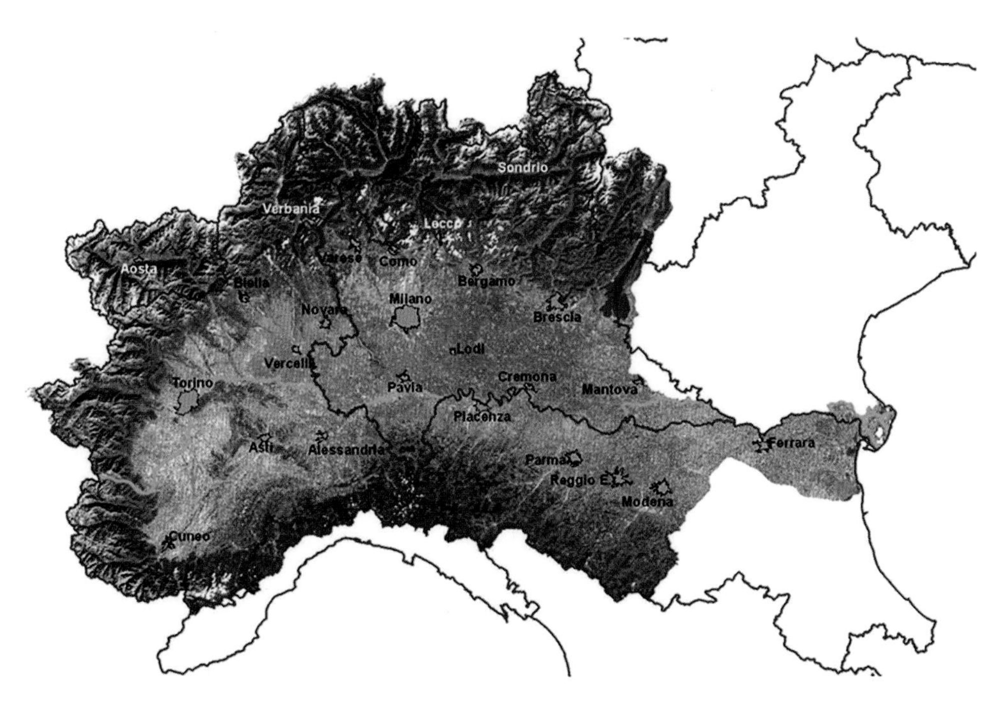

Figure 3.11.1 Po River Basin: regional and provincial administrative boundaries. *Source*: Po River Basin Authority.

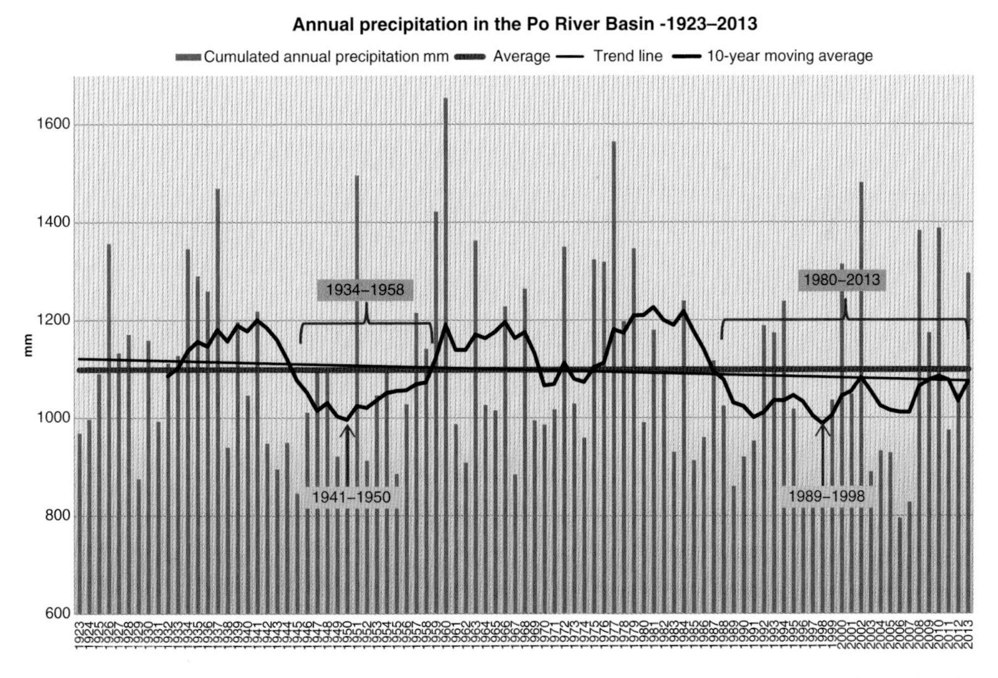

Figure 3.11.2 Annual precipitation on the Po River Basin (1923–2013). *Source*: Elaboration by Po River Basin Authority.

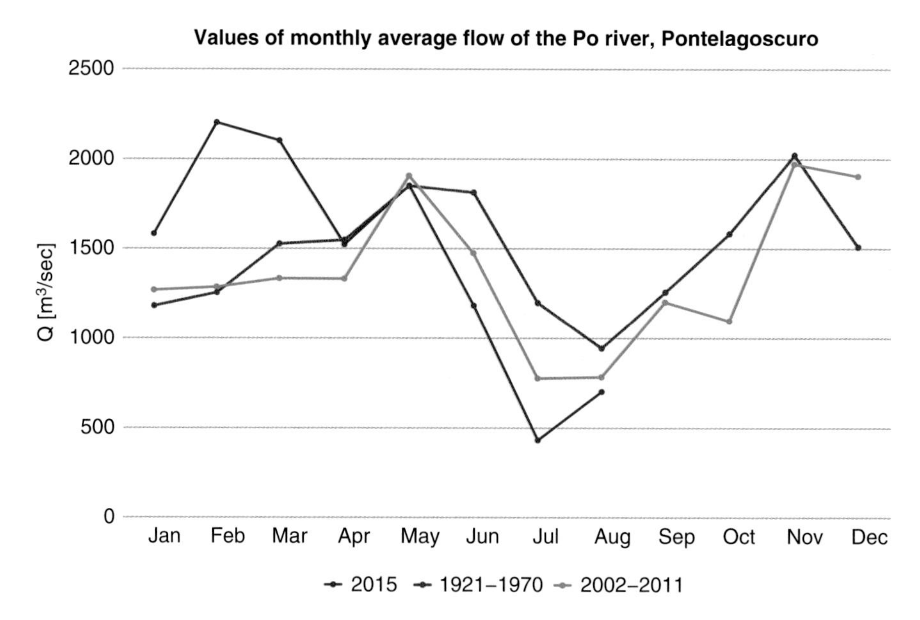

Figure 3.11.6 Average monthly flows of the Po river at Pontelagoscuro. *Source*: Elaboration by Po River Basin Authority.

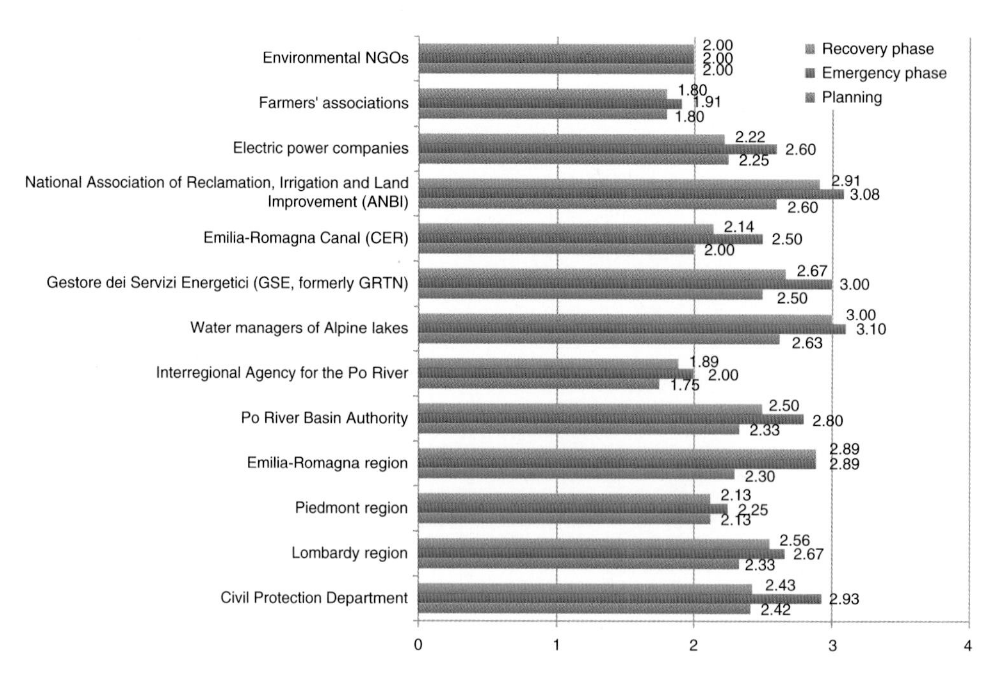

Figure 3.11.7 Average evaluation of the stakeholders' participation (rating: 4 = very adequate; 3 = adequate; 2 = not so adequate; 1 = not adequate) – survey conducted for the FP7 DROUGHT–R&SPI project.

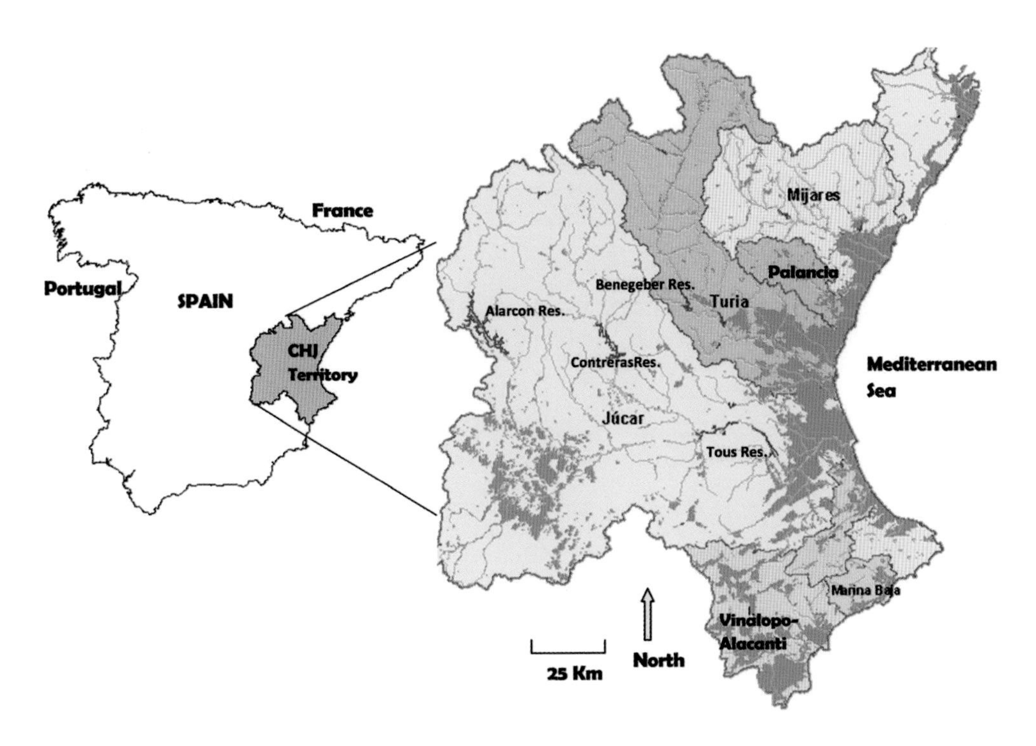

Figure 3.12.1 Location of the Jucar River Basin District, in Spain. *Source*: Self elaboration from Confederacion Hidrografica del Jucar originals, with permission.

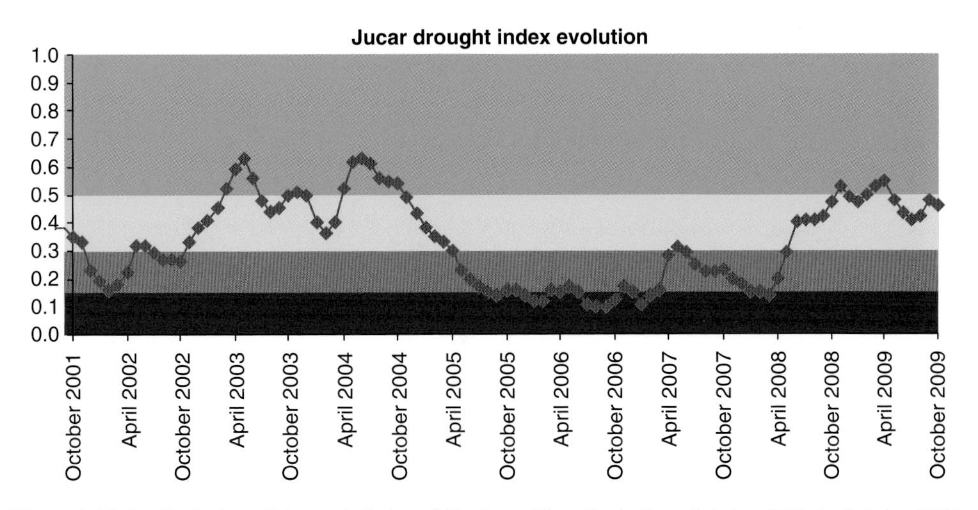

Figure 3.12.7 Evolution of the state index of the Jucar River Basin from October 2001 to October 2009. *Source*: Self-elaboration with data provided by CHJ.

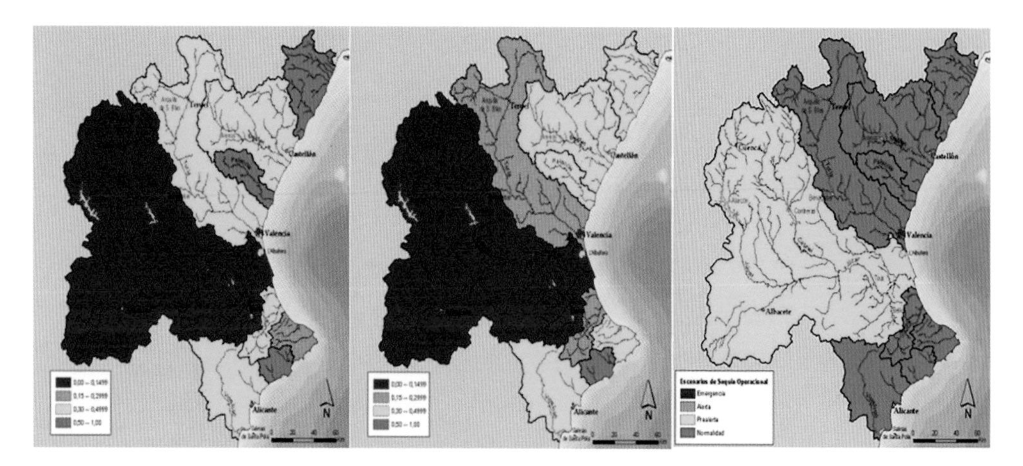

Figure 3.12.8 SODMIs for the CHJ basins corresponding to March 2006, January 2007, and March 2009 (from left to right). *Source*: Self elaboration from public domain information.

2.6 Indicators of Social Vulnerability to Drought

Gustavo Naumann[1], Hugo Carrão[2], and Paulo Barbosa[1]
[1]*European Commission, Joint Research Centre (JRC), Ispra, Italy*
[2]*Space4Environment, Niederanven, Luxemburg*

2.6.1 Introduction

Drought vulnerability is a complex concept that includes both biophysical and socio-economic drivers of drought impact that determine the capacity to cope with drought. The term 'vulnerability' is used here to convey the characteristics of a system or social group that makes it susceptible to suffer the consequences of drought. As with the definition of droughts, there is still a semantic debate on terminology, and hence 'vulnerability' may have different meanings when used in different disciplines and contexts (Brooks et al., 2005; Adger, 2006; Füssel, 2007; O'Brien et al., 2007). In the context of drought risk reduction, vulnerability depends on inadequate structures and management, on technological and economic limitations, and on environmental constraints. In many cases, social factors dominate (Turner et al., 2003). For example, although the direct impact of precipitation deficit may result in a reduction in crop yields, the underlying cause of this vulnerability to meteorological drought may be that the farmers did not use drought-resistant seeds – because their costs were too high, or because of some commitment to cultural beliefs. Another reason could be farm foreclosure related to drought. The underlying cause of this vulnerability could be manifold, such as small farm size because of historical land appropriation policies, lack of credit for diversification options, farming on marginal lands, limited knowledge of possible farming options, lack of local industry for off-farm supplemental income, or government policies.

Drought: Science and Policy, First Edition.
Edited by Ana Iglesias, Dionysis Assimacopoulos, and Henny A.J. Van Lanen.

2.6.2 Theoretical framework

Drought vulnerability can be understood as a starting point of the drought impacts, as a feature produced by multiple environmental and social processes (O'Brien et al., 2007; González-Tánago et al., 2016). In that sense, in order to assess drought vulnerability and then the risk for a certain location, the definition of vulnerability to drought should reflect the complex interactions between the socioeconomic systems and the physical environment. However, defining vulnerability to drought is complex, and, in any case, involves some measure of susceptibility, exposure, coping capacity, and adaptive capacity (Birkmann, 2007; Iglesias et al., 2009; Naumann et al., 2014).

In the context of disaster risk reduction, vulnerability has a multifaceted and multidimensional nature (Turner et al., 2003), and there is no single measure that can entirely represent its complexity. This multidimensional concept of vulnerability can be divided into different subgroups or components (biophysical, social, economic, and institutional). These components can be interdependent, and linkages may exist between them.

In the framework depicted by Naumann et al. (2014), the drought vulnerability index (DVI) is expressed as a function of four components that address the different aspects of vulnerability: renewable natural capital, economic capacity, human and civic resources, and infrastructure and technology. The definition of the components is based on the relevance of each indicator for policy development and the entire statistical structure of the dataset. An analytical approach was then used to explore whether the components were statistically well balanced in the composite indicator.

Each component is based on the aggregation of different factors or variables. Some variables are specific to drought (e.g. water infrastructure), while others are likely to influence vulnerability to droughts in an indirect way (e.g. poverty, population without access to improved water), as well as to influence the vulnerability of other hazards in diverse sociopolitical contexts (Cardona et al., 2012).

During the construction process of the composite indicator, it is desirable to account for the sources of uncertainty, while the inference process should be as objective and as simple as possible (OECD/JRC, 2008). As depicted in Naumann et al. (2014), the analysis could be conceptually divided into three main parts that are essential to any vulnerability assessment approach (Figure 2.6.1): (i) definition of the components of drought vulnerability, (ii) selection of variables and their normalisation, and (iii) model validation through a weighting and sensitivity analysis, and comparison with other indicators. A detailed analysis on the main decisions adopted (weighting scheme, aggregation, etc.), as well as a comparison with the impacts of previous drought disasters, may help the acceptance of the indicator by policymakers and end-users.

In Naumann et al. (2014), the set of variables that has been used to characterise the four components of socioeconomic vulnerability was compiled for African countries from the sources listed in Table 2.6.1. A sub-indicator for each of the four components may be computed as the weighted average of all the representative variables within the component.

In this case, the DVI is a composite indicator calculated by the weighted aggregation of 17 variables that represent the four components. The selection of the variables followed two criteria: they represent the concept to be explored (the theoretical framework) and are publicly available. Later on, this vulnerability index may be used

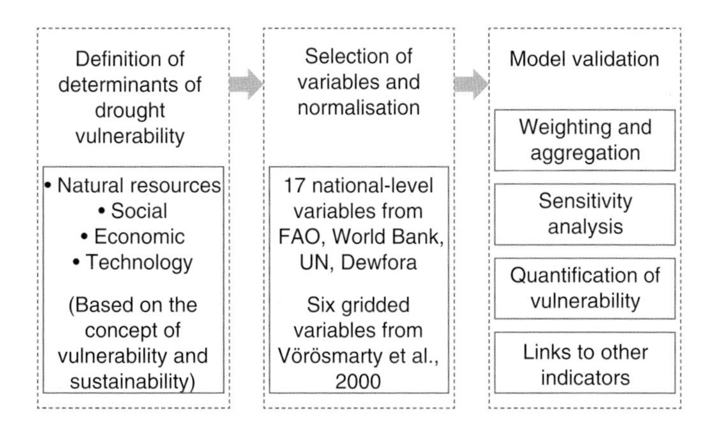

Figure 2.6.1 Summary of methodological framework proposed in Naumann et al. (2014).

Table 2.6.1 Vulnerability factors and components included in the DVI used in Naumann et al. (2014).

Component	Aspect relevant to drought management and type of influence	Indicator	Data source
1. Renewable natural capital	Water management, positive influence	Agricultural water use (% of total); irrigation water withdrawals (millions of m^3 $year^{-1}$ per grid cell)	Aquastat; World Water Assessment Program (World Water Development Report II: http://wwdrii.sr.unh.edu/index.html)
	Water management	Total water use (% of renewable)	FAO (Aquastat); CRU
	Water management	Irrigated area (% of cropland); irrigation-equipped area (km^2 per grid cell); agricultural area (km^2)	Aquastat, World Water Assessment Program (World Water Development Report II: http://wwdrii sr.unh.edu/index html)
		Rural population, year 2000 (people per grid cell) and total population, year 2000 (people per grid cell)	
	Water availability	Average precipitation: 61–90 mm $year^{-1}$	Aquastat; Global Precipitation Climatology Centre (GPCC) at Deutscher Wetterdienst
	Pressure on resources	Population density (inhabitants per km^2)	Aquastat; World Water Assessment Program (World Water Development Report II)
	Economic welfare	GDP per capita (in US$)	United Nations Development Programme (UNDP) Human Development Index; World Statistics Pocketbook (United Nations Statistics Division)

(Continued)

Table 2.6.1 (Continued)

Component	Aspect relevant to drought management and type of influence	Indicator	Data source
	Food security	Agricultural value added/ GDP %	Aquastat
2. Economic capacity	Economic welfare	Energy use (kg oil equivalent per capita)	World Bank; World Statistics Pocketbook (United Nations Statistics Division)
	Collective capacity	Population living below US$1.25 purchasing power parity per day	UNDP Human Development Index
	Human development (individual level)	Adult literacy rate (%)	UNDP Human Development Index
	Human development (individual level)	Life expectancy at birth (years)	UNDP Human Development Index
3. Human and civic resources	Collective capacity, institutional coordination	Government effectiveness (ranges from approximately − 2.5 (weak) to 2.5 (strong) governance performance)	World Bank
	Collective capacity, institutional coordination	Institutional capacity (0–1)	Drought Early Warning and Forecasting to Strengthen Preparedness and Adaptation to Droughts in Africa (DEWFORA)
	Collective capacity	Population without access to improved water (%)	World Bank
	Human displacement	Refugees (% of total population)	UNHCR
4. Infrastructure and technology	Development	Fertiliser consumption (kg per hectare of arable land)	World Bank; fertiliser consumption (total million tons) from FAOSTAT, arable land in kilohectares from Aquastat
	Water management potential	Water infrastructure (storage as proportion of total renewable water resources)	Aquastat

to understand the sensitivity of the system and to assist in the selection of measures to be adopted.

As highlighted by González-Tánago et al. (2016), data availability is still the major constraint in building sound and policy-relevant vulnerability assessments. However, depending on the availability of data, such indicators could be refined at higher resolution for a single component or for the full composite indicator. For instance, in Naumann et al. (2014), the analysis of the renewable natural capital component of the DVI was also carried out at sub-basin level, based on gridded data. Blauhut et al. (2016) explored the capability of frequently applied drought indices and vulnerability factors to predict the annual drought impact for different sectors and macro regions in Europe at different spatial resolutions. In a global assessment, Carrão et al. (2016) combined the economic and social indicators of vulnerability at the country level, with infrastructural vulnerability composed by three factors at the sub-national scale.

2.6.3 Selection of policy-relevant variables

Although drought impacts depend on local processes and conditions, a national-level analysis seems appropriate to be used by central governments and international organisations in the determination of drought policies. This index is developed based on demographic and other data sources to establish indicators that assess the capacity of the societies to cope with droughts. This indicator evaluates the societal capacity to develop knowledge and awareness to drought and their potential impacts, the natural capital including the availability and reliability of water resources, the efficiency to develop and use technologies, and the economic capacity for investment in food security and income stabilisation (Werner et al., 2015). The variables presented in Table 2.6.1 are mostly at country level; however, if data is available at finer scales, the analysis could be redefined accordingly. The socioeconomic vulnerability components and the related variables were selected on the following basis: the data is readily available, and the variables are geographically explicit and can be drought scenario dependent.

The five variables that compose the renewable natural capital component and are relevant to assess drought vulnerability are: (1) *Agricultural water use* – this is the amount of water used for agriculture as the percentage of the total water used in the country; it is a measure of the dependency of the agricultural sector on water availability. (2) *Total water use* – this is the total freshwater withdrawn in a given year, expressed in percentage of the actual total renewable water resources; it is an indication of the pressure on renewable water resources. (3) *Average precipitation* – this relates to the dependency of the country on the aridity level, and therefore the need for regulation of water sources. (4) *Irrigated area* – this is a share of the total agricultural area, which directly lowers the vulnerability to meteorological drought; however, mismanagement of irrigation allocation may result in increased urban or ecosystem vulnerability. (5) *Population density* – this is an indicator of the human pressure on water resources; a higher population density increases drought vulnerability.

It is clear that a higher resolution is preferred in order to characterise local disparities within countries. The renewable natural capital component was also characterised in Naumann et al. (2014) at higher resolution by using similar corresponding variables available from the University of New Hampshire datasets (Vörösmarty et al., 2000). From the variables available in this digital archive, the following variables were selected to obtain an index equivalent to the renewable natural capital: irrigation-equipped area, irrigation water withdrawals, agricultural area, rural population, and total population. Gridded normal precipitation from the Global Precipitation Climatology Centre dataset (Schneider et al., 2013) was also used. The indicators at the finer resolution level were then aggregated in the study at the sub-basin level, which can be of use for water basin management.

The four variables included to characterise the economic capacity component of the DVI, as well as their relevance for assessing drought vulnerability, is as follows: (1) *Gross domestic product (GDP) per capita* – this is the total economic output of a country divided by the number of people in the country; although an imperfect measure of well-being, it is widely used in sustainability and human development indicators as the main variable affecting a country's economic capacity, and is directly correlated to lower vulnerability. (2) *Agricultural value added per unit of GDP* – this is associated with the manufacturing processes that increase the value of primary agricultural production, and is directly correlated to lower vulnerability. (3) *Energy use* – this is the use of primary energy before transformation to other end-use fuels; it reflects economic capacity, and therefore also correlates positively with a lower vulnerability potential. (4) *Population living below poverty line* – this is the category of the population with purchasing power parity below US$1.25/day; in contrast to the other variables, this correlates with higher vulnerability levels, since poverty influences the capacity to cope and respond to drought impacts.

The selection of variables to characterise human and civic resources is more controversial, and data less readily available. Here, we have selected six variables that have been widely used in previous studies. *Adult literacy rate*, *life expectancy at birth*, and *population without access to improved water* are included in the human development index (HDI) of the United Nations. In addition, we have considered *institutional capacity* and *government effectiveness* to represent the management dimensions of drought vulnerability. Finally, a measure of the displaced population and refugees was included, since this is an important factor that reduces the coping capacity of population to drought. The relevance of these variables for assessing drought vulnerability is summarised in the following text.

Institutional capacity refers to the capacity of a country to cope with drought events; a higher institutional capacity implies lower drought vulnerability. *Government effectiveness* reflects perceptions of the quality of public services, the quality of the civil service and the degree of its independence from political pressures, the quality of policy formulation and implementation, and the credibility of a government's commitment to such policies. *Adult literacy rate* refers to the percentage of the population aged 15 years and older who can, with understanding, both read and write a short simple statement on their everyday lives. A higher literacy rate implies a higher capacity to deal with drought events. *Life expectancy at birth* is related to a population's vulnerability to extreme events, including drought, because the lack of sufficiently elderly people will prevent appropriate traditional knowledge transmission to younger generations. *Population without access* to improved water refers to the percentage of population without reasonable access to an adequate amount of water from an improved source. It is the most widely used indicator of drought damage in the most vulnerable areas, and has been a subject of the Millennium Development Goals. *Reasonable access* is defined as the availability of at least 20 litres of water per person a day from a source within 1 km of his or her dwelling; greater access to improved water reduces drought vulnerability. The number of refugees and the size of displaced population (as defined by the United Nations High Commissioner for Refugees, UNHCR) increase the drought vulnerability of a country, since this category is more likely to be exposed to natural hazards and less capable of coping with disasters.

The two variables selected for the infrastructure and technology components were fertiliser consumption and water infrastructure. Fertiliser consumption is a widely accepted measure of agricultural technology, and it is included as an indicator in most rural development studies. Water infrastructure measures the water stored as proportion of total renewable water resources, and reduces the vulnerability to drought.

It can be noticed that the components depicted previously are defined for a specific context and scope, but the combination of components to produce an overall vulnerability assessment may vary. The interdependences between the components are not clearly defined, as variations of the same factor could be used to characterise different components (González-Tánago et al., 2016). For instance, in Carrão et al. (2016), the factors representing population, assets, or other valuable elements present in an area in which drought events may occur (population, agricultural lands, livestock, and baseline water stress) were classified as the exposed elements, and represents a separate component of the drought risk. This approach follows the UNISDR (2004) definition of risk, depicted as the combination of hazard, exposure, and vulnerability as separated components.

2.6.3.1 Normalisation of variables to a common baseline

The drought vulnerability indicator is a way of summarising complex phenomena, and, as depicted previously, is composed of individual variables that have different measurement units. During the construction of the composite indicator, there is a need to normalise the variables in order to make them comparable. There are many possible normalisation methods, and the selection of the best procedure should be related to the data properties and the scope of the composition. Different data normalisation methods could lead to different outcomes. The effect of the normalisation, and therefore the robustness of the final scores, should be tested by employing different normalisation techniques. Normalisation techniques, such as the method of ranking and the distance to a reference value, are summarised and discussed in OECD/JRC (2008).

In Naumann et al. (2014), the variables in Table 2.6.1 were normalised between the different countries in order to be able to directly compare results. The normalisation has been made taking into account the maximum and minimum value of each variable across all countries in order to combine variables within the categories and to guarantee that variables have an identical range between 0 and 1. For variables with a positive correlation to the overall vulnerability, the normalised value is then calculated according to the general linear transformation, with

$$Z_i = \frac{\left(X_i - X_{min}\right)}{\left(X_{max} - X_{min}\right)}$$

Here, X_i represents the variable value for a generic country or administrative unit i, and X_{min} and X_{max} represent the respective minimum and maximum values across all administrative units i. In some cases, there is an inverse relationship between vulnerability and adaptive indicators (e.g. GDP per capita, adult literacy rate, or water infrastructure). For variables with negative correlation to the overall vulnerability, a transformation was applied to link the lowest variable values with the highest values of vulnerability:

$$Z_i = 1 - \frac{(X_i - X_{min})}{(X_{max} - X_{min})}$$

In this way, all normalised indicators (Z_i) have values between 0 (less vulnerable) and 1 (most vulnerable). Then, for each administrative unit, any of the k ($k = 1,\ldots, n$) components are computed as the arithmetic mean of the variables Z_i that define each component.

2.6.3.2 Composite indicator of drought vulnerability (weighting and aggregation)

Composite indicators of drought vulnerability are most useful to identify highly vulnerable countries for purposes of adaptation assistance, or as an entry point for case studies of systemic vulnerability (Brooks et al., 2005). Those interested in the determinants behind vulnerability for particular regions, countries, or groups of countries should examine the respective scores and rankings in the individual vulnerability dimensions and/or variables. Nevertheless, composite indicators can always be decomposed in order to measure the individual impact of each dimension and extend the analysis of country performance for improved policy guidelines (Naumann et al., 2014).

According to Kaufmann et al. (1999), Alkire and Santos (2014), and Naumann et al. (2014), to cite but a few, there are considerable benefits from aggregating related variables into a small number of composite indicators of vulnerability. For example, composite indicators span a much larger set of factors and can provide more precise measures of vulnerability than any individual variable, permitting comparisons of vulnerability across a broad set of regions. Moreover, it is possible to construct quantitative measures of both the composite vulnerability indicators and their dimensions and/or variables, allowing formal testing of hypotheses regarding cross-country differences in vulnerability.

Nevertheless, caution should be exercised when interpreting key vulnerability variables, or when using them to construct composite vulnerability indicators. First, there is a high degree of heterogeneity in the way datasets are collected by different organisations, and many times datasets are imperfect subjective proxies of unobservable variables (Brooks et al., 2005). The heterogeneity in the data means that these datasets can at best be used to place countries in groups, rather than compare governance between individual countries. Second, caution should also be exercised against methods for averaging across variables for a particular country and against methods for weighting variables into composite indicators of drought vulnerability (Kaufmann et al., 1999).

When used in a benchmarking framework, weights can have a significant effect on the overall composite indicator values and respective country rankings (OECD/JRC, 2008). Regardless of which method is used to derive variables' weights, these are essentially value judgements. While some analysts might choose weights based only on statistical methods, others might reward (or punish) components that are deemed more (or less) influential, depending on expert opinion, to better reflect policy priorities or theoretical factors (OECD/JRC, 2008). When there is insufficient knowledge of causal relationships or a lack of consensus on the alternative, composite indicators best rely on equal weighting (EW) – it does not mean 'no weights', but implicitly implies that all variables are given the same weight.

Most composite indicators rely on equal weighting. For example, HDI (UNDP, 2013) is a composite indicator that measures countries' achievements in key dimensions of human vulnerability: a long and healthy life, access to knowledge, and a decent standard of living. The HDI is the geometric aggregation of the three equally weighted normalised dimensions described in the preceding text, as follows (UNDP, 2013):

$$HDI = \left(V_{Health} \cdot V_{Education} \cdot V_{Income}\right)^{1/2}$$

The HDI assigns equal weight to all three dimension variables; the two education sub-variables are also weighted equally. The choice of equal weights is based on the normative assumption that all human beings value the three dimensions equally. Research papers that provide a statistical justification for this approach include Noorkbakhsh (1998) and Decancq and Lugo (2008).

The global multidimensional poverty index (MPI) is another equal-weighted composite indicator of vulnerability designed to capture the multiple deprivations that each poor person faces at the same time with respect to education, health, and other aspects of living standards (Alkire and Santos, 2014). Although MPI follows the same dimensions and weights as HDI, it is calculated by multiplying the incidence or headcount ratio of poverty (H) – the proportion of the population that is multidimensionally poor, by the average intensity of their poverty (A) – the average proportion of variables in which poor people are deprived, as follows (Alkire and Santos, 2014):

$$MPI = H \times A$$

A person is identified as 'poor' if he or she is deprived in at least one-third of the weighted variables, and as 'vulnerable to poverty' if he or she is deprived in 20–33.33% of the weighted variables.

Kim et al. (2013) also proposed an equally weighted composite indicator of vulnerability to describe the degree to which the socioeconomic system of a region and its physical assets are either susceptible or resilient to the impacts of a drought. The DVI proposed by Kim et al. (2013) (DVI_{kim}) is a relative measure among regions, and it is computed as the arithmetic average of quantifiable socioeconomic and physical variables, as follows:

$$DVI_{Kim} = \frac{IL \times AO \times CP \times PD \times MW \times IW \times AW}{7}$$

where IL, AO, CP, PD, MW, IW, and AW are the normalised values assigned to irrigated land, agricultural occupation, crop production, population density, municipal water, industrial water, and agricultural water, respectively.

However, if variables are grouped into dimensions, and those are further aggregated into the composite, then applying equal weighting to dimensions while grouping a different number of variables in each dimension could result in an unbalanced structure in the composite indicator (OECD/JRC, 2008). Naumann et al. (2014) studied the stability and robustness of different weighting schema applied to a composite indicator of vulnerability to drought. Their DVI ($DVI_{Naumann}$) is calculated as the weighted arithmetic average of the four dimensions of socioeconomic vulnerability: (1) renewable natural capital, (2) economic capacity, (3) human and civic resources, and (4) infrastructure and technology, as similar as for the Human Development Index.:

$$DVI_{Naumann} = \sum_{d=1}^{4} W_d \times AA_d$$

where W_d is the weight assigned to dimension d (with $\sum W_d = 1$), and AA_d is the statistical mean of normalised vulnerability variables within each dimension. The final weighting scheme was selected by taking into account the relative importance of each dimension of the DVI. The influence of different weights on $DVI_{Naumann}$ values was tested using three schema: equal weights (*EW*), a weighting scheme according the number of variables in each dimension (proportional weights, *PW*), and random weights (Montecarlo with 1000 simulations, *RW*). The results of a sensitivity analysis performed by the authors (see Section 2.6.3.3) show that the proportional weights lead to the lowest dispersion in countries' vulnerability ranks, adding value as compared to the composite indicator with random weights and reducing the extreme behaviour of the composite indicators with equal weights. Moreover, the results attained by Naumann et al. (2014) suggest that the composite indicator derived with the proportional weights was the more robust option for aggregating vulnerability dimensions in their study area – and regarding the selected input variables.

Besides weighting schemes, we showed that aggregation schema of vulnerability dimensions also vary for the different composite indicators. The averaging of dimensions into a composite indicator of drought vulnerability can be performed by compensatory or non-compensatory aggregation schema. The composite indicators presented in the preceding text are based on compensatory aggregation schema. The decision of whether to use a compensatory or non-compensatory approach depends on whether the values of one variable can be traded for values of another variable in the composite indicator or not. By following a compensatory approach, drought vulnerability is estimated by considering all of the indicators' values, and by trading off the region's high value on one or more indicators, with its lower values on other indicators. While linear aggregation schema (sum or product) reward dimensions proportionally to the weights, geometric aggregations reward those countries with higher scores (OECD/JRC, 2008). Indeed, poor performance in any dimension is directly reflected in the geometric aggregation (UNDP, 2013). That is to say, a low achievement in one dimension is not anymore linearly compensated by a high achievement in another dimension. Thus, as a basis for comparisons of achievements, the geometric aggregation is also more respectful of the intrinsic differences across the dimensions than the arithmetic aggregation (UNDP, 2013). Nevertheless, in both linear and geometric aggregations, weights express trade-offs between dimensions, and a deficit in one dimension can thus be offset (compensated) by a surplus in another.

To ensure that weights remain a measure of importance, other aggregation methods should be used – particularly methods that do not allow compensability. Among different non-compensatory methods, data envelopment analysis (DEA) (Lovell and Pastor, 1999; Cook et al., 2014) is a non-parametric approach that can be used to average variables of drought vulnerability into a relative measure of drought vulnerability for a specific region or country. This statistical method has been used, for example, by Ramos and Silber (2005), and Anderson et al. (2011), to derive the relative socioeconomic welfare of countries and their rankings from multidimensional sets of variables. The following assumptions are made for DEA benchmarking and ranking (OECD/JRC, 2008):

- The higher the value of a given variable, the more vulnerable it is for the corresponding region
- Non-discrimination of regions that are the most vulnerable in a single variable, thus ranking them equally

DEA is a linear programming-based method that estimates a normalised distance function to an observed multidimensional frontier of maximum drought exposure characterised by several variables. The relative exposure of each region to drought is determined by the statistical position of the region and its multidimensional distance to the benchmark frontier. Both issues are represented in Figure 2.6.2 for the simple case of six regions ($R_1, R_2, \ldots, R_6$) and two generic exposure indicators (y_1 and y_2). The line connecting the observed indicators' values for countries R_1, R_2, R_3, and R_4 (that has been notionally extended to the axes by the lines '$R_1y'_2$' and '$R_4y'_1$' to enclose the entire dataset) constitutes the performance frontier (i.e. maximum vulnerability among the regions or countries represented in the sample dataset) and the benchmark for regions R_5 and R_6, which lie below that frontier. The regions supporting the frontier are classified as the most exposed according to their values in one or both indicators. The most exposed regions will have a performance score of 1, while regions R_5 and R_6, which are within this envelope, are less vulnerable than the others, and will have score values between 0 and 1.

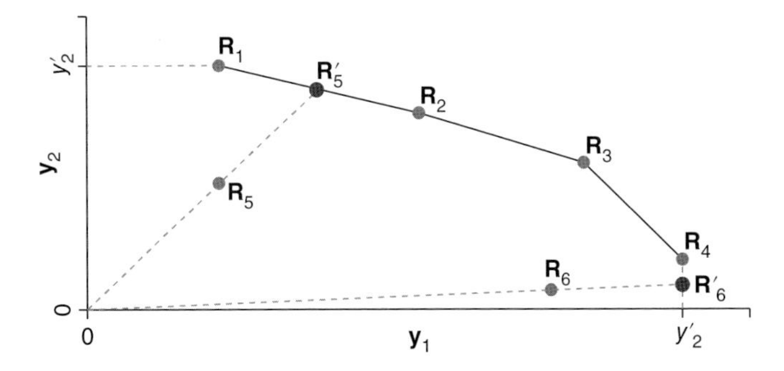

Figure 2.6.2 Computation of a performance frontier in a simulated DEA for six regions and two variables. Red: observed indicators' values; and blue: projected indicators' values in the frontier of maximum exposure. *Source:* From Carrão et al. (2016).

The non-compensatory drought vulnerability values can be computed with DEA for regions R_5 and R_6, as follows (OECD/JRC, 2008):

$$dv_i = \overline{0R_i} \Big/ \overline{0R_i'}$$

where $\overline{0R_i}$ is the multivariate distance between the origin and the actual observed indicators' values for region i, and $\overline{0R_i'}$ is the distance between the origin and the projected regional values in the frontier of maximum vulnerability. The vulnerability value for each region depends on its statistical position with respect to the frontier, while the benchmark corresponds to the worst situation according to the values of the other regions measured in the set of the same indicators.

2.6.3.3 Sensitivity analysis and model validation

Uncertainty analysis focuses on how uncertainty in the input factors (variables included in the model, data normalisation, weighting, and aggregation) propagates through the overall construction of the DVI composite indicator. *Sensitivity analysis* evaluates the contribution of each individual source of uncertainty to the overall variance. The iterative use of uncertainty and sensitivity analyses during the development of a DVI could improve its structure (Gall, 2007). Likewise, sensitivity analysis can help to increase the transparency and to identify the regions that are favoured or weakened under the different assumptions. The approach taken to assess uncertainties could include the following steps (OECD/JRC, 2008):

- Inclusion and exclusion of individual variables or factors
- Using alternative editing schemes – for example, single or multiple imputation
- Using alternative data normalisation schemes, such as Min–Max, standardisation, use of rankings, etc.
- Using different aggregation systems – for example, linear, geometric mean of unscaled variables
- Using different weighting schemes and plausible values for the weights

For instance, Naumann et al. (2014) performed a sensitivity analysis to assess the robustness of the DVI. This examination was designed as different Monte Carlo experiments to assess the contribution of any individual source of uncertainty to the output variance. This methodology is based on multiple evaluations of the model, with three weighting and two aggregation schemes that generate different probabilistic density functions of model outputs. Similarly, Carrão et al. (2016) compared the regional vulnerability ranks derived with the model proposed to alternative composite statistics and aggregation schema. The criterion to evaluate the robustness of the different models is internal and based on the distance between the respective regional rankings and the median regional ranking of the ensemble set defined by the outputs of all possible models.

The uncertainties from the input factors are then expressed as the resulting probability distribution functions (PDFs) of the DVI and shifts in rankings between administrative units. The uncertainty bounds associated with the DVI values are also useful for communicating to the end-users all the plausible values that the DVI can reach for each administrative unit.

2.6.4 Application: Drought risk assessment in Latin America

This methodology was applied to formulate and validate a drought risk assessment in Latin America. The maps were derived from the combination of global drought hazard, exposure, and vulnerability determinants as defined in Carrão et al. (2016). In this case study, we propose the data-driven approach for computing each determinant of risk, and derive a final global map based on the theoretical formulation proposed by the UNISDR (2004).

The overall drought risk map is computed at the sub-national administrative level (Figure 2.6.3) to facilitate the management of layers and their associations on a fixed and common minimum mapping unit, following the approach of Carrão et al. (2016). Moreover, it is easy for stakeholders and policy-makers to use and manipulate the products delivered at the level of administrative regions. Each determinant of risk is formulated independently of each other, and validated on its own.

A major limitation of the proposed approach is the use of national information for computing the model – it might be criticised for the lack of spatial detail. This solution is the best compromise, since harmonised databases with socioeconomic information at the sub-national administrative level were not available for most users. Nevertheless, it is important to highlight that, in case this information would be available, other difficulties on their harmonisation would arise. For example, the intra-national management of regional disparities is not easily understood when a hazard such as drought strikes: some countries might provide immediate national support to their poorer regions, whereas others may not have this capacity. The richest region in some poor country might not respond to a hazard in the same manner as the poorest region in some rich country. Therefore, the use of socioeconomic factors at the national level could be a limitation in this study, but it is simultaneously a good compromise between unknown

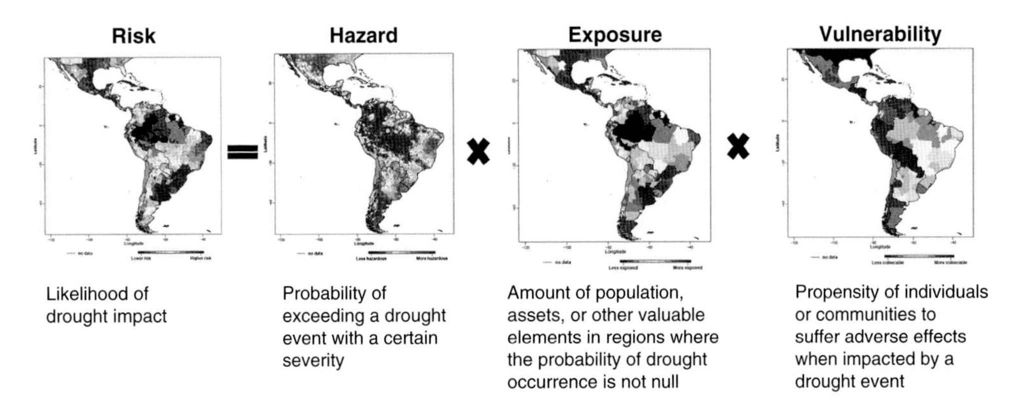

Figure 2.6.3 Latin American map of drought risk by combining hazard, exposure, and vulnerability.

multi-scale relationships (that would increase the bias in the model and lower thematic accuracy) and output knowledge of high spatial resolution.

Another limitation relates to the selection of the model. This model is based on an internal validation procedure that chooses the best model as the one giving regional vulnerability ranks that approximate the median of the ensemble of all models tested. A neater solution could be tested, but the absence of reference data for performing an independent validation reduces the lack of effective testing options.

Finally, regarding the drought risk in Latin America, attained results show that drought occurrence is higher in southeast and northeast Brazil, as well as between the northeast and southwest of Argentina, and the whole southern part of Argentina. Exposure is higher in southeast Brazil and northeast Argentina, as well as in the Central American Dry Corridor (CADC); vulnerability is higher in the CADC region. Overall, and due to regional differences in the characteristics of its determinants, drought risk is more prominent in Central America and the southeast of South America. Since the determinants of risk vary in these areas, we suggest that there are no single best-management approaches to drought, and that mitigation or adaptation measures should be evaluated depending on each specific case following a bottom-up approach.

References

Alkire, s. and Santos, M. (2014). Measuring acute poverty in the developing world: robustness and scope of the multidimensional poverty index. *World Development* 59: 251–274.

Adger, W.N. (2006). Vulnerability. *Global Environmental Change* 16: 268–281.

Anderson, g., Crawford, i., and Leicester, A. (2011). Welfare rankings from multivariate data, a nonparametric approach. *Journal of Public Economics* 95: 247–252.

Birkmann, J. (2007). Risk and vulnerability indicators at different scales: applicability, usefulness and policy implications. *Environmental Hazards* 7: 20–31.

Blauhut, V., Stahl, K., Stagge, J.H., Tallaksen, L.M., De Stefano, L., and Vogt, J. (2016). Estimating drought risk across Europe from reported drought impacts, drought indices, and vulnerability factors. *Hydrology and Earth System Sciences* 20 (7): 2779–2800.

Brooks, N., Adger, W.N., and Kelly, P.M. (2005). The determinants of vulnerability and adaptive capacity at the national level and the implications for adaptation. *Global Environmental Change* 15: 151–163.

Carrão, H., Naumann, G., and Barbosa, P. (2016). Mapping global patterns of drought risk: an empirical framework based on sub-national estimates of hazard, exposure and vulnerability. *Global and Environmental Change* 39: 108–124.

Cardona, O., Van Aalst, M., Birkmann, J., Fordham, M., Mcgregor, G., Perez, R., Pulwarty, R., Schipper, E., and Sinh, B. (2012). Determinants of risk: exposure and vulnerability. In: *Managing the Risks of Extreme Events and Disasters to Advance Climate Change Adaptation. A Special Report of Working Groups I and II of the Intergovernmental Panel on Climate Change (IPCC)*, pp. 65–108. Cambridge, UK, and New York, NY, USA: Cambridge University Press.

Cook, W., Tone, K., and Zhu, J. (2014). Data envelopment analysis: prior to choosing a model. *Omega* 44: 1–4.

Decancq, k. and Lugo, M.A. (2008). Setting Weights in Multidimensional Indices of Well-Being. *OPHI Working Paper 18*, University of Oxford.

Füssel, H.M. (2007). Vulnerability: a generally applicable conceptual framework for climate change research. *Global Environmental Change* 17: 155–167.

Gall, M. (2007). Indices of social vulnerability to natural hazards: a comparative evaluation. PhD dissertation, Department of Geography, University of South Carolina.

González-Tánago, I.G., Urquijo, J., Blauhut, V., Villarroya, F., and De Stefano, L. (2016). Learning from experience: a systematic review of assessments of vulnerability to drought. *Natural Hazards* 80 (2): 951–973.

Iglesias, A., Garrote, L., Cancelliere, A., Cubillo, F., and Wilhite, D.A. (2009). Coping with drought risk in agriculture and water supply systems. In: *Drought Management and Policy Development in the Mediterranean, Series: Advances in Natural and Technological Hazards Research*. The Netherlands: Springer.

Kaufmann, d., Kraay, a., and Zoido-Lobato, p. (1999). Aggregating Governance Indicators. *Policy Research Working Paper 2195*, World Bank, Washington, DC.

Kim, h., Park, j., Yoo, j., and Kim, t. (2013). Assessment of drought hazard, vulnerability, and risk: a case study for administrative districts in South Korea. *Journal of Hydro-Environment Research* 9: 28–35.

Lovell, c. and Pastor, j.t. (1999). Radial DEA models without inputs or without outputs. *European Journal of Operational Research* 118: 46–51.

Naumann, g., Barbosa, p., Garrote, l., Iglesias, a., and Vogt, J. (2014). Exploring drought vulnerability in Africa: an indicator based analysis to be used in early warning systems. *Hydrology and Earth System Sciences* 18: 1591–1604.

Noorkbakhsh, j. (1998). The human development index: some technical issues and alternative indices. *Journal of International Development* 10: 589–605.

O'Brien, K., Eriksen, S., Nygaard, L.P., and Schjolden, A. (2007). Why different interpretations of vulnerability matter in climate change discourses. *Climate Policy* 7 (1): 73–88.

OECD/JRC. (2008). *Handbook on Constructing Composite Indicators. Methodology and User Guide, Social Policies and Data Series*. Paris: OECD Publisher.

Ramos, x. and Silber, J. (2005). On the application of efficiency analysis to the study of the dimensions of human development. *The Review of Income and Wealth* 51: 285–309.

Schneider, U., Becker, A., Finger, P., Meyer-Christoffer, A., Ziese, M., and Rudolf, B. (2013). GPCC's new land surface precipitation climatology based on quality-controlled in situ data and its role in quantifying the global water cycle. *Theoretical and Applied Climatology* 115: 1–26.

Turner, B.L., Kasperson, R.E., Matson, P.A., McCarthy, J.J., Corell, R.W., Christensen, L., Eckley, N., Kasperson, J.X., Luers, A., Martello, M.L., Polsky, C., Pulsipher, A., and Schiller, A. (2003). A framework for vulnerability analysis in sustainability science. *Proceedings of the National Academy of Sciences* 100 (14): 8074–8079.

UNISDR. (2004). *Living with Risk: A Global Review of Disaster Reduction Initiatives, Review Volume 1*. New York and Geneva: United Nations International Strategy for Disaster Reduction.

UNDP. (2013). The Rise of the South: Human Progress in a Diverse World. *Human Development Reports 19902013 23*, United Nations Development Programme, New York, USA.

Vörösmarty, C.J., Green, P., Salisbury, J., and Lammers, R. (2000). Global water resources: vulnerability from climate change and population growth. *Science* 289: 284–288.

Werner, M., Vermooten, S., Iglesias, A., Maia, R., Vogt, J., and Naumann, G. (2015). *Developing a Framework for Drought Forecasting and Warning: Results of the DEWFORA Project. Drought: Research and Science–Policy Interfacing, 01/2015: chapter 41*, pp. 279–285. New York: CRC Press, Taylor and Francis Group.

2.7

Drought Vulnerability Under Climate Change: A Case Study in La Plata Basin

Alvaro Sordo-Ward[1], María D. Bejarano[1], Luis Garrote[1], Victor Asenjo[2], and Paola Bianucci[3]

[1]*Universidad Politécnica de Madrid, Madrid, Spain*
[2]*GIS and Environmental Sciences Consultant, Spain*
[3]*AQUATEC (Suez Group), Dep. Basin Management, Madrid, Spain*

2.7.1 Introduction

Disaster risk reduction and climate change adaptation policies include the idea of reducing vulnerability and enhancing resilience. Drought policies are being adopted by most South American countries at a rapid pace. This apparent transformation in policy has recently accelerated, especially since United Nations International Strategy for Disaster Reduction's Hyogo Framework for Action was adopted (HFA, 2005), and the United Nations Development Programme included disaster management as one of the adaptation strategies to climate change. This chapter seeks to understand the roles of precipitation and evapotranspiration as drivers of drought vulnerability. We studied the spatiotemporal characterisation of the droughts in La Plata Basin by using the SPEI index for the period 1961–2100 at a macro-basin scale.

La Plata Basin is spread over five countries in the centre–south region of South America and comprises 3 174 229 km². It is one of the largest fresh water reserves on the planet. Despite the significant impact of droughts on agriculture, cattle rearing, urban water supply, natural water courses, and wetlands, droughts are still difficult to predict in the region, both in time and space. However, this information is needed in order to accurately define specific measures and plans for drought prevention and mitigation. Drought and water scarcity greatly influence areas with temperate grasslands and all crops throughout the region, as well as the critical production of hydropower. The Intergovernmental Panel on Climate Change clearly shows that an increased frequency

Drought: Science and Policy, First Edition.
Edited by Ana Iglesias, Dionysis Assimacopoulos, and Henny A.J. Van Lanen.

of extreme events is likely to have larger impacts than changes in mean precipitation (IPCC, 2014).

Despite the high uncertainty regarding the evaluation of the factors that cause droughts and the difficulties on predicting drought events, there are means and methods aimed at fighting the damage resulting from droughts, such as: definition of tolerable droughts and degree of losses based on indicators of drought level; prevention methods aimed at supply and demand or minimisation of impacts; application of instruments of damage reduction such as soil improvement or crop changes towards more drought-tolerant varieties; and the organisation and coordination of the agents involved, among others (Spanish Ministry of Environmental Affairs, 2007). Therefore, the phenomenon of drought and its implications for sustainable management of water resources depend fundamentally on the interplay of three factors: (a) hydrometeorological conditions of the basin; (b) main water uses in the geographical area; and (c) water infrastructure and the capacity for regulation available in the basin.

Droughts in La Plata Basin, similar to other large basins, has been little studied. This chapter is focused on the spatiotemporal analysis of hydrometeorological conditions in the sub-basins for both the current and future scenarios covering the period 1961–2100. A study carried out by Brazil's National Institute for Space Research (INPE) on the current (1961–1990) and future rainfall variability in La Plata Basin (Marengo et al., 2014; Ferraz, 2014a, 2014b; among others) states that the projection of the daily average precipitation in December, January, and February (DJF) indicates: an increase of up to 1 mm/day in the south, and a decrease of up to 3 mm/day in the north of the basin for the period 2011–2040; an increase of up to 1 mm/day in the south, southwest, and west, and a decrease of up to 1 mm/day in the north and northeast of the basin for the period 2041–2070; and an increase in the south/southwest of up to 2 mm/day, and a decrease in the north/northeast of up to 1 mm/day for the period 2071–2099. On the other hand, the average temperature for the entire trimester (DJF) is projected to increase for the analysed periods all over the basin. The greatest differences are found between the latitudes 10°S and 23°S, with a DJF average temperature increase of up to 3°C for 2011–2040, up to 3.5°C for 2041–2070, and up to 4°C for 2071–2099. However, when rainfall and temperature are analysed together, drought projections appear to be contrasting: droughts would be favoured by longer periods of consecutive non-rainy days (Donat et al., 2013) and future scenarios characterised by higher temperatures, while they would be disfavoured by the stabilisation or increase of the in-between-droughts period and the annual runoff described by some authors for current and future scenarios (note that this pattern is not as clear for some areas of the basin). Summing up, it is not possible to derive any strong conclusion on the future patterns of drought in La Plata Basin from these previous studies.

2.7.2 Methods

2.7.2.1 Framework

We used P, PET, and SPEI to characterise droughts. PET and P were obtained for all 10 × 10 km–sized cells within the basin. Cell-to-cell information was integrated into a

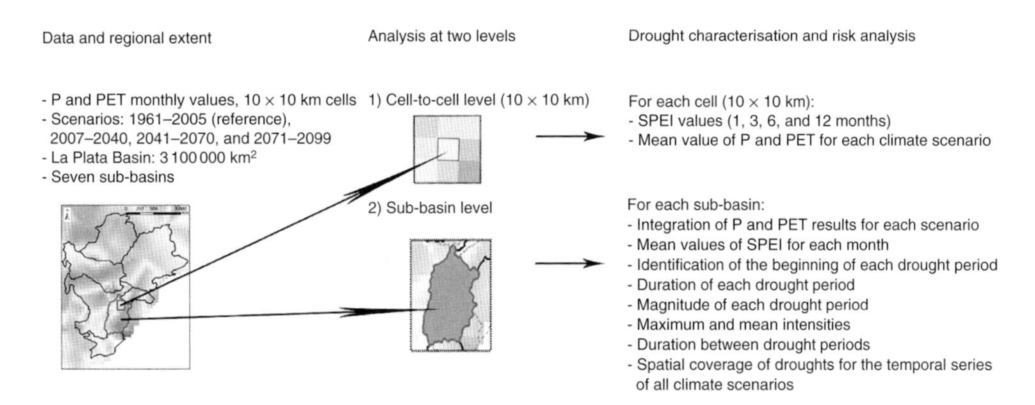

Figure 2.7.1 General conceptual framework of the applied methodology.

sub-basin level (seven sub-basins were defined). For each sub-basin, climate scenario, and temporal scale of SPEI (1, 3, 6, and 12 months), we showed and analysed the results. Additionally, for each SPEI temporal scale and sub-basin, we described the spatial coverage of droughts for the temporal series of all climate scenarios. Finally, we carried out a drought risk analysis at a sub-basin level. A general scheme of the proposed methodology is shown in Figure 2.7.1.

2.7.2.2 Data and regional extent

The identification and characterisation of dry and wet periods for the current and future scenarios was carried out on a monthly timescale and on a distributed spatial scale, and by using SPEI (Vicente-Serrano et al., 2010a, 2010b; Beguería et al., 2010). The SPEI calculation was based on the monthly series of precipitation (P) and potential evapotranspiration (PET) provided by INPE. These were estimated using the Eta model (Ferraz, 2014a; Bustamante et al., 2002, 2006; Alves et al., 2004; Chou et al., 2005) for different climate scenarios. The information provided by INPE is spatially distributed in 10 × 10 km cells and temporarily divided into four climate scenarios: 1961–2005, 2007–2040, 2041–2070, and 2070–2099. We used the 10 × 10 km cells as calculation units, and the seven sub-basins that comprise La Plata Basin, as defined by the Intergovernmental Coordinating Committee of the Countries of La Plata Basin (CIC) (Figure 2.7.2), as data analysis and result presentation areas. The sub-basins were named as: Upper Paraguay (UpPy), Upper Paraná (UpPn), Lower Paraguay (LoPy), Lower Paraná (LoPn), Upper Uruguay (UpUy), Lower Uruguay (LoUy), and Río de la Plata (PlBn). The annual mean values of P and PET for each sub-basin and climate scenario is shown in Table 2.7.1. Additionally, we used observed rainfall data from 46 stations located in the basin from University of East Anglia's Climatic Research Unit Time Series (CRU TS) database (version 3.22) (CRU, 2014). In some stations, there were available rainfall series from 1851.

In Figure 2.7.3, for each sub-basin and climate scenario, we show the temporal evolution of the monthly mean P and PET. Cycles characterised by high and low values, intensities, and durations of P and PET can be distinguished.

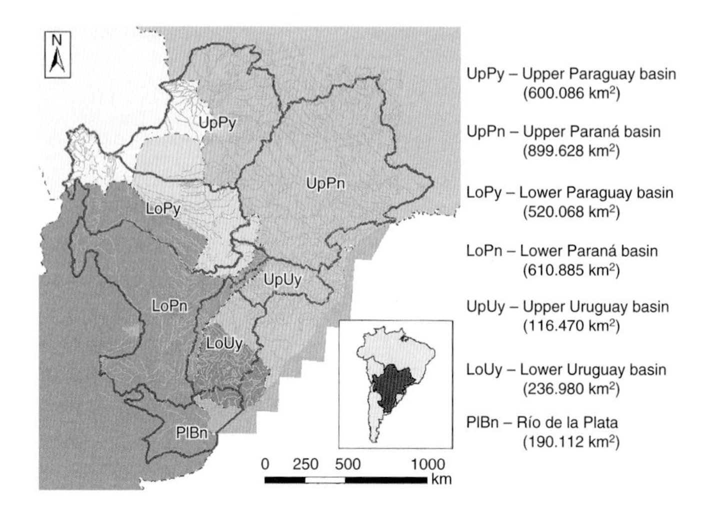

Figure 2.7.2 Sub-basins that constitute La Plata Basin. Continuous red line represents borders between sub-basins; discontinuous line represents international borders; coloured backgrounds represent different country areas; and light grey line represents the main fluvial network from La Plata Basin. Acronyms of each sub-basin are defined.

Table 2.7.1 Annual mean precipitation and evapotranspiration (mm) for each sub-basin and climate scenario.

			Mean annual P (mm/year)				
Time frame	UpPy	UpPn	LoPy	LoPn	UpUy	LoUy	PlBn
1961–2005	1201	1792	1005	1032	2254	1609	1138
2007–2040	1036	1477	975	1098	2452	1787	1303
2041–2070	1167	1780	1105	1220	2781	1878	1279
2071–2099	1226	1848	1170	1270	2914	1961	1294
			Mean annual PET (mm/year)				
Time frame	UpPy	UpPn	LoPy	LoPn	UpUy	LoUy	PlBn
1961–2005	2579	2169	2243	2113	1825	1820	1780
2007–2040	2923	2518	2399	2089	1907	1794	1676
2041–2070	3008	2513	2466	2158	1962	1874	1809
2071–2099	2983	2527	2407	2126	1952	1863	1797

2.7.2.3 Data analysis

First, we carried out an analysis of the current scenario at the annual timescale using historical rainfall data from the basin. When the rainfall in a year was lower than 80% of the mean annual rainfall of the whole dataset recorded in a station, we defined this year as a *dry year* (or a year characterised by a major deficit in a rainfall station) (Tucci, 2013). For this aim, we used monthly rainfall series from 46 stations located in the basin from the CRU TS database (version 3.22; CRU, 2014). We identified the dry years from the period without missing data from each station. To evaluate a possible change in the current climate pattern regarding the complete series, we also performed the calculations for the period 1961–2005.

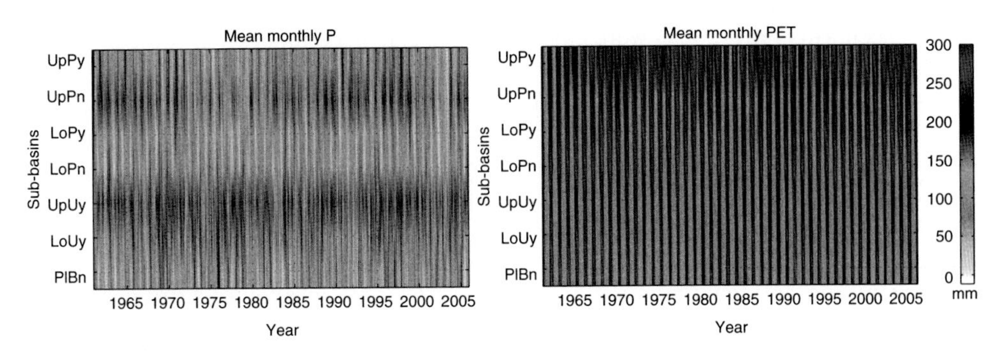

Figure 2.7.3 Temporal distribution of the mean monthly P and PET in mm for the different sub-basins and climate scenarios.

Later, we identified and characterised the dry and wet periods for the current and future scenarios at a monthly timescale, spatially distributed by using SPEI (Vicente-Serrano et al., 2010a, 2010b; Beguería et al., 2010). The consecutive months starting when monthly SPEI < –1 and ending when it is > 0 were defined as a *dry period*. A 1-month SPEI provides short-term information. When SPEI is lower than –1 for an isolated month, although it is theoretically considered a dry period (or a period characterised by water deficit), it hardly implies significant negative impacts on any activity related to water resources or society. Consequently, it cannot not be considered a drought. Only when such a deficit is repeated for several consecutive months will some activities (e.g. farming) be severely affected. A 3-month SPEI provides seasonal information. A prolonged deficit at this temporal scale would result in significant impacts – for example, on crops due to insufficient available soil moisture. A 6-month SPEI indicates medium-term patterns (and, depending on regions, it would provide seasonal information). A prolonged deficit at this temporal scale would also result in negative impacts on agriculture and other economic sectors. Finally, a 12-month SPEI provides medium- and long-term patterns. This index, in addition to critical impacts on agriculture, would also impact on other sectors, owing to decreases in the water stored in reservoirs and aquifers and changes in the hydrological regimes of large basins. At this temporal scale, it is possible to identify long dry periods that would indicate dry medium-term cycles. In this study, we used 1-, 3-, 6-, and 12-month SPEIs for each defined sub-basin and climate scenario.

We used the 10 × 10 km cells as calculation units, and the seven sub-basins that comprise La Plata Basin as data analysis and result presentation areas. Each scenario has a different temporal length, so results should be analysed with caution, avoiding comparisons with absolute values (Table 2.7.1). However, all scenarios are sufficiently long (over 30 years) to be representative of the climate of that period. First, we defined the range of cells that belongs to each sub-basin, and their corresponding P and PET were assigned using a geographic information system (GIS). In order to assess the effects of climate change on the PET and P in each sub-basin, we analysed the variations of both variables in the different climate scenarios. Therefore, in each cell belonging to each sub-basin, the mean monthly PET and P were calculated for the entire period of each scenario considered. Then, the variations of each scenario's PET and P relative to the control scenario (1961–2005) were calculated per cell. Second, in each cell comprising La Plata Basin, we calculated the SPEI values corresponding to the different timescales (1, 3, 6, and 12 months) and scenarios relative to the control (1961–2005). For each sub-basin, we identified each dry period and characterised its average patterns. For each identified dry period per timescale and scenario, we calculated the total duration, magnitude, maximum and mean intensities, and the durations of in-between dry periods and start moments. For each scenario, we also compared and obtained patterns, recorded the total number of dry periods and the averages and standard deviations for the duration, magnitude, and maximum and mean intensities. The *duration* of a dry period was measured as the number of months under water deficit, the *magnitude* as the sum of the SPEI values during that period, the *maximum intensity* as the SPEI value corresponding to the month with the lowest value, and the *average intensity* as the mean SPEI for that period.

On the other hand, we determined the proportion of area (i.e. percentage from the total cells of the sub-basin) that was affected by dry periods per each month of the analysed time series for each SPEI timescale, each scenario, and each sub-basin. The interest of this analysis lies in the fact that, from many unidentified dry periods for the whole sub-basin, a significant part of the area is in fact suffering from water deficit (although it is not on average).

2.7.2.4 Drought indicator used

In Table 2.7.2, the analysed variables are shown for each scenario.

2.7.2.5 Limitations of the methodology

While the information used in this study (mainly P and PET) is based on scientifically rigorous studies (Ferraz, 2014a; Bustamante et al., 2002, 2006; Alves et al., 2004; Chou et al., 2005), they are the result of the application of models (both current and future situations), and the conclusions from this study are thus inextricably affected by the models' uncertainties. Additionally, the aggregation of results from the 10 × 10 km cells into large sub-basins, in some cases, leads to a significant percentage of the area suffering from water deficit (but not on average) in unidentified dry periods for the whole sub-basin. The same is true for periods characterised by water excess. If the study was

Table 2.7.2 List of proposed variables for drought characterisation.

Variable	Definition
Total number of dry periods	Number of dry periods identified (similar and different durations)
Total number of dry periods per year	Ratio of the number of identified dry periods and the length of the series from the analysed scenario in years
Average duration of dry periods	Average duration of the identified dry periods in months
Average magnitude of dry periods	Average for each scenario and SPEI timescale of the sum of SPEIs from each month of the dry period
Average mean intensity of dry periods	Average for each scenario and SPEI timescale of the mean SPEIs from each dry period

carried out on a smaller spatial scale, dry or wet periods in certain locations could change in comparison to those identified at the sub-basin scale. Furthermore, if we consider the land uses and existing hydraulic infrastructures in those particular areas, we could have problems in water resources management undetectable by this study due to its working scale.

2.7.3 Results and discussion

2.7.3.1 Drought characterisation: Changes in rainfall

Figure 2.7.4 shows results from the characterisation of the dry periods at the sub-basin level for the current scenario, at the annual timescale and based on rainfall historical data. Figure 2.7.4A shows the average time between dry years using the whole series from the rainfall stations. Figure 2.7.4B shows the average time between dry years using the series from 1961 to 2005. Figure 2.7.4C shows the ratio of the average time between dry years using the series from 1961 to 2005, and the average time between dry years using the whole series. The average time between dry years for the whole basin within the whole dataset is 6 years, whereas for 1961–2005 it is 9 years. Results show that, in the UpPy sub-basin, the average time between dry years hardly decreases during 1961–2005, and that, in the LoPy sub-basin, it fluctuates depending on the analysed rainfall station. In the remaining sub-basins, the average time between dry years increases during 1961–2005. It is noted that variation is minimal in areas where the average time of occurrence of dry years is high, while the time increases considerably in the short series in areas where it is lower. This would be positive for water-resource-dependent activities.

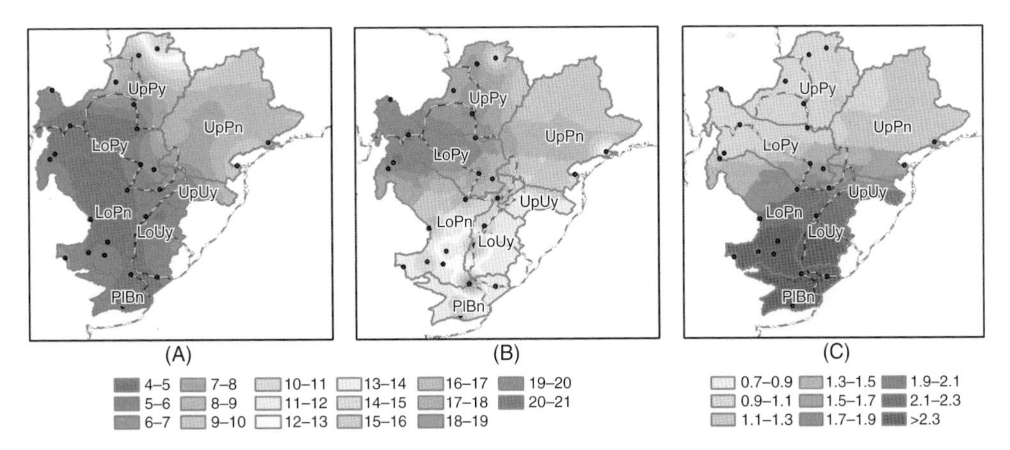

Figure 2.7.4 Characterisation of dry periods for the current scenario at the annual timescale and based on rainfall historical data. Figure 2.7.4A shows the average time between dry years using the whole series from the rainfall stations; Figure 2.7.4B shows the average time between dry years using the series from 1961–2005; and Figure 2.7.4C shows the ratio of the average time (years) between dry years, considering the 1961–2005 period and the whole series for each station. Dots show the locations of the rainfall stations used for this analysis.

2.7.3.2 Drought characterisation: Changes in SPEI drought indicator

Figure 2.7.5 shows the SPEI values when comparing with the series from 1961 to 2005, for the 12-month SPEI timescale, and for each sub-basin and climate scenario. Interesting results were obtained from the analysis of the 12-month SPEI. In the 2007–2040 scenario, the effect compared with the control scenario is considerable in some sub-basins. In the UpPy and UpPn sub-basins, we observed an initial period from 2008 to 2015 characterised by SPEIs higher than the average of the 1961–2005 series, and then, between 2015 and 2040, a dry period. It is likely that, once the projections are confirmed, the implementation of adaptation measures is required in these areas. Although attenuated, the situation was similar in the LoPy sub-basin as well. However, we found the opposite situation in the UpPy, LoUy, and PlBn sub-basins, which showed in general SPEIs higher than the average for the 1961–2005 series. In these cases, a wetter climate than the current one might also need adaptation measures, such as adjustments of the types of crops to the new climate reality.

In the 2041–2070 scenario, the UpPy, UpPn, and LoPy sub-basins presented a general water deficit. On the contrary, LoPn, UpUy, and LoUy showed water excess. The PlBn sub-basin showed an intermediate pattern. In the 2071–2099 scenario, the UpPy sub-basin remained water deficit most of the time, although the deficit was of lower intensity. It also showed short periods characterised by a normal situation in terms of water availability and even a few of water excess. Dry and wet periods alternated in the UpPn sub-basin, but still with a predominance of the latter. The LoPn, UpUy, and

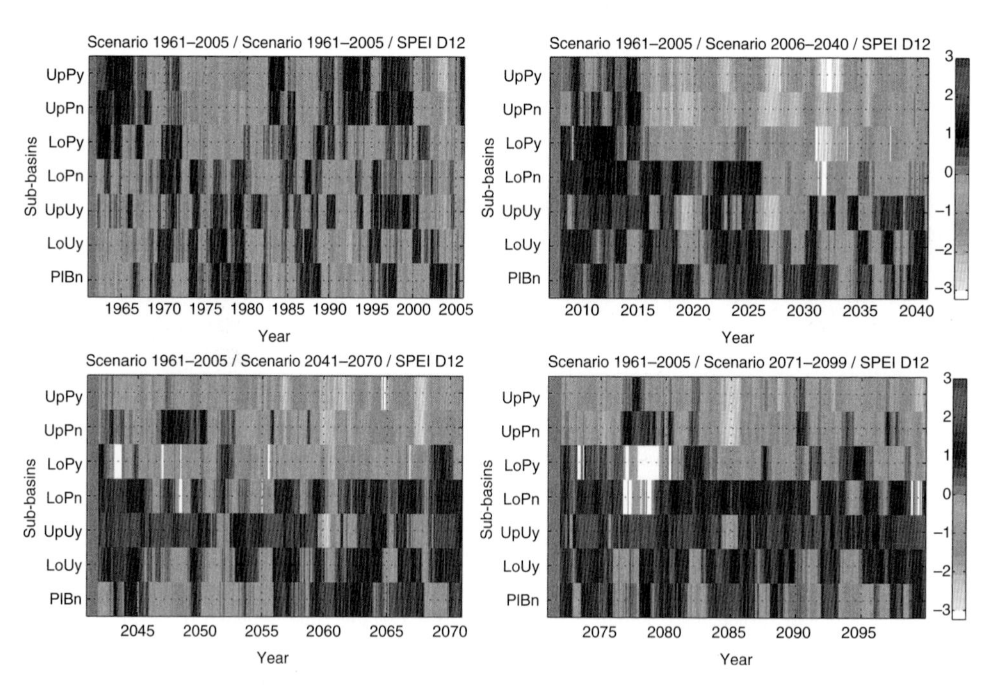

Figure 2.7.5 Comparison of SPEI – 12 months for all sub-basins and scenarios. (*See colour plate section for the colour representation of this figure.*)

LoUy sub-basins presented more wet periods, as did the PlBn sub-basin, except for a few dry periods. In conclusion, it appears that, in the 2007–2040 scenario, the sub-basins located towards the north of La Plata Basin showed a predominance of dry periods, whereas those located towards the south showed a predominance of wet periods alternating with short dry periods. As more distant future scenarios are discussed, the wetter climate tends to spread to the northern sub-basins by increasing the number of wet periods or attenuating the dry periods.

Figure 2.7.6 shows the percentage of each total sub-basin area that suffers water deficit each month, for the 12-month SPEI timescale and climate scenario. Interesting results were obtained from the analysis of the 12-month SPEI. In the 2007–2040 scenario, for the UpPy, UpPn, and LoPy sub-basins, we observed a significant increase of the area affected by dry periods (not as marked in LoPy). LoPn and UpUy remained stable, although dry and wet periods distributed differently in time compared to the control scenario. LoUy and PlBn showed a significant decrease of the area affected by dry periods as compared to the control scenario. In the 2041–2070 scenario, the sub-basins UpPy, UpPn, and LoPy showed large areas, although lower than in the previous scenario, under water deficit. The same happened in the LoPn sub-basin, where there was a smaller area surface suffering water scarcity than in the 2007–2040 scenario, but areas affected by drought still predominated. The UpUy and LoUy sub-basins still showed a small area under water deficit, and PlBn increased the area affected by droughts as compared to the 2007–2040 scenario, with alternating dry and wet periods. Finally, during the 2071–2099 scenario, the UpPy, UpPn, and LoPy sub-basins showed a

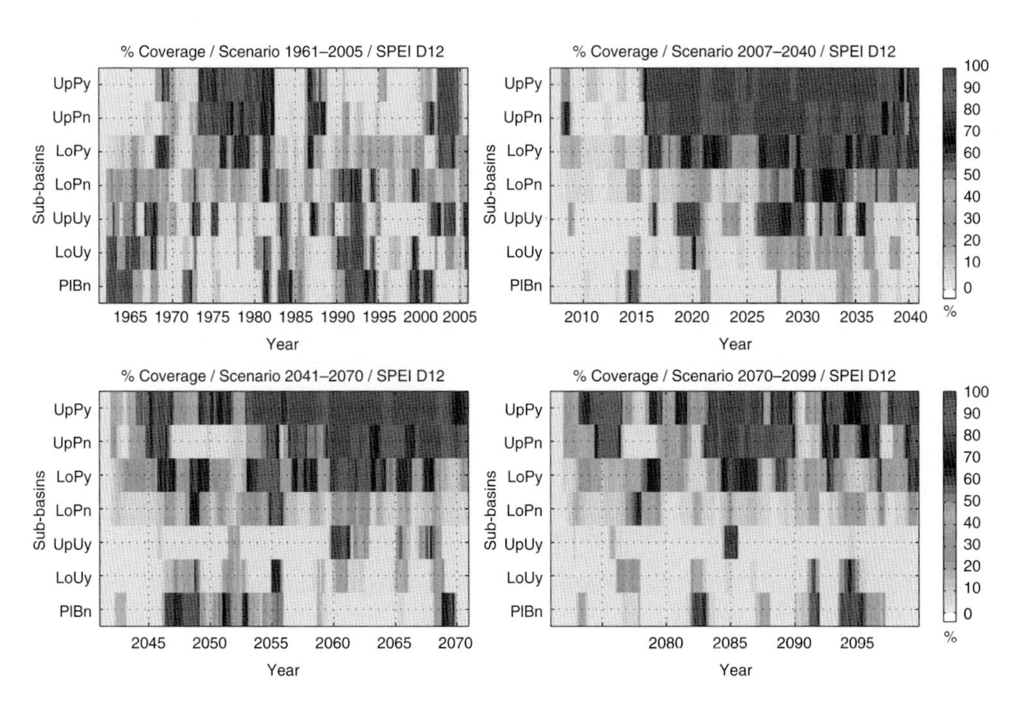

Figure 2.7.6 Percentage from the total cells of each sub-basin that was affected by dry periods (% coverage) per each month of the analysed time series for SPEI – 12 months, for all scenarios. (*See colour plate section for the colour representation of this figure.*)

decrease in the area affected by water deficit, although dry climate was still predominant. In LoPn, UpUy, LoUy, and PlBn, except for short periods, the area affected by water deficit during this scenario was small. As more distant future scenarios are discussed, droughts in the sub-basins tend to decrease. The situation in terms of water availability improves from south to north as time passes.

Table 2.7.3 and Figures 2.7.7 and 2.7.8 show the values of the selected variables to characterise the dry periods, comparing for each sub-basin the different climate scenarios for an SPEI 12-month timescale. In UpPy and UpPn sub-basins, we observed an increase in the frequency, duration, magnitude, and intensity of the dry periods. The worst situation in terms of water availability is shown in the 2007–2040 scenario. In the LoPy sub-basin, the situation is the same as that for the 2007–2040 scenario, but without reaching the levels that characterise the UpPy and UpPn sub-basins. In the control scenario, the LoPn, UpUy, and LoUy sub-basins showed a normal to slightly wet climate, showing only short, low-intensity dry periods. In the future scenario, the climate was gradually getting wetter for the future scenarios, decreasing the frequency, magnitude, intensity, and spatial coverage of the dry periods. Finally, in the PlBn sub-basin, the control scenario was characterised by alternating dry and wet periods. In the 2007–2040 scenario, we observed a strong decrease in number, duration, magnitude, and spatial coverage of the dry periods. However, unlike other sub-basins, and despite the predominance of normal to wet climate, there were punctual dry periods. These dry periods, however, were less severe than the ones observed in the control scenario.

For UpPy and UpPn in the 2007–2040 scenario, the entire basin showed positive patterns for PET, and most of the basin showed negative patterns for P. We observed an increase in the frequency, duration, magnitude, and mean intensity of dry periods, and of the affected surface. This pattern represented the worst situation due to the subsequent decrease in water resources in the sub-basins. The situation improved for the 2041–2070 and 2071–2099 scenarios, mainly P, but without reaching the levels of the control scenario. Also, a wide range was observed in P and PET values. For the same scenario, there were cells where the positive variation was moderate for PET but significant for P, which would result in an increase of water resources. However, for the same scenario, there were also cells with a significant positive variation of PET and a negative variation of P, which would result in a decrease in water resources. This spatial heterogeneity across sub-basins in the patterns of PET and P values for the different time scenarios highlights that results aggregated by sub-basins should be used with caution.

For LoPy, in the 2007–2040 scenario, almost the entire basin showed positive patterns for PET, whereas there were no marked patterns for P that depended on each analysed cell. However, for the remaining scenarios, the positive variations for PET stabilised or occasionally decreased, and there were positive patterns for P. The duration, magnitude, and affected area by dry periods increased significantly during 2007–2040, although without reaching the levels of the other sub-basins. The situation improves from the 2041–2070 scenario, but without reaching similar characteristics of the control scenario.

In the control scenario, the LoPn sub-basin showed a normal to slightly humid climate, with specific low-intensity dry periods. Climate gradually became wetter with the farthest scenarios over time, decreasing the frequency, magnitude, intensity, and spatial coverage of dry periods. Consequently, it is expected that future scenarios will present

Table 2.7.3 Values of the selected variables to characterise the dry periods, comparing the different climate scenarios for each SPEI 12-month timescale for the current climate and 2071–2099 timescale.

Variable	Time frame	Upper Paraguay	Upper Parana	Lower Paraguay	Lower Parana	Upper Uruguay	Lower Uruguay	Río de la Plata
Total number of dry periods	1961–2005	4	3	4	4	10	8	6
Total number of dry periods	2071–2099	7	5	4	0	1	1	3
Total number of dry periods /year	1961–2005	0.1	0.1	0.1	0.1	0.2	0.2	0.1
Total number of dry periods /year	2071–2099	0.2	0.2	0.1	–	0	0	0.1
Average duration of dry periods (months)	1961–2005	46.3	57	17.8	21	13.4	14.4	26
Average duration of dry periods (months)	2071–2099	31.9	29.2	17.3	–	12	11	16.3
Average magnitude of the dry periods (–)	1961–2005	41.6	54.6	15.7	15.5	13.2	14.9	24.3
Average magnitude of the dry periods (–)	2071–2099	42.1	41	14.6	–	17.6	8.4	16.5
Average of the mean intensity of the dry periods (–)	1961–2005	–0.9	–1	–0.8	–0.7	–0.9	–0.9	–0.9
Average of the mean intensity of the dry periods (–)	2071–2099	–1.3	–1.2	–0.8	–	–1.5	–0.8	–1

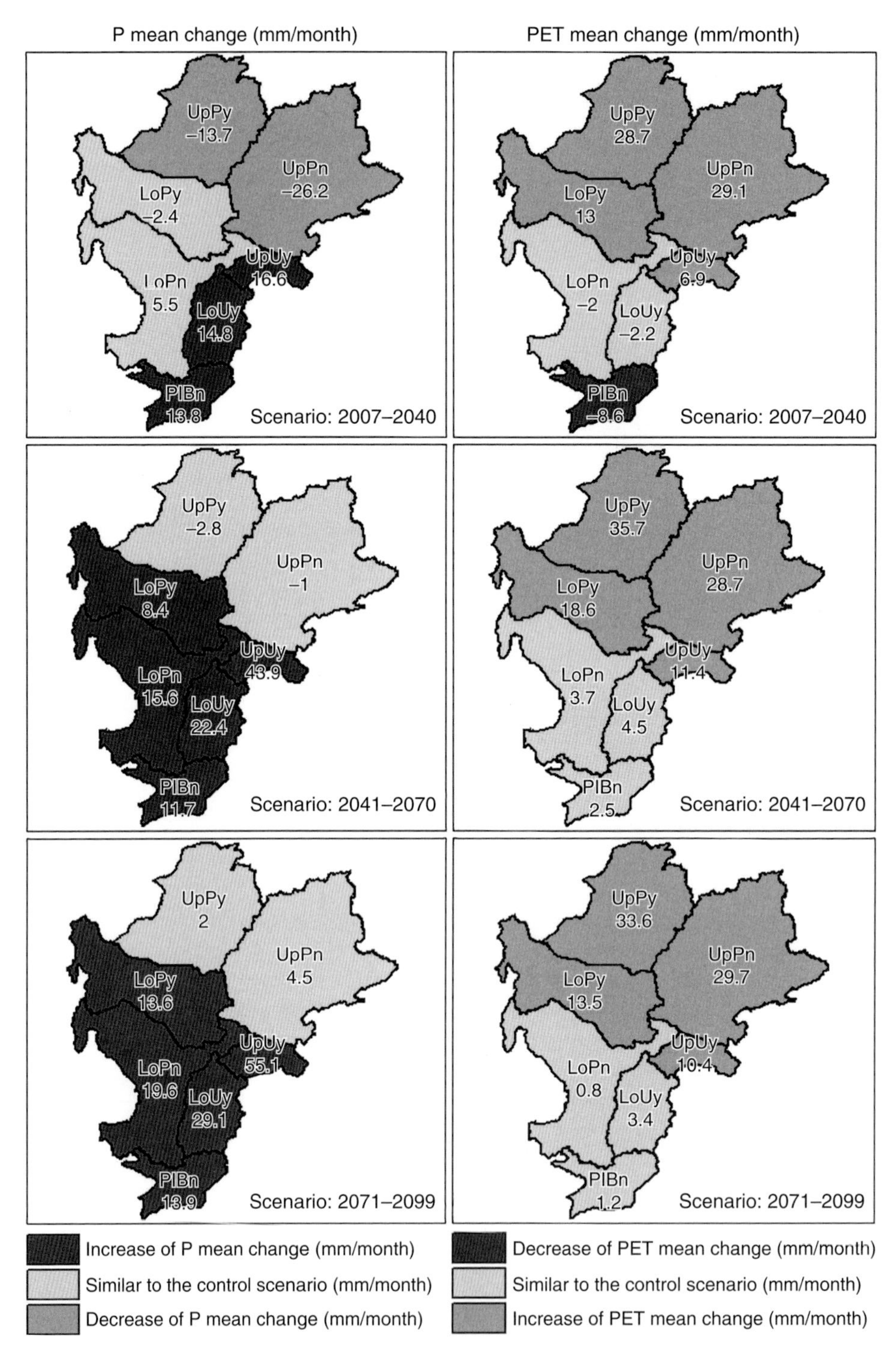

Figure 2.7.7 P and PET changes compared to the control scenario. The values show the mean changes of P and PET in mm for the whole sub-basin and the total time series for each scenario. (*See colour plate section for the colour representation of this figure.*)

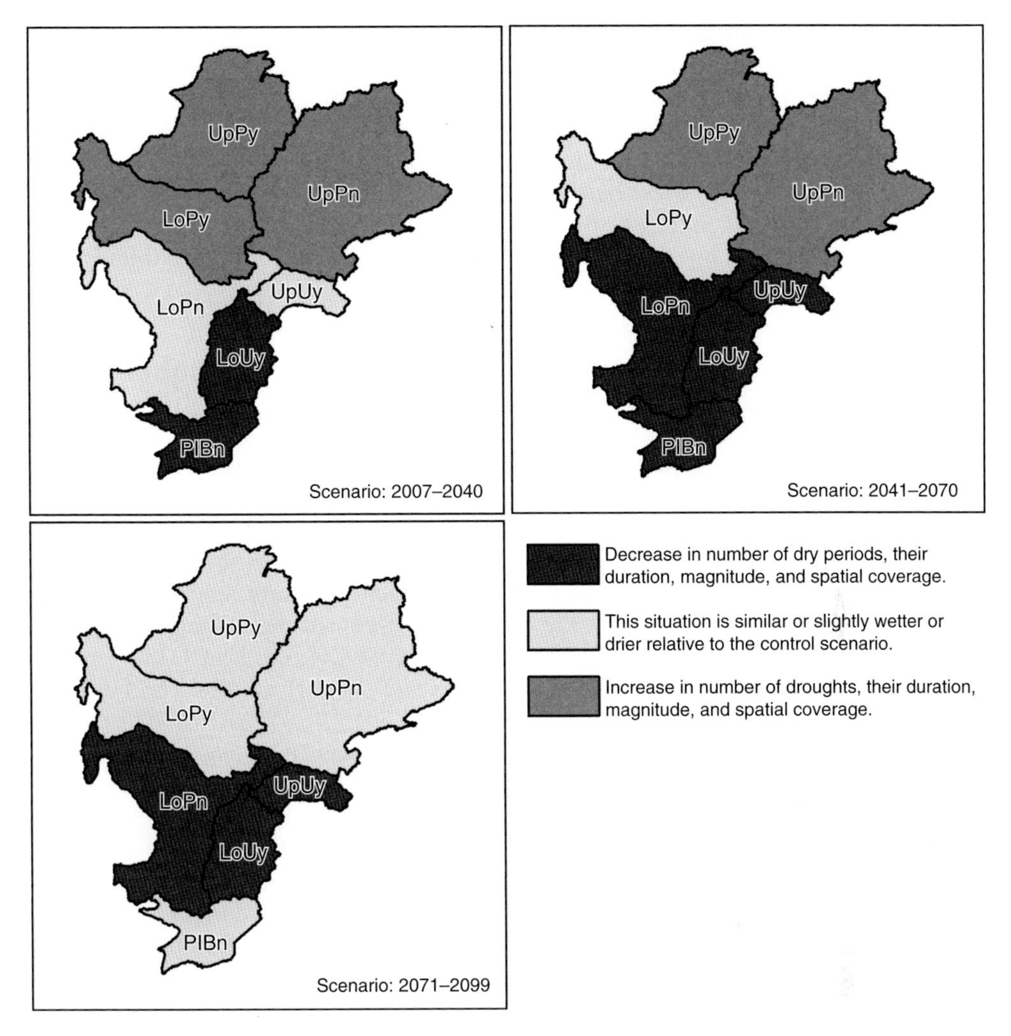

Figure 2.7.8 Variations in the situation of dry periods in the different sub-basins and time scenarios as compared to the control scenario (1961–2005). (*See colour plate section for the colour representation of this figure.*)

higher water resources than in the control situation. Nevertheless, we also found cells in the sub-basin that were characterised by either positive or negative variations of PET and P.

The UpUy sub-basin showed an alternation of dry and wet periods from 1961 to 2005. We observed a small positive variation of PET that remained similar during the future scenarios, and a strong positive variation of P as the more distant future scenarios were analysed. During 2007 to 2040 scenario, we observed less dry periods, although they were longer and more intense. From the 2041 to the 2070 scenario, wet climate prevails, decreasing the frequency, duration, intensity, and spatial coverage of dry periods. Consequently, water resources availability improves significantly in the future in this sub-basin.

The LoUy sub-basin showed a stable pattern of PET and a progressive positive pattern of P towards the farthest future scenarios. The combined behaviour of these two variables assures water resources availability in the future in this sub-basin. Finally, the PlBn sub-basin in the 2007–2040 scenario showed a negative pattern of PET and a positive pattern of P, although variations were moderate on average. The combined behaviour of these two variables benefits water availability. In the other scenarios, PET remained similar to the values of the control scenario, and the positive pattern of P remained stable. We did not observe the progressive increase of P that was found in the other sub-basins.

2.7.3.3 Graphical user interface as a tool for drought management

The present work generated a large amount of information, both at 10 × 10 km cell and sub-basin levels. We designed a tool with a graphical user interface (GUI) in order to facilitate the visualisation of results in general and specific ways by the meteorological services, water services, national and international institutions associated with watershed management, stakeholders, policymakers, and general users, among others. The devised tool allows queries of the study results at different scales, taking into account the needs of the users. The tool was developed using MATLAB, Java, GIS tools, and MapServer programming, among other programs and support utilities. The GUI tool has a home page with the available content (Figure 2.7.9).

It is possible to query at three levels: (1) La Plata Basin, (2) sub-basin, and (3) 10 × 10 km cell. At the level of the entire basin and sub-basin, the SPEI values can be consulted at different timescales (1, 3, 6, and 12 months) and climate scenarios for a time interval selected by the user (e.g. whole basin, SPEI 1 month, February 2000; see Figure 2.7.10). The program offers the option to display an animation of the temporal evolution of the selected index. Also, queries can be related to a database where the data with their corresponding coordinates can be obtained. At the sub-basin level, in addition to SPEI, it is possible to visualise information derived from the study – such as PET and P values, PET and P changes between climate scenarios, and percentages of spatial drought coverage. At the cell scale, it is possible to query by clicking in a cell on the map and connect to the database of SPEI for the different timescales and according to the different climate scenarios. Results are shown as graphic or data files. Figure 2.7.11 shows the output format of the information available in each selected cell, for each climate scenario and each SPEI analysed (1, 3, 6, and 12 months). Graphic information or a data file is obtained by clicking in a cell.

2.7.4 Conclusions

The analysis of historical data (rainfall stations have records since 1851) on an annual scale reveals that the average time between dry years in La Plata Basin is 6 years. This value increases to 9 years when considering only the period 1961–2005 (representative period of the current climate). The duration of the period between dry years shows significant heterogeneity when the analysis is carried out at the sub-basin level, being between 6 and 16 years. The average time between dry periods hardly decreases for the

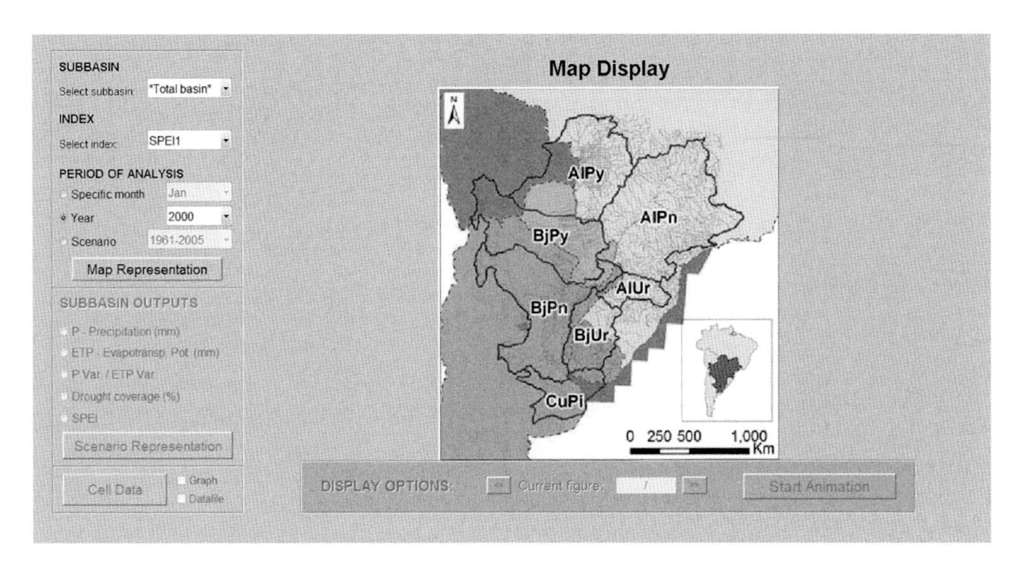

Figure 2.7.9 GUI tool main page.

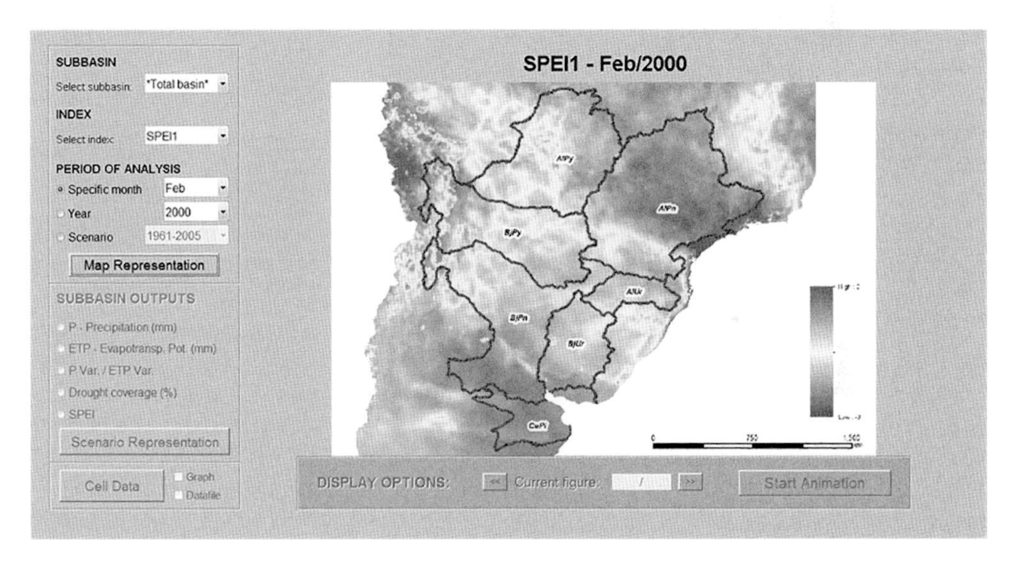

Figure 2.7.10 Output: SPEI – 1 month (February 2000) for the whole basin.

period 1961–2005 in the UpPy sub-basin (as compared to the complete historical series), and, in the LoPy sub-basin, it varies according to the season that is analysed. In the remaining sub-basins, the average time between dry periods increases for the period 1961–2005. In the 2007–2040 scenario, we noted that northern sub-basins show a predominance of dry periods, while the ones in the south show a predominance of wet periods with short dry periods in between. With scenarios further in the future, wet climate is expected to expand to the northern sub-basins, increasing the frequency of wet periods and decreasing the magnitude and duration of the dry periods in northern locations. Similarly, the area of each sub-basin that was affected by drought tended, in

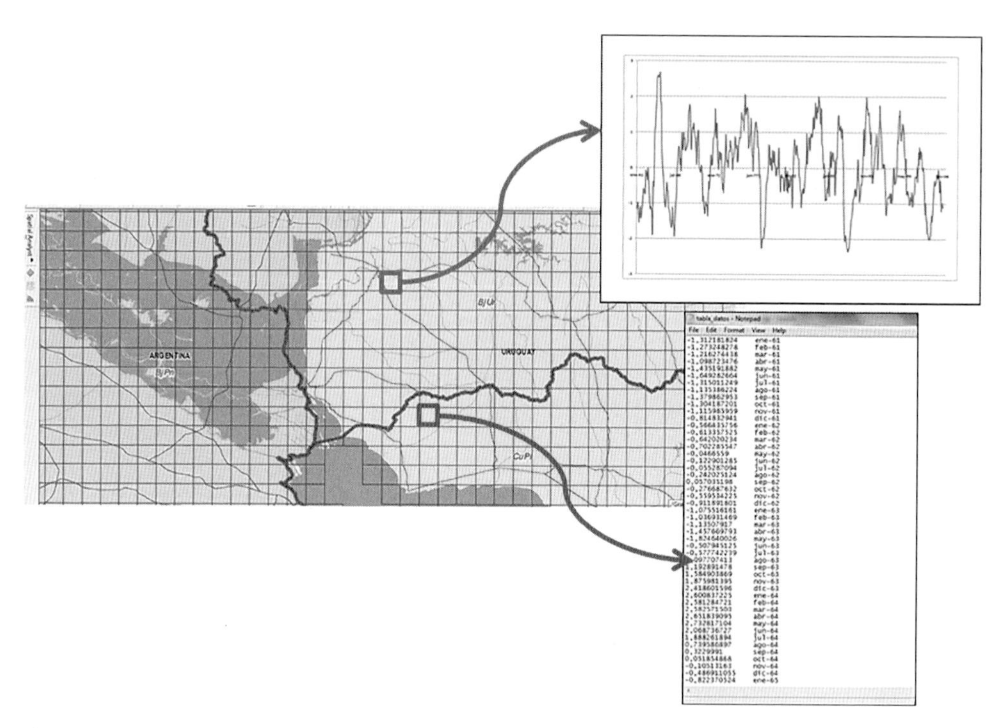

Figure 2.7.11 Interactive map interface. Graphic information or a data file can be obtained by clicking in a cell on the map.

general, to decrease with scenarios further in the future. If the climate projections become a reality, it is expected a significant heterogeneity between sub-basins in relation to the existence of dry, normal, and wet periods.

Although the drought problem in La Plata Basin is widespread, drought management is not uniform across the five countries of the region. While there are, in some cases, drought monitoring and warning tools, droughts are undervalued in others. The analysis of the hydroclimatic conditions of a basin is key for characterising the climate and its variations; for identifying and characterising particularly dry areas and dry periods as compared to the average conditions of the basin; and in general for assessing the situation of the basin. However, the evaluation and management of drought involves the study of other factors that should be included in future studies, such as the evaluation and analysis of water uses in each basin and water infrastructure, and the regulatory capacity available in the basin. Hence, there is the short-term need to relate the areas most likely to suffer dry periods to the analysis of the satisfaction of demand; to evaluate the existing infrastructure, and the effect of new infrastructures, on the satisfaction of demand; and to assess the regulation capacity of sub-basins, all for current and future scenarios. A basin that shows a high probability of suffering prolonged dry periods will be more or less vulnerable to droughts, depending on the existing uses, the resulting impacts, and its regulation capacity. It is an important future task to quantitatively evaluate the degree of satisfaction of the current and future water demands, and to analyse the regulation requirements to meet different levels of demand. In a drought management plan, it would be essential to have a primary regional vision of the current situation in the sub-basins, and of the need for new facilities based on different future scenarios.

Given the large size of the analysed sub-basins and the aggregation of results, significant areas within a sub-basin may suffer drought when the average SPEI value obtained for the entire sub-basin informs about a normal situation. If we also consider the economic activities, land uses, and existing infrastructures in these areas, there could be water resources management problems that the working scale of this study would be neglecting. Therefore, we propose to select the most sensitive areas and study them in detail.

Additionally, this study shows the heterogeneity of dry periods within sub-basins and climate scenarios. Having information on the characteristics of droughts (duration, intensity, frequency, spatial coverage) and their impacts, areas within each sub-basin that are critical or highly vulnerable to drought could be identified both for past (based on historic data series) and future climate scenarios. It would also be possible to identify such areas in real time to prioritise measures during the occurrence and development of each drought.

One of the main medium-term objectives should be the integrated management of drought by using sub-basins as management units. Given that sub-basins are currently shared by several countries, different legal and institutional frameworks would be involved in drought prevention and mitigation actions. Having a single legal and institutional framework aimed at drought management in the entire sub-basin where all stakeholders are involved would be key to defining, coordinating, and managing actions before (prevention), during (management), and after (restoration) droughts. Furthermore, to facilitate drought management at the sub-basin level, it would be necessary to define, by common agreement of all countries, indexes and their thresholds that characterise, as an example, normal, pre-alert, alert, and emergency situations in relation to droughts. Each of these situations would be associated to structural and non-structural measures.

Despite being a recurring phenomenon in La Plata Basin and, in the light of this study, of importance in the future, droughts tend to be managed as crises and solved with emergency measures that rarely belong to an already drafted plan. These measures are usually aimed at guaranteeing water resources availability by pumping underground water, implementing temporal water supply restrictions, and interconnecting water systems, which are often accompanied by environmental, functional, and economic costs. It is therefore proposed herewith to establish the necessary mechanisms that can help implement a drought management plan at the sub-basin level. The main objective of this plan would be the early detection of water scarcity situations based on a set of indicators and already established thresholds that enable the characterisation of the situations in relation to drought. The plan also includes actions to be implemented in any of the identified situations in order to prevent, mitigate, or solve drought effects. Measures are classified as *strategic*, *tactical*, and *emergency measures* (Spanish Ministry of Environmental Affairs, Spain, 2007).

Strategic measures are long-term, institutional actions that include the development of infrastructures as part of the hydrological planning (water storage and regulation structures, legislation, and water management). They require long-term implantation, big budgets, political negotiations, social acceptance, and, eventually, legislative changes.

Tactical measures are short-term actions that need to be planned and validated in advance under the drought management plan. They must be devised on historical situations and adopted once the drought is observed by the indicator system. Finally, *emergency measures* are adopted late in the drought and vary depending on its severity, extent, or degree of affection to the basin.

Moreover, climate and hydrological information and drought indices are key components used to monitor drought at the basin scale. It would be very useful to have a tool that shows this information on maps (e.g. with weekly or monthly time steps), and which centralises the information from each sub-basin. This tool will enable decision-makers and scientists to monitor and evaluate the evolution of the drought and compare it to past events. In an observatory, all stakeholders from the public bodies should be involved in order to establish a centre of knowledge, prediction, mitigation, and monitoring of the effects of drought in La Plata Basin.

Summing up, the proposal for future activities according to the problems that have been identified are: (a) to identify critical areas that require further and detailed analysis; (b) to analyse the maximum water demand that can be supplied according to the current infrastructure at a more detailed spatial scale, and quantify alternatives aimed at increasing water supply; (c) to establish sectoral and sub-basin methodologies / protocols in order to objectively quantify the economic losses resulting from droughts; (d) to identify and assess the impacts of historical droughts; (e) to develop drought management plans at the sub-basin level; (f) to create a drought observatory for La Plata Basin; and (g) to develop programs of cross-cooperation among countries, encompassing technical support for the selection, commissioning, and operation of new technologies, training and mobility activities, and other activities aimed at institutional strengthening.

Acknowledgements

The authors are grateful for the support from 'The Framework Program for the Sustainable Management of La Plata Basin's Water Resources with respect to the effects of climate variability and change'. They also thank the CIC Plata team for providing data and comments during the development of this study, the INPE team for providing P and PET data, and David J. Vicente for his collaboration in the GUI tool's development.

References

Alves, L.F., Marengo, J.A., and Chou, S.C. (2004). Avaliação das previsões de chuvas sazonais do modelo Eta climático sobre o Brasil. In: CongressoBrasiliero de Meteorologia, 13. 29 Agosto–13 setembro, 2004, Fortaleza, (CE).

Beguería, S., Vicente-Serrano, S.M., and Angulo, M. (2010). A multi-scalar global drought data set: the SPEIbase: a new gridded product for the analysis of drought variability and impacts. *Bulletin of the American Meteorological Society* 91: 1351–1354.

Bustamante, J.F., Gomes, J.L., and Chou, S.C. (2002). Influência da temperatura da superfície do mar sobre as previsões climáticas sazonais do modelo regional Eta. In: CongressoBrasiliero de Meteorologia, 12. 4–9 agosto 2002, Foz do Iguaçu (PR), pp. 2145–2152.

Bustamante, J.F., Gomes, J.L., and Chou, S.C. (2006). 5-year Eta model seasonal forecast climatology over South America. In: *International Conference on Southern Hemisphere Meteorology and Oceanography*, 08. 24–28 April 2006, Foz do Iguaçu (PR), pp. 503–506.

Chou, S.C., Fonseca, J.F.B., and Gomes, J.L. (2005). Evaluation of Eta model seasonal precipitation forecasts over South America. *Nonlinear Processes in Geophysics* 12 (4): 537–555.

Donat, M.G., Alexander, L.V., Yang, H., Durre, I., Vose, R., Dunn, R.J.H., Willett, K.M., Aguilar, E., Brunet, M., Caesar, J., Hewitson, B., Jack, C., Klein Tank, A.M.G., Kruger, A.C., Marengo, J., Peterson, T.C., Renom, M., Oria Rojas, C., Rusticucci, M., Salinger, J., Elrayah, A.S., Sekele, S.S., Srivastava, A.K., Trewin, B., Villarroel, C., Vincent, L.A., Zhai, P., Zhang, X., and Kitching, S. (2013). Updated analyses of temperature and precipitation extreme indices since the beginning of the twentieth century: the HadEX2 dataset. *Journal of Geophysical Research* 20: 1–16. doi: 10.1002/jgrd.50150.

Ferraz, C. (2014a). Producto 1: Relatório contendo a análise das simulações do modelo Eta-20 km para a região da Bacia do Prata, utilizando as condições do HadGEM2-ES RCP 4.5, para o período de 1961–2100. Organización de Estado Americanos, Brasil, 28 pp.

Ferraz, C. (2014b). Producto 2: Relatóriocontendo a análise das simulações do modelo Eta-10 km para a região da Bacia do Prata, utilizando as condições do HadGEM2-ES RCP 4.5, para o período de 1961–2100. Organización de Estado Americanos, Brasil, 28 pp.

HFA. (2005). Hyogo Framework for Action 2005–2015: Building the Resilience of Nations and Communities to Disasters. World Conference on Disaster Reduction, Hyogo, Kobe Japan, 18–22 Jan 2005.

IPCC. (2014). *Climate Change 2014: Synthesis Report. Contribution of Working Groups I, II and III to the Fifth Assessment Report of the Intergovernmental Panel on Climate Change* (Core Writing Team, ed. R.K. Pachauri and L.A. Meyer). Geneva, Switzerland: IPCC, 151 pp.

Marengo, J., Chou, S., and Alves, L. (2014). Clima, variabilidadehidro- climática e cenários futuros de clima na Bacia do Prata: Umarevisãogeral. INPE, Brasil, 31 pp.

Spanish Ministry of Environmental Affairs (Ministerio de Medio Ambiente), Spain (2007). Plan Especial de actuación en situaciones de alerta y eventual sequía de la Cuenca Hidrográfica del Tajo. Programa AGUA, Ministerio de Medio Ambiente, Madrid, España, 146 pp.

Tucci, C. (2013). Avaliacao e Estratégias para os Recursos Hídricos no Río Grande Do Sul. RhamaConsultoriaAmbiental, Brasil, 25 pp.

University of East Anglia Climatic Research Unit (CRU). (2014). CRU TS version 3.22. Climatic Research Unit (CRU) time-series datasets of variations in climate with variations in other phenomena. Climatic Research Unit, NCAS British Atmospheric Data Centre, 2014.

Vicente-Serrano, S.M., Santiago Beguería, and Juan I. López-Moreno (2010a). A multi-scalar drought index sensitive to global warming: the standardized precipitation evapotranspiration index – SPEI. *Journal of Climate* 23: 1696–1718.

Vicente-Serrano, S.M., Beguería, S., López-Moreno, J.I., Angulo, M., and El Kenawy, A. (2010b). A new global 0.5° gridded dataset (1901–2006) of a multiscalar drought index: comparison with current drought index datasets based on the Palmer Drought Severity Index. *Journal of Hydrometeorology* 11: 1033–1043.

2.8
Drought Insurance

Teresa Maestro, Alberto Garrido, and María Bielza
Universidad Politécnica de Madrid, Madrid, Spain

2.8.1 Introduction

Drought is a complex natural hazard that impacts ecosystems and society (Van Loon, 2015). Estimates of drought impacts in recent years indicate that drought-related losses are increasing (Stahl et al., 2012).

One of the main adaptation tools to face drought in arid zones is irrigation. However, drought not only affects rainfed agriculture, but also the availability of water for irrigation. The term *meteorological drought* is exclusively used to refer to the lack of precipitation for a long time, usually affecting rainfed agriculture in large extensions of land. But when meteorological drought reduces the water levels in streams, rivers, lakes, and reservoirs, then a hydrological drought sets in, affecting water availability for irrigation (and for other uses).

Many water supply systems are currently under pressure, because human water use has more than doubled during the past decades, while available freshwater resources are finite (Wada et al., 2011). Environmental policies, aiming to protect endangered species, pose more restrictions for irrigation water, intensifying water shortage problems for irrigated agriculture (Johansson et al., 2002; Buchholz and Musshoff, 2014).

Besides, climate change projections indicate that droughts will intensify in the twenty-first century in southern Europe, the Mediterranean region, central Europe, central North America, Central America, Mexico, northeast Brazil, and southern Africa (IPCC, 2014). Multi-model experiments show that climate change is likely to exacerbate regional and global water scarcity considerably (Schewe et al., 2014), and the global severity of hydrological drought at the end of the twenty-first century is likely to increase (Prudhomme et al., 2014). Milly et al. (2005) project a 10–30% decrease in runoff in southern Africa, southern Europe, the Middle East, and mid-latitude western North America by the year 2050.

In order to cope with drought risks in agriculture, some of the adaptation tools suggested by the IPCC are crop insurance, subsidised drought assistance, and water trading

Drought: Science and Policy, First Edition.
Edited by Ana Iglesias, Dionysis Assimacopoulos, and Henny A.J. Van Lanen.

(IPCC, 2014). Crop insurance as an adaptation tool against climate change was already proposed by the European Commission (2007).

Agricultural insurance is one of the best tools that farmers can be offered to manage risks that are too large for them to manage on their own (World Bank, 2011). Agricultural insurance can monetise the expected increase in the vulnerability of the agricultural sector and enables the producer to transfer his uncertainty to third parties, thus reducing the variability and risk associated with the loss of income (Pérez Blanco et al., 2011).

Whereas rainfed systems are fairly protected against drought by insurance schemes, irrigated systems remain mostly unprotected. In the following sections, we provide a description of the main difficulties and challenges that drought insurance poses to insurance developers and the different types of drought insurance in both rainfed systems and irrigated systems that are implemented or in study. Weather index insurance is of particular importance for protecting irrigated crops against drought.

2.8.2 Main difficulties and challenges in developing drought insurance

Agricultural insurance is among the most difficult insurance schemes to develop. There are several important reasons why agricultural insurance markets may fail, such as (1) spatially correlated risk, (2) moral hazard, (3) adverse selection, and (4) high administrative costs (World Bank, 2005).

Drought is a phenomenon that is particularly difficult to insure against, representing the third risk in importance in agricultural systems, after hail and frost (Agroseguro, 2015a) (see Figure 2.8.1). Its effects are prolonged in time and can be extended to more than one growing season, and it may contribute to the occurrence of other problems or aggravate them (i.e. weakening plants and making them more susceptible

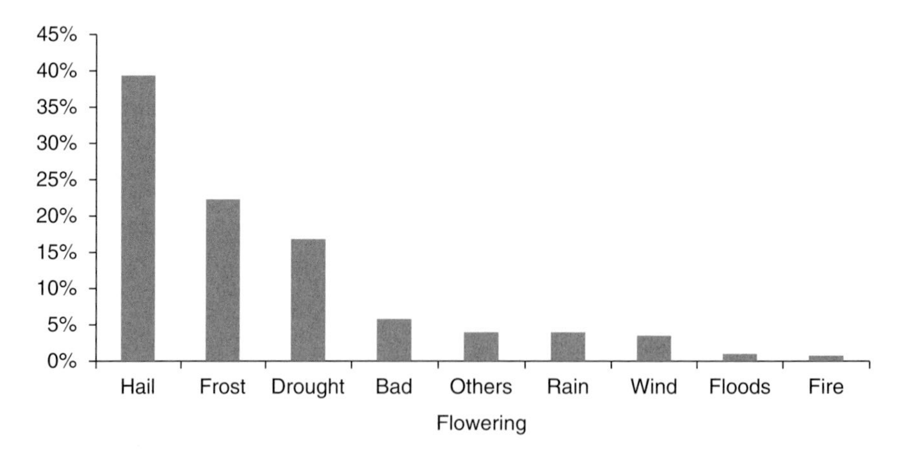

Figure 2.8.1 Payoff distribution by peril for the period 1980–2014 in Spain. *Source*: Own elaboration from Agroseguro (2015a).

to diseases). Besides, drought can be predicted in advance in some regions, leading to intertemporal adverse selection. Intertemporal adverse selection comes from the fact that preseason weather information can influence in crop insurance decisions (Carriquiry and Osgood, 2012), as farmers may use this information to purchase insurance only in years with enhanced drought risk and probability of payout (Luo et al., 1994).

The probability and severity of droughts can also be influenced by climate change. Climate change represents a major factor of uncertainty in drought risks (Bielza Diaz-Caneja et al., 2008). Increasing drought risk resulting from climate change affects the price of insurance in two ways. First, ambiguity and catastrophe loads increase, because uncertainty associated with future climate change impacts leads insurers to plan for the worst likely scenario when establishing these loads. Historical return periods may not be valid since they might underestimate the likelihood of agricultural losses in the future. Second, increasing drought risk changes the pure risk (Collier et al., 2009). In consequence, insurance parameters have to be adjusted over time to effectively hedge future weather risk (Kapphan et al., 2012).

In irrigated agriculture, difficulties augment, since there is a high uncertainty that stems from institutional decisions about water availability; moral hazard increases due to the water management and crop decisions at the farm; and farmers can have access to a wide selection of water resources.

2.8.3 Types of drought insurance

Drought insurance can be classified into two groups – depending on the way the loss adjustment is performed: *indemnity-based* or *index-based insurance*; and depending on the crops insured: *rainfed* or *irrigated crops*. In Table 2.8.1, some examples of drought insurance implemented are listed, and further explained in the following sections. Whereas the risk of drought in rainfed systems is sufficiently managed (at least in developed countries), the risk of drought in irrigated systems is still insufficiently managed.

Table 2.8.1 Types of drought insurance and implemented examples. *Source*: Own elaboration.

		Rainfed crops	Irrigated crops
Indemnity-based		Multiple-peril crop insurance (MPCI) programmes in Spain, Cyprus, Argentina, Chile, Peru, and the United States	Only the MPCI programme in the United States
Index-based	Area–yield	The United States, Sweden, Canada, and India	–
	Meteorological index	Rainfall index insurance in Mexico and India	Vietnam (not offered anymore)
	Remote sensing index	Canada, Spain, and The United States	–
	Hydrological index	–	–

2.8.4 Drought indemnity–based insurance

In indemnity-based insurance, the damage is measured in the field. The most common indemnity-based insurance is the MPCI, or yield insurance. It typically protects against multiple perils –that is, it covers many different causes of yield loss. For MPCI, administrative costs are particularly high, owing to the need for establishing expected yields, verifying realised yields, categorising potential insureds into appropriate risk pools, and monitoring to make sure that insureds employ appropriate production inputs and practices (Coble and Barnett, 2013).

There are many countries that insure rainfed crops against drought under this type of policy: Spain (ENESA, 2012), Cyprus (Tsiourtis, 2005), Argentina (Oficina del Riesgo Agropecuario, 2010), Chile (Magallanes, 2015), Peru (La Positiva, 2015), and the United States (Risk Management Agency, 2015).

MPCI insurance covering irrigated crops is, to our knowledge, only implemented in the United States. In an irrigated practice, farmers can choose to only insure the acreage that would be fully irrigated, considering the expected water availability at the beginning of the crop season (when insurance attaches). The expected water availability is based on the water available in reservoirs, soil moisture levels, snow pack storage levels (if applicable), and the precipitation that would normally be received during the crop season. To be covered by MPCI, any failure in irrigation water supply must be due to a naturally occurring event. Decreased water allocations due to diversion of water for environmental reasons, compact compliance, or other non-natural causes are not covered (Risk Management Agency – Topeka, 2015).

Although not implemented, several researchers have proposed other insurance schemes covering water shortage risk. Indemnity-based insurance schemes, that is, using crop production functions or crop simulation models to assess the economic impact of drought, have been proposed in some irrigated regions of Spain (Quiroga et al., 2011; Pérez Blanco and Gómez Gómez, 2013; Ruiz et al., 2013, 2015).

2.8.5 Drought index–based insurance

Index insurance indemnifies the insured based on the observed value of a specified index. Ideally, an index is a random variable that is objectively observable, reliably measurable, highly correlated with the losses of the insured, and, additionally, not influenced by the actions of the insured (Miranda and Farrin, 2012).

Index insurance is uncommon in high-income countries, which are dominated by markets with high uptake of MPCI insurance (World Bank, 2011). Index products are useful for systemic risk, and are best suited for homogeneous areas (Bielza Diaz-Caneja et al., 2008).

We can differentiate between area–yield (or revenue) index insurance, in which indemnities are paid based on a shortfall in area average yield (or revenue) (Miranda, 1991; Skees et al., 1997; Coble and Barnett, 2008), and weather index insurance (Martin et al., 2001; Turvey, 2001; Vedenov and Barnett, 2004; Barnett and Mahul, 2007; Collier et al., 2009; Skees, 2010; Ritter et al., 2012). Area–yield (or revenue) insurance is based on the returns that have historically been produced in a homogeneous geographical area, such that, if zonal yields fall below a certain value, all the insured farmers in that area receive compensation, regardless of who has actually suffered losses or who has not (the United States, Sweden, Canada, and India, among others, offer this type of insurance). For its application on drought insurance, this section will focus on weather index insurance.

With weather index insurance, indemnities are paid based on realisations of a specific weather parameter (e.g. rainfall or temperature), measured over a pre-specified period of time, at a particular point of measurement. Weather index insurance schemes are sometimes referred to as 'derivative products' (Hess and Syroka, 2005). Although the risk transfer characteristics and benefits are similar, the two instruments feature different regulatory, accounting, tax, and legal issues. Derivative products are not necessarily associated with any physical loss (World Bank, 2011). The Chicago Mercantile Exchange (CME) started trading weather derivatives in 1999, and standardised rainfall derivatives for 10 selected US cities in 2011 (Ritter et al., 2012).

Some of the advantages of index insurance as compared to MPCI schemes are:

- Reduced risk of adverse selection, because index instruments do not require insurers to classify potential purchasers according to their risk exposures. Weather index instruments can, however, be susceptible to intertemporal adverse selection (Barnett et al., 2008).
- Reduced moral hazard, because farmers cannot influence the claim. Besides, since farmers have incentives to continue to produce or to try to save their crops and livestock even in the face of bad weather events, index insurance should provide for more efficient allocation of resources (World Bank, 2005).
- Field loss assessment is eliminated, decreasing administrative costs.
- Reduced information requirements and bureaucracy, decreasing administrative costs.
- Transparency and facilitation of reinsurance, because it is based on independently measured weather events.

The selection of the index on which the insurance should be based has been widely discussed in the literature (Bielza Diaz-Caneja et al., 2008; Leiva and Skees, 2008; World Bank, 2011). The two basic prerequisites of an appropriate index are: (1) high correlation with the potential loss; and (2) fulfilment of the quality standards of the insurance industry (observable, easily measured, objective, transparent, independently verifiable, reported in a timely manner, consistent over time, experienced over a wide area, and not a subject of manipulation).

There are different types of drought indicators, depending on the variables on which they are based, that can be used as underlying indices in drought insurance. They are used for drought monitoring and drought management purposes (Estrela and Vargas, 2012).

In Table 2.8.2, some examples of meteorological, agronomic, hydrologic, and remote sensing drought indicators are listed, and some of their applications in drought management and drought insurance are mentioned.

Depending on the underlying index, we can classify weather-based drought index insurance in several categories: meteorological drought index insurance, remote sensing drought index insurance, and hydrological drought index insurance (see Table 2.8.1).

A major obstacle for the use of index-based insurance in practice is the existence of basis risk (Vedenov and Barnett, 2004; Woodard and Garcia, 2008). Basis risk is the

Table 2.8.2 Types of drought indicators, examples, variables used, and applications.

Type of indicator	Name and reference	Based on	Application
Meteorological	Deciles (Gibbs and Maher, 1967)	Rainfall	Used for drought monitoring in Australia (Australian Bureau of Meteorology, 2015)
	Standardised precipitation index (McKee and Doesken, 1993)	Rainfall, evapotranspiration, and temperature	Used for drought monitoring in the Unites States (National Drought Mitigation Center, 2015)
	Various cumulative precipitation indices	Rainfall	Used in India (Agriculture Insurance Company of India Limited, 2014a, 2014b) and Canada (AGRICORP, 2015; AFSC, 2015a) as triggers for drought insurance
Agronomic	Palmer drought severity index (PDSI) (Palmer, 1965)	Rainfall, evapotranspiration, soil water content, and water balances	Used for drought monitoring in the United States (US Drought Portal, 2015b)
	Crop moisture index (CMI) (Palmer, 1968)	Rainfall, evapotranspiration, soil water content, and water balances	Used for drought monitoring in the United States (US Drought Portal, 2015a)
	FCDD indexes (Hartell et al., 2006; Bielza Diaz-Caneja et al., 2008)	Temperature, humidity, and precipitation	Used as a trigger in weather risk transfer for the Agroasemex Agricultural Portofolio (Hartell et al., 2006; Bielza Diaz-Caneja et al., 2008)
Hydrological	Status indicators (Confederación Hidrográfica del Júcar (CHJ), 2007)	Streamflow data, reservoir volumes, aquifers, and snow	Drought management plans in Spain (Estrela and Vargas, 2012)
	Sacramento Valley index (Bureau of Reclamation, 2008)	Streamflow data	Used by the California Department of Water Resources and the United States Bureau of Reclamation Use (Bureau of Reclamation, 2008)
Remote sensing	Normalised difference vegetation index (NDVI) (Rouse et al., 1974)	Satellite images	Used for the triggering of pasture insurance against drought in Spain (Agroseguro, 2015b) and Alberta, Canada (AFSC, 2015b)

[1] For a more detailed list of drought indicators, readers may refer to Byun and Wilhite (1999); Heim (2002); Keyantash (2002); Niemeyer (2008); Sivakumar et al. (2011); and Zargar et al. (2011).

deviation between actual losses and insurance payoff. Basis risk reduces the hedging effectiveness of index insurance and lowers the willingness to pay for these instruments (Ritter et al., 2012). Due to the fact that index insurance indemnities are triggered not by farm-level losses but rather by the value of an independent measure (the index), a policyholder can experience a loss and yet receive no indemnity. Conversely, the policyholder may not experience a loss and yet receive an indemnity. The effectiveness of index insurance as a risk management tool depends on how positively farm-level losses are correlated with the underlying index.

In addition to defining the index, specifying the weather station from which the weather variables taken to construct the index were measured and recorded, declaring the buyer/seller information (names, crop, and surface insured), and specifying the premium and risk protection period of the contract, an index-based weather insurance contract must also include the following information (Hartell et al., 2006):

- *Strike, trigger, or upper threshold*, which is the index level at which weather protection is triggered. The trigger that signals for payment owing to crop loss is very important in the pricing of index insurance products, and therefore it is desirable to obtain an optimal trigger for the indication of crop loss (Choudhury et al., 2015). The strike determines the level of risk retention of the insured. A trigger very close to the mean of the index indicates a low level of risk retention by the end user and a contract that will pay out with high probability (Hartell et al., 2006).
- *Tick size*, which is the financial compensation per unit index deviation above or below the trigger(s), and which converts the index outcome into a monetary amount (Ritter et al., 2012; World Bank, 2011).
- *Liability*, which is the level of protection of the index insurance (Zeuli and Skees, 2005). It corresponds to the maximum indemnity that an insured may receive (Miranda and Farrin, 2012).

Index insurance usually follows the scheme shown in Figure 2.8.2, where the variable used for the contract design is the cumulative rainfall. The policy pays zero if accumulated rainfall during the crop season exceeds the upper threshold; otherwise, the policy pays an indemnity (the tick) for each millimetre of rainfall deficiency relative to the strike, until

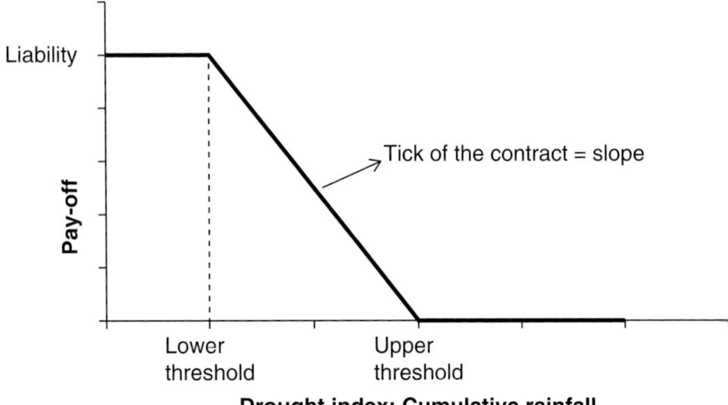

Figure 2.8.2 Structure of index contract (own elaboration).

the lower threshold is reached. If rainfall is below the lower threshold value, the policy pays a fixed, higher indemnity (Martin et al., 2001; Vedenov and Barnett, 2004; World Bank, 2011; Giné et al., 2007). Appropriate thresholds will therefore need to be set for each risk and each country (Hess and Syroka, 2005).

The following sections provide some examples of drought index insurance implemented for rainfed and irrigated crops (see Table 2.8.1), and some research advances in designing drought index insurance (weather based) for irrigated crops.

2.8.5.1 Meteorological-based index insurance

The use of meteorological-based index insurance to cover rainfed crops against drought is widespread among middle- and lower-income countries.

Mexico and India currently have the most developed rainfall index insurance programmes. In both countries, the products were first introduced in 2003, focusing primarily on rainfall deficiency (drought) (Barnett and Mahul, 2007). In Mexico, there is an insurance to protect maize, corn, beans, sorghum, and barley growers against catastrophic losses due to droughts and floods. This insurance is based on a rainfall index that sets two triggers (to determine if there has been a drought or a flooding). These trigger levels differ according to the crop, region, and crop growth stages (i.e. sowing, flowering, and harvesting) (Hazell et al., 2010). The rainfall index insurance scheme named 'Varsha Bima' in India covers anticipated yield losses on account of shortfall in the actual rainfall within a specific location and period. It provides for two options: 'sowing failure' (SF) and 'rainfall distribution index' (RDI). SF covers the risk of prevented / failed sowing, while RDI covers the entire season's rainfall requirement with due respect to moisture requirement at critical stages of crop growth. Rainfall index is created by (i) giving weightage to rainfall during critical periods, and (ii) by capping for excess rainfall. Varsha Bima has been designed for popular and widely grown field crops such as rice, pearl millet, maize, sorghum, groundnut, etc., in the country's drought-prone areas (Agriculture Insurance Company of India Limited, 2014b).

Pilot projects have been started in several countries (mostly supported by the World Bank), including Thailand, Indonesia, Malawi, Kenya, and Nicaragua (World Bank, 2011; Instituto Internacional de Investigación para el Clima y la Sociedad, 2010).

Instead of covering yield losses, some insurance schemes protect against additional irrigation costs in a period of drought. From an economic point of view, a loss due to a shortfall in yields is not different from a loss due to increased costs of irrigation (Mafoua and Turvey, 2003). A rainfall-based drought index insurance scheme for individual coffee farmers in Vietnam aimed to compensate for additional costs incurred during a period of drought. Generally, coffee is irrigated three times a year. In the event of drought, a fourth irrigation is needed. This insurance was studied by GlobalAgRisk (2009) and implemented by the Bao Minh Company at a premium rate of 10% (Bao Minh, 2015). However, owing to losses, the insurer had to stop offering drought insurance for coffee in Daklak (Mai and Hung, 2013).

Although not yet implemented, several researchers have investigated the potential of meteorological-based index insurance to cover irrigated crops against drought:

- Mafoua and Turvey (2003) provided a starting point for examining how rainfall can be used to mitigate excessive irrigation costs for farmers. For this, they developed an economic model of irrigation cost insurance to illustrate the relationship between a weather variable (rainfall), crop yields, costs of irrigation, and profits.

- Zeuli and Skees (2005) also proposed a financial product called 'cumulative rainfall insurance contract' in a New South Wales irrigation district in Australia, to manage water supply risk. For its implementation, historic rainfall levels need to be highly and positively correlated to historic reservoir water levels. The coverage level and liability are chosen by the insured, and are independent of the crops in the farm.
- Thompson et al. (2009) analysed the potential for using a rainfall derivative to manage the annual risk associated with entering the crop year with a less-than-normal irrigation supply.
- Buchholz and Musshoff (2014) investigated the potential of index-based weather insurance (with underlying precipitation and temperature indices) to cope with the economic disadvantages for farmers resulting from reductions in water quotas and increased water prices.

2.8.5.2 Remote sensing–based index insurance

Remote sensing–based index insurance provides an inexpensive method of hedging against drought risk in rainfed crops: pasture, rangeland, and large monocultures. Recent advances in satellite remote sensing technology now permit accurate measurements at particular spatial scales and spectral bandwidths that allow dynamic monitoring of environmental conditions such as vegetation cover. Remote sensing has proven a powerful tool for evaluating crop-growing conditions and drought (Peters et al., 2002).

NDVI-based insurance programmes currently exist in the Canada, Spain, and the United States (Turvey and Mclaurin, 2012). In 2001, Alberta (Canada) launched a pilot project using satellite imagery to define the historical benchmark production and assess annual pasture production (Hartell et al., 2006; World Bank, 2005). In Spain too, the product has been offered since 2001 for all the farms performing extensive livestock production, specifically on the following species: cattle, sheep, horses, and goats. The insurance product is designed to cover farmers experiencing more than 30 dry days (definition of 'dry days' based on the average historical information on pasture) (Agroseguro, 2015b). In the United States, there has been a vegetation index–based index insurance since 2007 to cover apiculture, pasture, rangeland, and forage (Risk Management Agency, 2009).

Differences between payments and actual losses (i.e. basis risk) might decrease over time, since improvements in this field are very quick, with regular increases in imagery resolution and rapid emergence of new technologies (Leblois and Quirion, 2010).

2.8.5.3 Hydrological drought index insurance

Although not commercially offered, several hydrological drought index insurance mechanisms have been proposed in the literature, seeking to cover financial risks for water managers (Brown and Carriquiry, 2007; Zeff and Characklis, 2013) and offering covers for agricultural production (Leiva and Skees, 2008). Brown and Carriquiry (2007) used seasonal inflows to develop an index insurance that compensates urban water suppliers for the increased costs associated with acquiring water from irrigators during dry periods via option contracts. Zeff and Characklis (2013) proposed an index insurance based on streamflows and withdrawals from a multi-reservoir system that generally refills each year. They proposed a contract that is purchased once per year at the beginning of the irrigation season, when reservoir levels are typically full. Leiva and Skees (2008) proposed an index insurance presenting two types of contracts, based on 12- or 18-month river flow accumulation in the whole Rio Mayo irrigation system in northwestern Mexico. Cumulated river flows are highly correlated to annual plantings. When the accumulation of reservoir inflows falls below a predetermined threshold, the insured district obtains an indemnity based on the district average expected income per hectare.

In the works presented, it is mentioned that intertemporal adverse selection might occur in the presence of skilful contracts. In such cases, Brown and Carriquiry (2007) suggested that contracts should be signed before forecasts are available; otherwise, insureds could adjust premiums in accordance with forecasts. However, none of the cited studies adapt the insurance scheme to deal with intertemporal adverse selection.

Table 2.8.3 includes the main characteristics (risk covered, policyholder, index used, tick of the insurance, method for premium calculation, duration of contract, contract conditions, and the study case analysed) of the three index insurance schemes proposed in the literature. None of them have been implemented, but they show potential in protecting irrigators or water utilities against hydrological drought economic losses.

Table 2.8.3 Hydrological drought index insurance for irrigation.

Insurance name	Third-party index insurance contracts (Zeff and Characklis, 2013)	Irrigation insurance (Leiva and Skees, 2008)	Reservoir index insurance (linked to option contracts) (Brown and Carriquiry, 2007)
Risk covered	Water utility financial risks	Irrigation district income stabilisation	Water utility financial risks
Policyholder	Water utility	Irrigation district	Urban water supplier
Index	Streamflow index (compared to withdrawals from a reservoir)	Streamflow index (linked to irrigated plantings)	Streamflow index (inflows to a reservoir)
Tick of the insurance	US$1 contracts; insured could buy the number of contracts desired	Based on hectare-equivalent income	Based on water prices
Premium calculation	Synthetic inflows and Monte Carlo simulation	Simulation model taking into account reservoir operation policies	Monte Carlo simulation
Duration of contract	Annual	12-month or 18-month	Annual
Contract conditions (if mentioned)	Single-purpose reservoir; reservoir refills each year	–	Insurance contracts signed before forecasts are available
Study case	Durham water utility, North Carolina, United States	Rio Mayo irrigation district	Water system in Metro Manila, Philippines

2.8.6 Conclusions

There has been significant work done in attempting to develop insurance policies with drought coverage for farmers. So far, it seems clear that the risk of drought in rainfed crop production is adequately covered with insurance policies. It might be expensive, and require significant efforts to combat the risks of asymmetric information (adverse selection and moral hazard). However, there are many examples of effective drought insurance policies implemented for rainfed farmers around the world.

However, while irrigated farms do also face the risks of droughts, resulting in reduced supply or complete interruption of water deliveries for irrigation, there are very few experiences of insurance schemes covering this consequence. There are various reasons explaining the lack of such experiences: (a) suffering water unavailability is, in most cases, a second-order consequence of a natural event, which means that it is not a purely natural and random event; (b) this requires that insurance be based on indices that are in turn conditional on natural and random variables, easily and objectively measurable; (c) this might result in the indices, based on which an insurance policy is structured, not being perfectly correlated with the volume of water that an irrigator can use; and (d) there is a risk of intertemporal adverse selection. These reasons explain the technical difficulties in developing these types of policies, and the doubtful demand that such products might have among farmers.

In closing, we would like to stress that, in some circumstances, drought coverage might be hedged with insurance mechanisms. Hence, there is a need to do more research on the topic and see more practical applications of the theoretical and numerical analyses.

Acknowledgements

Funding for this study was provided by the Spanish Ministry of Economy, Industry and Competitiveness (MINECO). The authors thank all data providers, and the national authorities for the insurance records.

References

AFSC (2015a). Moisture Deficiency Insurance. https://www.afsc.ca/doc.aspx?id=7859.

AFSC (2015b). Satellite Yield Insurance. http://www.afsc.ca/doc.aspx?id=3360.

AGRICORP (2015). Production Insurance Program for Forage Rainfall.

Agroseguro (2015a). *El Sistema Español de Seguros Agrarios En Cifras 1980/2014.*

Agroseguro (2015b). *Condiciones Del Seguro de Compensación Por Pérdida de Pastos.*

Agriculture Insurance Company of India Limited (2014a). Rainfall Insurance Scheme – Coffee. http://www.aicofindia.com/AICEng/General_Documents/Product_Profiles/COFFEE-RISC-ENG.pdf.

Agriculture Insurance Company of India Limited (2014b). Rainfall Insurance Scheme – Varsha Bima (Rainfall Insurance). http://www.aicofindia.com/AICEng/General_Documents/Product_Profiles/VARSHA-RAINFALL-ENG.pdf.

Australian Bureau of Meteorology (2015). Australian Rainfall Deciles. http://www.bom.gov.au/climate/enso/d6a1941.shtml.

Bao, M. (2015). Insurance Products. http://www.baominh.com.vn/vi-vn/chuyenmuc-948-bao-hiem-nong-nghiep.aspx.

Barnett, B.J., Barnett, C.H., and Skees, J.R. (2008). Poverty traps and index-based risk transfer products. *World Development* (Elsevier Ltd) 36 (10): 1766–1785. doi:10.1016/j.worlddev.2007.10.016.

Barnett, B.J. and Mahul, O. (2007). Weather index insurance for agriculture and rural areas in lower-income countries. *American Journal of Agricultural Economics* 89 (5): 1241–1247. doi:10.1111/j.1467-8276.2007.01091.x.

Bielza Diaz-Caneja, M., Conte, C.J., Catenaro, R., and Gallego Pinilla, J. (2008). *Agricultural Insurance Schemes II Index Insurances*. Italy: Ispra.

Brown, C. and Carriquiry, M. (2007). Managing hydroclimatological risk to water supply with option contracts and reservoir index insurance. *Water Resources Research* (American Geophysical Union) 43 (11): W11423+. doi:10.1029/2007WR006093.

Buchholz, M. and Musshoff, O. (2014). The role of weather derivatives and portfolio effects in agricultural water management. *Agricultural Water Management* (Elsevier B.V.) 146 (February): 34–44. doi:10.1016/j.agwat.2014.07.011.

Bureau of Reclamation (2008). *Central Valley Project and State Water Project Operations Criteria and Plan Biological Assessment.*

Byun, H.R. and Wilhite, D.A. (1999). Objective quantification of drought severity and duration. *Journal of Climate* 12 (9): 2747–2756. doi:10.1175/1520-0442(1999)012<2747:OQODSA>2.0.CO;2.

Carriquiry, M.A. and Osgood, D.E. (2012). Index insurance, probabilistic climate forecasts, and production. *Journal of Risk and Insurance* 79 (0): 287–300. doi:10.1111/j.1539-6975.2011.01422.x.

Choudhury, A., Jones, J., Okine, A., and Choudhary, R. (2015). *Drought Triggered Index Insurance Using Cluster Analysis of Rainfall Affected by Climate Change.* Normal, Illinois: Illinois State University. doi:10.1017/CBO9781107415324.004.

Coble, K.H. and Barnett, B.J. (2008). Implications of integrated commodity programs and crop insurance. *Journal of Agricultural and Applied Economics* 2 (August): 431–442.

Coble, K.H. and Barnett, B.J. (2013). Why do we subsidize crop insurance? *American Journal of Agricultural Economics* 95 (2): 498–504. doi:10.1093/ajae/aas093.

Collier, B., Skees, J., and Barnett, B. (2009). Weather index insurance and climate change: opportunities and challenges in lower income countries. *The Geneva Papers on Risk and Insurance Issues and Practice* 34 (3): 401–424. doi:10.1057/gpp.2009.11.

Confederación Hidrográfica del Júcar (CHJ) (2007). Plan Especial de Alerta Y Eventual Sequía En La Confederación Hidrográfica Del Júcar.

ENESA (2012). La Sequía, Un Riesgo Incluido En Los Seguros Agrarios. *Noticias Del Seguro Agrario* 82: 3–4.

Estrela, T. and Vargas, E. (2012). Drought management plans in the European Union. The case of Spain. *Water Resources Management* 26 (6): 1537–1553. doi:10.1007/s11269-011-9971-2.

European Commission (2007). *Green Paper from the Commission to the Council, the European Parliament, the European Economic and Social Committee and the Committee of the Regions: Adapting to Climate Change in Europe – Options for EU Action.* Brussels.

Gibbs, W.J. and Maher, J.V. (1967). Rainfall deciles as drought indicators. *Bureau of Meteorology Bulletin, Commonwealth of Australia, Melbourne No. 48.*

Giné, X., Townsend, R., and Vickery, J. (2007). Statistical analysis of rainfall insurance payouts in Southern India. *American Journal of Agricultural Economics* 89 (5): 1248–1254. doi:10.1111/j.1467-8276.2007.01092.x.

GlobalAgRisk (2009). *Designing Agricultural Index Insurance in Developing Countries: A GlobalAgRisk Market Development Model Handbook for Policy and Decision Makers.* Lexington, KY: GlobalAgRisk, Inc.

Hartell, J., Ibarra, H., Skees, J., and Syroka, J. (2006). *Risk Management in Agriculture for Natural Hazards.* Rome: ISMEA.

Hazell, P., Anderson, J., Balzer, N., Hastruo Clemensen, A., Hess, U., and Rispoli, F. (2010). *Potential for Scale and Sustainability in Weather Index Insurance for Agriculture and Rural*

Livelihoods. Rome: Rural Finance. http://www.ruralfinance.org/servlet/CDSServlet?status=ND0xODA1LjcyMzQ1JjY9ZW4mMzM9ZG9jdW1lbnRzJjM3PWluZm8~.

Heim, R.R. (2002). A review of twentieth-century drought indices used in the United States. *Bulletin of the American Meteorological Society* (August): 1149–1166. http://www.engr.colostate.edu/~jsalas/classes/ce624/Handouts/20th century drought-Heim-02.pdf.

Hess, U. and Syroka, J. (2005). *Weather-Based Insurance in Southern Africa The Case of Malawi*. 13. *Agriculture and Rural Development Discussion Paper*. Agriculture and Rural Development Discussion Paper.

Instituto Internacional de Investigación para el Clima y la Sociedad (2010). *Seguros En Base a índices Climáticos Y Riesgo Climático: Perspectivas Para El Desarrollo Y La Gestión de Desastres*. Clima y sociedad, no. 2.

IPCC (2014). *Climate Change 2014: Synthesis Report. Contribution of Working Groups I, II and III to the Fifth Assessment Report of the Intergovernmental Panel on Climate Change. IPCC*.

Johansson, R.C., Tsur, Y., Roe, T.L., Doukkali, R., and Dinar, A. (2002). Pricing irrigation water: a review of theory and practice. *Water Policy* 4: 173–199. http://www.sciencedirect.com/science/article/pii/S1366701702000260.

Kapphan, I., Calanca, P., and Holzkaemper, A. (2012). Climate change, weather insurance design and hedging effectiveness. *The Geneva Papers on Risk and Insurance Issues and Practice* (Nature Publishing Group) 37 (2): 286–317. doi:10.1057/gpp.2012.8.

Keyantash, J. (2002). The quantification of drought: an evaluation of drought indices. *The American Meteorological Society* (August). http://adsabs.harvard.edu/abs/2002BAMS...83.1167K.

La Positiva (2015). Seguro agrícola catastrófico. https:// https://www.lapositiva.com.pe/principal/seguros/seguro-agricola-catastrofico/1023/c-1023.

Leblois, A. and Quirion, P. (2010). *Agricultural Insurances Based on Meteorological Indices: Realizations, Methods and Research Agenda*. 71. *Sustainable Development Series*.

Leiva, A. and Skees, J. (2008). Using irrigation insurance to improve water usage of the Rio Mayo irrigation system in Northwestern Mexico. *World Development (*Elsevier Ltd) 36 (12): 2663–2678. doi:10.1016/j.worlddev.2007.12.004.

Luo, H., Skees, J.R., and Marchant, M.A. (1994). Weather information and the potential for intertemporal adverse selection in crop insurance. *Review of Agricultural Economics* 16: 441–451.

Mafoua, E. and Turvey, C.G. (2003). Weather insurance to protect specialty crops against costs of irrigation in drought years. In: *American Agricultural Economics Association Annual Meeting*. Montreal, Canada. http://ageconsearch.umn.edu/bitstream/21922/1/sp03ma13.pdf.

Magallanes, A. (2015). Seguro contra riesgos climáticos. https:// http://www.magallanes.cl/MagallanesWebNeo/index.aspx?channel=8102.

Mai, B. and Hung, N. (2013). Sale of drought insurance for coffee suspended. http://english.thesaigontimes.vn/Home/business/financial-markets/28426/.

Martin, S.W., Barnett, B.J., and Coble, K.H. (2001). Developing and pricing precipitation insurance. *Journal of Agricultural and Resource Economics* 26 (1): 261–274.

Mckee, T.B. and Doesken, N.J. (1993). The relationship of drought frequency and duration to time scales. In: *8th Conference on Applied Climatology*. http://ccc.atmos.colostate.edu/relationshipofdroughtfrequency.pdf.

Milly, P.C.D., Dunne, K.A., and Vecchia, A.V. (2005). Global pattern of trends in streamflow and water availability in a changing climate. *Nature* 438 (November): 347–350. doi:10.1038/nature04312.

Miranda, M.J. (1991). Area-yield crop insurance reconsidered. *American Journal of Agricultural Economics* 73 (2): 233–242. doi:10.2307/1242708.

Miranda, M.J. and Farrin, K. (2012). Index insurance for developing countries. *Applied Economic Perspectives and Policy* 34 (3): 391–427. doi:10.1093/aepp/pps031.

National Drought Mitigation Center (2015). *Daily Gridded SPI*. http://drought.unl.edu/MonitoringTools/DailyGriddedSPI.aspx.

Niemeyer, S. (2008). New drought indices. *Water Management Options Méditerranéennes. Série A*. 80: 267–274. http://194.204.215.84/index.php/fre/content/download/8820/130859/file/New_drought_indices.pdf.

Oficina del Riesgo Agropecuario (2010). Gestión de Riesgos Y La Experiencia En El Aseguramiento En La Argentina. In *Fourth Conference on Agricultural Risk and Insurance*. Buenos Aires, Argentina.

Palmer, W.C. (1965). *Meteorologic Drought. Research Paper No. 45*, 58 p. NOAA, Washington DC, USA.

Palmer, W.C. (1968). *The abnormally dry weather of 1961–1966 in the Northeastern United States*. In: (ed. J. Spar) *Proceeding of the Conference on the Drought in Northeastern United States, New York University Geophysical Research Laboratory Report TR-68-3*: 32–56.

Pérez Blanco, C.D. and Gómez Gómez, C.M. (2013). Designing optimum insurance schemes to reduce water overexploitation during drought events: a case study of La Campiña, Guadalquivir River Basin, Spain. *Journal of Environmental Economics and Policy* 2 (1): 1–15. doi:10.1080/21606544.2012.745232.

Pérez Blanco, C.D., Gómez Gómez, C.M., and Del Villar García, A. (2011). El Riesgo de Disponibilidad de Agua En La Agricultura: Una Aplicación a Las Cuencas Del Guadalquivir Y Del Segura/Water availability risk in agriculture: an application to Guadalquivir and Segura River Basins. *Estudios de Econom\ia Aplicada* 29: 333–358. http://ideas.repec.org/a/lrk/eeaart/29_1_9.html.

Peters, A.J., Walter-Shea, E.A., Lel, J.I., Viña, A., Hayes, M., and Svoboda, M.D. (2002). Drought monitoring with NDVI-based standardized vegetation index. *Photogrammetric Engineering & Remote Sensing* 68 (1): 71–75. http://www.asprs.org/a/publications/pers/2002journal/january/2002_jan_71-75.pdf.

Prudhomme, C., Giuntoli, I., Robinson, E.L., Clark, D.B., Arnell, N.W., Dankers, R., Fekete, B.M., Franssen, W., Gerten, D., Gosling, S.N., Hagemann, S., Hannah, D.M., Kim, H., Masaki, Y., Satoh, Y., Stacke, T., Wada, Y., and Wisser, D. (2014). Hydrological droughts in the 21st century, hotspots and uncertainties from a global multimodel ensemble experiment. *Proceedings of the National Academy of Sciences of the United States of America* 111 (9): 3262–3267. doi:10.1073/pnas.1222473110.

Quiroga, S., Garrote, M L., Fernandez-Haddad, Z., and Iglesias, A. (2011). Valuing drought information for irrigation farmers: potential development of a hydrological risk insurance in Spain. *Spanish Journal of Agricultural Research* 9 (4): 1059–1075. doi:10.5424/https://doi.org/10.5424/sjar/20110904-063-11.

Risk Management Agency (2009). *Vegetation Index Insurance Standards Handbook*. Washington DC, USA: USDA.

Risk Management Agency (2015). *2015 Crop Policies and Pilots*. Washington DC, USA: USDA. http://www.rma.usda.gov/policies/2015policy.html.

Risk Management Agency – Topeka (2015). Risk management agency update. In: *Water Symposium*. Topeka, Kansas, USA: Risk Management Symposium 2015. http://www.farmranchwater.org/2015/Presentations/RMA.pdf.

Ritter, M., Musshoff, O., and Odening, M. (2012). Minimizing geographical basis risk of weather derivatives using a multi-site rainfall model. In: *Price Volatility and Farm Income Stabilisation*. Dublin. doi:10.1007/s10614-013-9410-y.

Rouse, J.W., Haas, R.H., Schell, J.A., Deering, D.W., and Harlan, C.J. (1974). Monitoring the vernal advancement and retrogadation (greenwave effect) of natural vegetation. *Texas A&M University Remote Sensing Center*, no. September 1972.

Ruiz, J., Bielza, M., Garrido, A., and Iglesias, A. (2013). Managing drought economic effects through insurance schemes based on local water availability. *Proceedings of the 8th*

International Conference of European Water Resources Association, Porto, Portugal, 26–29 June 2013.

Ruiz, J., Bielza, M., Garrido, A., and Iglesias, A. (2015). Dealing with drought in irrigated agriculture through insurance schemes: an application to an irrigation district in southern Spain. *Spanish Journal of Agricultural Research* 13 (4): 15 p.

Schewe, J., Heinke, J., Gerten, D., Haddeland, I., Arnell, N.W., Clark, D.B., Dankers, R., Eisner, S., Fekete, B.M., Colón-González, F.J., Gosling, S.N., Kim, H., Liu, X., Masaki, Y., Portmann, F.T., Satoh, Y., Stacke, T., Tang, Q., Wada, Y., Wisser, D., Albrecht, T., Frieler, K., Piontek, F., Warszawski, L., and Kabatt, P. (2014). Multimodel assessment of water scarcity under climate change. *Proceedings of the National Academy of Sciences of the United States of America* 111 (9): 3245–3250. doi:10.1073/pnas.1222460110.

Sivakumar, M.V.K., Motha, R.P., Wilhite, D.A., and Wood, D.A. (Eds.) (2011). Agricultural drought indices proceedings of an expert meeting. In: *Agricultural Drought Indices 2–4 June 2010*, 205.

Skees, J.R. (2010). *State of Knowledge Report – Data Requirements for the Design of Weather Index Insurance*. Lexington, KY.

Skees, J.R., Black, R., and Barnett, B.J. (1997). Designing and rating an area yield crop insurance contract. *American Journal of Agricultural Economics* 79 (2): 430–438. doi:10.2307/1244141.

Stahl, K., Blauhut, V., Kohn, I., and Acácio, V. (2012). A European Drought Impact Report Inventory (EDII): Design and Test for Selected Recent Droughts in Europe, *Technical Report No. 3*.

Thompson, C.L., Supalla, R., Martin, D.L., Neely, B.J., and Mcmullen, B.P. (2009). Weather derivatives as a potential risk management tool for irrigators. In: *AWRA Summer Specialty Conference*: 1–6. http://digitalcommons.unl.edu/agecon_cornhusker/456/.

Tsiourtis, N. (2005). Cyprus. *Options Méditerranéennes, Series B* 51: 25–47.

Turvey, C.G. (2001). Weather derivatives for specific event risks in agriculture. *Review of Agricultural Economics* 23 (2): 333–351. doi:10.1111/1467-9353.00065.

Turvey, C.G. and Mclaurin, M.K. (2012). Applicability of the normalized difference vegetation index (NDVI) in index-based crop insurance design. *Weather, Climate, and Society* 4 (4): 271–284. doi:10.1175/WCAS-D-11-00059.1.

US Drought Portal (2015a). Crop Moisture Index. https://www.drought.gov/drought/content/products-current-drought-and-monitoring-drought-indicators/crop-moisture-index.

US Drought Portal (2015b). Palmer Drought Severity Index. https://www.drought.gov/drought/content/products-current-drought-and-monitoring-drought-indicators/palmer-drought-severity-index.

Van Loon, A.F. (2015). Hydrological drought explained. *Wiley Interdisciplinary Reviews: Water* 2 (August): 359–392. doi:10.1002/wat2.1085.

Vedenov, D.V. and Barnett, B.J. (2004). Efficiency of weather derivatives as primary crop insurance instruments. *Journal of Agricultural and Resource Economics* 29 (3): 387–403.

Wada, Y., Van Beek, L.P.H., and Bierkens, M.F.P. (2011). Modelling global water stress of the recent past: on the relative importance of trends in water demand and climate variability. *Hydrology and Earth System Sciences* 15 (12): 3785–3808. doi:10.5194/hess-15-3785-2011.

Woodard, J.D. and Garcia, P. (2008). Basis risk and weather hedging effectiveness. *Agricultural Finance Review* 68: 99–117. doi:10.1108/00214660880001221.

World Bank (2005). *Managing Agricultural Production Risk Managing Agricultural Production Risk*. Washington, DC.

World Bank (2011). *Weather Index Insurance for Agriculture: Guidance for Development Practitioners. Agriculture and Rural Development Discussion Paper*. Washington, DC.

Zargar, A. Sadiq, R., Naser, B., and Khan, F.I. (2011). A review of drought indices. *Environmental Reviews* 19: 333–349. doi:10.1139/a11-013.

Zeff, H.B. and Characklis, G.W. (2013). Managing water utility financial risks through third-party index insurance contracts. *Water Resources Research* 49: 4939–4951. http://onlinelibrary.wiley.com/doi/10.1002/wrcr.20364/full.

Zeuli, K.A. and Skees, J.B. (2005). Rainfall insurance: a promising tool for drought management. *International Journal of Water Resources Development* 21 (4): 663–675. doi:10.1080/07900620500258414.

Part Three

Drought Management Experiences and Links to Stakeholders

3.9
Drought and Water Management in The Netherlands

Wouter Wolters[1], Henny A.J. Van Lanen[2], and Francien van Luijn[3]

[1]*Wageningen Environmental Research (Alterra), Wageningen, The Netherlands*
[2]*Wageningen University, Wageningen, The Netherlands*
[3]*Rijkswaterstaat Water, Traffic and Environment, Lelystad, The Netherlands*

3.9.1 General context

3.9.1.1 Physical and socioeconomic systems

Physical system Drought management in the Netherlands is not a separate activity; it is integrated into a comprehensive, multi-faceted water management regime covering both floods and droughts (i.e. ensuring freshwater supply and availability) and water quantity and quality, particularly to avoid too-high chloride (Cl) concentrations. In large parts of western Netherlands, that is, the part below mean sea level, water availability is less of a problem than its Cl concentration that renders the water unusable.

Long-term average precipitation (1981–2010, reference period) in the Netherlands falls in the range of 725–950 mm/year, and excess precipitation (i.e. precipitation minus potential evaporation) in the range of 150–350 mm/year (KNMI, 2015a). On average, the excess precipitation occurs in the October–March period. In the summer months, a precipitation deficit is observed (Figure 3.9.1).

The highest maximum potential precipitation deficit in the reference period 1981–2012 is about 230 mm, while the lowest deficit is 125 mm (Figure 3.9.1, left). These numbers are higher in dry years – for example, in 5 out of 100 years, the maximum deficit during the summer varies from about 200 to 320 mm.

Rain supplies freshwater to the Netherlands, but the main freshwater sources are the rivers Rhine and Meuse. Over the centuries, an extensive water distribution network was developed. This system is primarily used to drain excess water from the Netherlands, but also to supply river water to vast areas in the south, west, and north of the country.

Drought: Science and Policy, First Edition.
Edited by Ana Iglesias, Dionysis Assimacopoulos, and Henny A.J. Van Lanen.

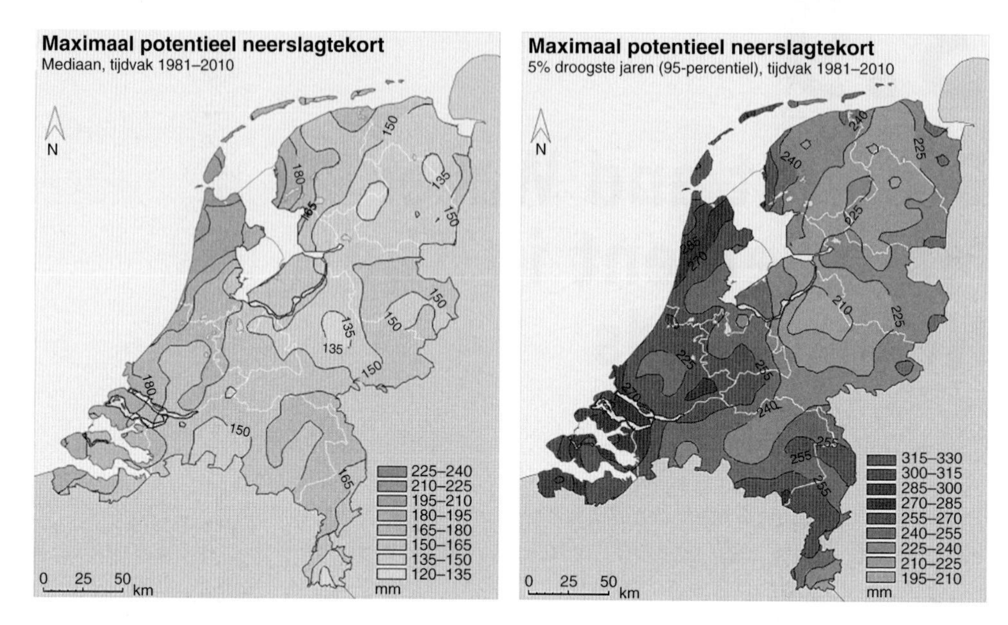

Figure 3.9.1 Maximum potential precipitation deficit in summer months over the period 1981–2010. Left: median; right: 95th percentile. *Source*: KNMI (2015a). (*See colour plate section for the colour representation of this figure.*)

Only some parts of the south and east of the Netherlands cannot be supplied with freshwater (so-called 'free-draining' areas). Freshwater is required, for example, to prevent salt seawater inflow in the river mouths (sea water intrusion), to supply freshwater to the polders to mix with the salty or brackish groundwater to avoid too-high Cl concentrations, and to protect water-based ecosystems. Furthermore, old peat dikes can shrink under dry weather, which reduces dike stability and can lead to the flooding of polder land. Freshwater is also required to feed drinking water reservoirs in the western Netherlands, and to infiltrate treated river water in the dunes to provide the major cities with drinking water. Vast volumes of river water are used for cooling energy-producing plants. To protect river ecosystems, river water temperature cannot be allowed to increase too much. River water is also supplied for agricultural (supplementary irrigation), industrial (process water, cooling, and cleaning), and recreational (e.g. bathing water quality) needs, and is needed to maintain adequate water levels in the rivers for waterborne transportation.

Socioeconomic system The Netherlands owes much of its prosperity and welfare to its water supply. The country has a large variety of special water-dependent ecosystems, as well as landscapes that depend on the availability of water. A considerable part of the economy depends on freshwater – namely, the agriculture and horticulture industries, the food and chemical industries, the energy sector, the recreation sector, inland fisheries, and navigation. About 8.5% of the working population works in a water-related sector. Water is an important factor in the national economy, with a total production value of about €183.5 billion (i.e. 16% of direct production value).

3.9.1.2 Scenarios

Van den Hurk et al. (2014) distinguishes four climate scenarios for the Netherlands that combine possible values for the global temperature increase and changes in the air circulation pattern, which together span the likely changes in the climate. Annual average temperature is projected to increase by 1.3–3.7°C in 2085, which results in a potential evaporation increase of 2.5–10%. Annual average precipitation is also expected to increase (5–7%). These changes lead to a projected increase in the maximum potential precipitation deficit of up to 17–19% (low temperature increase scenario) and 40–50% (high temperature increase scenario).

The socioeconomic outlook for the future, which determines all kinds of strategic decisions with respect to land uses and water (Janssen et al., 2006), was taken up by three national institutions – the Netherlands Environmental Assessment Agency (MNP), the Netherlands Bureau for Economic Policy Analysis (CPB), and the National Institute for Spatial Research (RPB). Its purpose is to visualise the challenges that the Dutch government may expect when continuing with 'business as usual'. Structural measures in the physical environment usually take a long time to be realised, may be costly, and have long-term consequences. Therefore, the government has to balance short-term costs, benefits, and risks with possible developments in the future. In addition, there is a need to compare the effects for different generations, social groups, and regions.

It is well understood that long-term trends – such as a decreasing household size, an ageing population, international migration, economic growth, and increasing personal welfare – will change the Dutch natural and built environment significantly. Therefore, the scenarios described in the following paragraphs analyse the combined impact of these trends on various aspects of the Dutch urban and rural landscape, including residential and industrial land use, traffic and transport, energy, agriculture, nature and landscape, water safety, and environment and health.

The scenario development process was based on an assessment of the long-term effects of current policy, given the international economic and demographic context of the Netherlands. Its qualitative and quantitative results should serve as a reference for policymakers involved in spatial planning, housing, natural resources, infrastructure, and the environment. By exploring how land use and various aspects of the living environment may develop on the long run (2040), the study shows when current policy objectives may be difficult to meet, and which new issues may emerge.

Two extremes of this socioeconomic scenario analysis have been linked with climate change scenarios as developed for the Delta Programme (Delta Programme, 2011). Two extreme climate scenarios (moderate and warm; i.e. without and with changed air circulation over Europe) were combined with two 'extreme' socioeconomic scenarios (one scenario with high socioeconomic growth, and the other with some socioeconomic decline). These, combined, lead to a broad width of conditions between 'the worst that can happen' and 'the least that can happen'. The following scenarios are distinguished (Delta Programme, 2011; Figure 3.9.2):

- *Quiet*: moderate climate change 'with less pressure on space'
- *Full*: moderate climate change 'with more pressure on space'
- *Warm*: fast climate change 'with less pressure on space'
- *Steam*: fast climate change 'with more pressure on space'

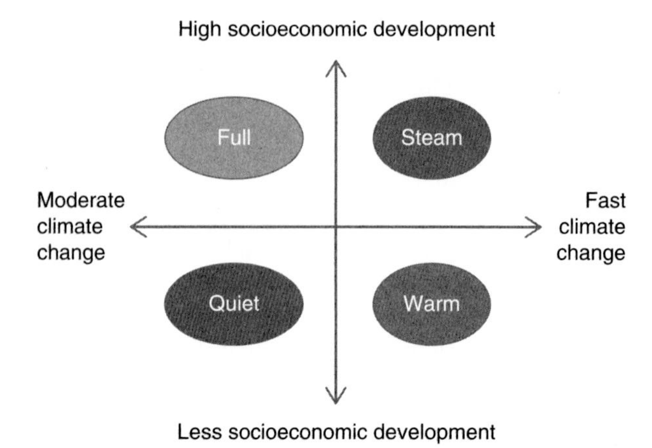

Figure 3.9.2 The four scenarios developed after considering climate change and socioeconomic development. *Source*: Delta Programme (2011).

In the 'full' and 'steam' scenarios, the population increases considerably until 2050, there is considerable economic growth, and the urban areas expand at the cost of land for agriculture and horticulture. These driving forces continue after the year 2050. In combination with higher temperatures, this also implies a larger per-capita demand for drinking and industrial water. In addition, the need for cooling water for energy production increases, and the water available in urban areas will need to fulfil higher quality standards.

For the 'quiet' and 'warm' scenarios, the increase in population is smaller, and the population size eventually decreases. Economic growth is limited, resulting in less demand for drinking and industrial water, but water demand for agriculture and nature will increase, especially with higher temperatures.

However, in the national 'bottlenecks' analysis, the following effects are yet to be considered: (i) upstream countries will likely divert more water in the future in order to cope with drought events in their territory; and (ii) there will be adaptation to climate change – for example, through technological advances in water use.

3.9.1.3 Drought characteristics: Frequency and severity

Drought has become more frequent in the Netherlands since 1951 (KNMI & PBL, 2015). Since little historic flow data is available, another approach for investigating droughts was adopted in a European-wide study, which included the Netherlands. This study focussed on the trends in low flow and drought characteristics derived from the spatially distributed multi-model mean total runoff, which was simulated with nine large-scale hydrological models for the period 1963–2001 (Alderlieste and Van Lanen, 2012). Downwards trends were found for most of the summer months, with negative values varying from –8 to –21%. The 7- and 30-day minimum flow changed by about –15%. The average flow deficit volume increased by about 20%, while drought duration increased by 15%. The number of drought events hardly changed.

Alderlieste and Van Lanen (2013) and Alderlieste et al. (2014) show that, for the most extreme IPCC climate scenario (A2), the median annual flow in the Netherlands is projected to increase by the end of the twenty-first century, whereas, in the period June–August, the median flow will decrease (20–60%) according to the large-scale models, which were forced with downscaled and bias-corrected CNRM GCM data. The decrease in the summer months is lower when the models are forced with ECHAM GCM data (25–30%) or IPSL GCM data (about 20%). The number of droughts is projected to decrease (about 30%), but the average duration of the droughts will double and the average flow deficit volume will triple (forcing CNRM GCM data). Projected changes in duration and deficit volume were lower when ECHAM GCM data were used (about 80% and 125%, respectively) or IPSL GCM data (about 90% and 60%, respectively). It should be mentioned that the Netherlands does not suffer from multi-year droughts.

3.9.1.4 Current management framework

With the present water system, freshwater availability and quality status is well managed in normal situations, and impacts from droughts of low frequency and intensity are accepted at the policy level. For instance, the current practice of dividing the flow of the Rhine over the main branches of the river will remain as is until 2050. However, in view of possible development in the future, the current strategy will no longer hold in the long term. The new strategy, to be taken up in the Delta Programme, entails: (i) solving current and future bottlenecks, and (ii) exploiting opportunities. In addition to the existing approach of reacting to shortages (e.g. the priority ranking for water supply and fighting desiccation), there is a need for a new policy towards 'preventing and being prepared for' water shortage and salinisation. All this is taken care of as part of the Delta Programme (with nine components, of which one is about a freshwater programme that includes drought management).

The Delta Programme is a broadly supported implementation programme, to which all water users commit. Parties will take up the relevant measures in their plans and reserve the required financial means in their budgets. The investments required for the Delta Plan on Freshwater Supply for the period 2015–2028 are estimated at €1.5–2 billion, with the government part about €550 million. The entire fund for the implementation of the Delta Programme for that period is €16 billion (i.e. about €1 billion annually).

A reliable water supply is essential for maintaining suitable liveability (quality of life) and health conditions, as well as for the economic development of companies and sectors that depend on freshwater. The aim is to prevent and/or reduce damages in the future, and to seize opportunities to continue benefitting from the advantageous location in the delta. With the Delta Programme, the government has taken the initiative, together with all users and sectors, to investigate the areas where the parties involved (i.e. national and regional governments and water use stakeholders) can actively contribute to a more robust system – by working on 'service levels'. *Service levels* are defined as 'the availability of freshwater in normal and drought conditions in an area' (with the annotation that there is debate over the exact meaning of 'normal' and 'drought' in the definition). Governments and users jointly specify, in the form of region-specific as well as related national agreements, the government's responsibilities and obligations in terms of ensuring water availability, and the users' responsibilities and remaining risks. This concerns both surface and groundwater, and covers both quality and quantity. Moreover, the agreements have repercussions not only for the water system, but also for spatial planning and for the users.

Until 2021, risks of water shortage will be monitored, to support the dialogue between system managers and users. This will eventually clarify the basis on which users will be able to make technological or entrepreneurial adjustments.

3.9.2 Drought risk and mitigation

3.9.2.1 Vulnerability to drought

The main source of freshwater in the Netherlands is not precipitation, but the freshwater inflows of the rivers Rhine and Meuse. With the changes in both temperature and air circulation, the average low flow of the river Rhine is projected to decrease from about 1700 m^3/sec (average 1968–1998) to 1000–1250 m^3/sec (De Wit et al., 2008). Water balance modelling (Klijn et al., 2010; Kwadijk et al., 2008) using the previous set of climate scenarios (KNMI'06 scenarios) showed that, in general, there is enough water in the Netherlands (even in 2050!), but that it is not available at the right time, the right place, and in the right quality. This leads to bottlenecks in dry years, both now and in the future.

Droughts will increase in frequency, intensity, and duration, and this will lead to water shortages, water quality issues, and salinization. Sea level rise will contribute to salt water intrusion. Potential crop yields will increase, with a longer growing season and higher CO_2 concentrations, but changes in precipitation and the prevalence of extreme events could threaten harvests. Precipitation-dependent ecosystems are at risk, and there is an increased risk of natural fires.

The analysis of the aforesaid bottlenecks has clarified that climate change is a factor in this, and also that the reactions of water users to climate change, as well as societal and technological developments, are very important factors in determining the size of the bottlenecks.

The 'bottlenecks in dry years' have consequences for both *location-related water users* (agriculture, nature, urban area) and *network-related water users* (navigation, cooling water for the energy producers, and drinking water). It appears difficult to make a good estimation of the economic effects of water shortage, but the following effects are anticipated:

- The economic impacts seem to be highest in agriculture and navigation. In the Netherlands, there is a feeling of having a relative advantage, as droughts in the Netherlands are less pronounced than in the more drought-prone areas elsewhere in Europe.
- Industries will experience damages due to disturbances or failures in water supply, and may be required to spend more on systems for processing water, cooling, etc. Shortage of water may lead to a reduced reserve energy capacity, which in turn can lead to a higher cost of energy for users.
- Water intake points for drinking water or industry may need investments to secure the supply of good-quality water.
- In urban areas, water shortage will lead to damages to urban green, buildings, and infrastructures (both aboveground and underground). With the increasing heat

related to drought conditions, liveability or quality of life will deteriorate (e.g. heat stress), as will water related recreation (e.g. due to blue-green algal blooms).

As yet, the exact amount of water demand and availability cannot be predicted with the current state of investigations and modelling.

3.9.2.2 Existing framework for drought management

The Water Management Centre for the Netherlands (WMCN), under the Ministry of Infrastructure and Environment, is the information centre for the national Dutch water system. It bundles all information about monitoring and modelling products and services concerning water, including national drought management. WMCN provides daily information on extreme situations, among others, including water shortages, below-normal water states, and fluxes, and provides forecasts to the national and regional water authorities. The National Water Distribution Committee (LCW), which is part of WMCN, is responsible for dealing with actual drought.

Drought monitoring is relatively new in the Netherlands, although monitoring of the national, regional, and local water systems has always been important. The term 'drought' in the Netherlands is very closely related to freshwater supply to avoid too-high Cl concentrations in the surface water system. The 1976 drought gave rise to the first 'drought policy' – that is, a priority ranking of water allocation in times of drought. Several in-depth hydrological and impact studies (so-called 'drought studies') were carried out to improve the understanding of drought management. The 2003 drought boosted further development of water management under dry conditions. Figure 3.9.3 presents the national priority ranking of water allocation that was developed based on the experiences of 1976.

Under drought conditions, surface water is first supplied to the sectors dealing with safety and the prevention of irreversible damage (Figure 3.9.3, 'Category 1'). Reduction of (old peat) dike stability, prevention of soil compaction (peat soils), and impacts to wet vulnerable nature receives the highest priority. Then, water is allocated to drinking water supply and energy production (sufficient river water for cooling purposes) ('Category 2'). Navigation, agriculture, recreation, and fisheries belong to the sectors with the lowest priorities ('Category 4'). In the regions, priority ranking may deviate from national ranking. Priority ranking is the result of a political (importance of sectors) and technical (water supply infrastructure) decision-making process, which is concluded before any drought emerges and implemented during an upcoming or ongoing drought.

LCW is pro-active and takes timely measures to optimally distribute surface water to reduce drought impact. The Netherlands is divided into six regions (Figure 3.9.4) that are supplied with water from the rivers Rhine and Meuse. Different surface water supply options exist for these regions. LCW has meetings with representatives from these regions.

LCW starts to have regular meetings when water levels fall low in the Rhine or the Meuse. Another reason to convene such a meet is high river water temperatures. The following conditions are distinguished to call for LCW's attention:

- Flow of Rhine that is expected to last longer than 3 days (Figure 3.9.5):
    - <1400 m^3/sec in May
    - <1300 m^3/sec in June
    - <1200 m^3/sec in July

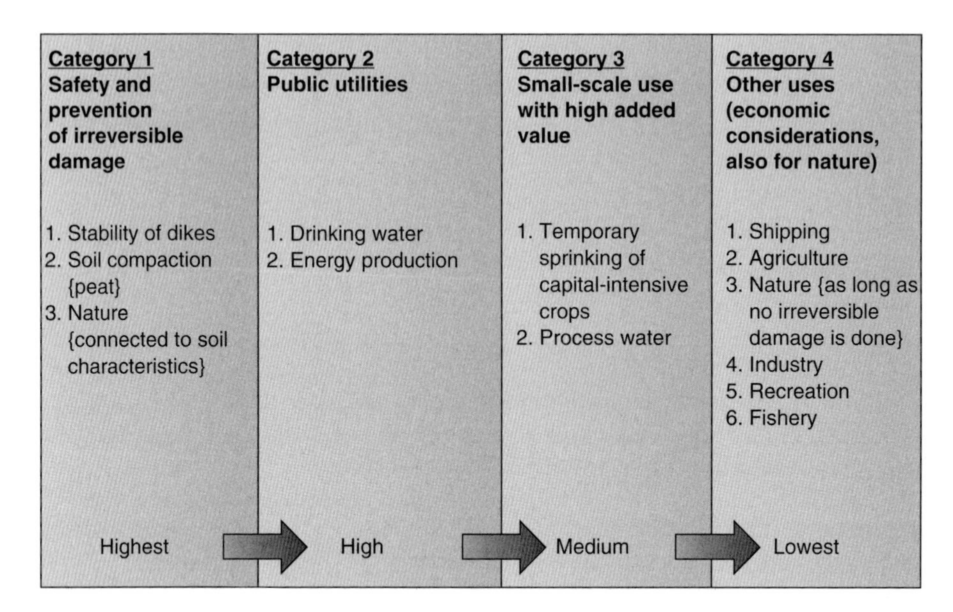

Figure 3.9.3 Priority ranking of water allocation at the national scale in the Netherlands under drought conditions. *Source*: Andreu et al. (2015).

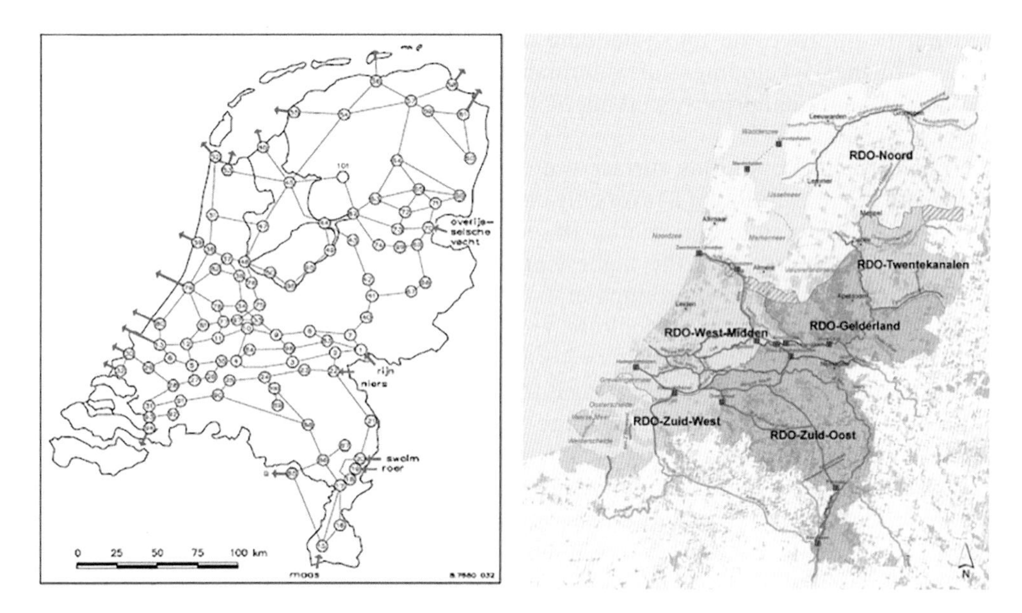

Figure 3.9.4 Surface water system that is used to supply freshwater in the Netherlands (left), and the six regions for regional drought platforms (right). *Source*: Andreu et al. (2015). (*See colour plate section for the colour representation of this figure.*)

  - <1100 m^3/sec in August
  - <1000 m^3/sec in September and October
- Flow of Meuse lower than 25 m^3/sec

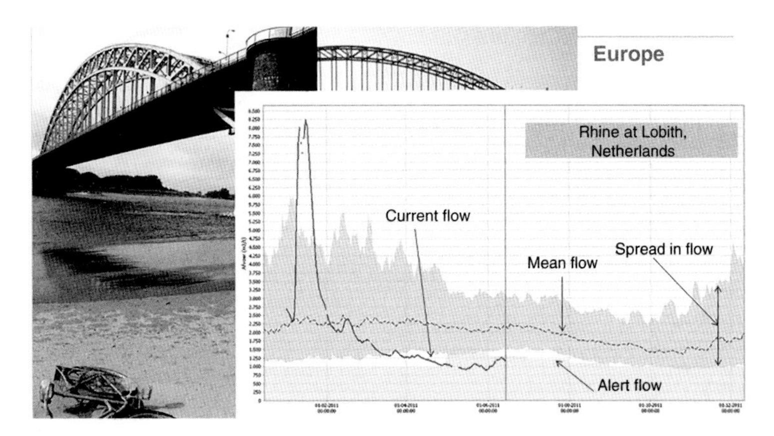

Figure 3.9.5 Flow of the Rhine and the threshold that triggers LCW meetings. *Source*: Andreu et al. (2015).

- Temperature of the Rhine at Lobith (near where the river enters the country from Germany) above 23°C

The indicators and indices used by LCW in this process are simple and relatively straightforward. When the river flow is below the specified thresholds, or the river water temperature above the specified limit, negative impacts of drought may occur. No standardised indices, such as SPI (Standardised Precipitation Index), SGI (Standardised Groundwater level Index), or SRI (Standardized Runoff Index), are used by LCW.

LCW does not issue a 'drought declaration', but distinguishes four different alert levels:

- Normal water conditions (surface water, groundwater, soil water)
- Potential water shortages (regional, temporal)
- Actual water shortages (like in 2003 or 2011)
- (Potential) emergency situation due to water shortages

When a (potential) drought is emerging, LCW, through the regional consultations: (i) collects and communicates information; (ii) coordinates and fine-tunes regional measures; and (iii) implements the national and regional priority ranking. LCW, as part of WMCN, consists of national and regional departments of the Directorate General of Public Works and Water Management (Rijkswaterstaat), Association of Regional Water Authorities (water boards), National Coordination Centre Crises Management (different ministries responsible for water supply and impacts due to water shortages), Interprovincial Consultation Committee, and the Royal Netherlands Meteorological Institute (KNMI). LCW operates under the authority of the Ministry of Infrastructure and Environment. If LCW, with the support of the technical system, is unable to identify appropriate measures, the minister of infrastructure and environment has the final responsibility for taking decisions under emergency situations arising from water shortages.

The ongoing drought conditions (precipitation deficit, groundwater levels, river flow, salt concentration, and river water temperature) in the Netherlands are discussed during LCW meetings, and the impacts are reported to LCW participants (Figure 3.9.6).

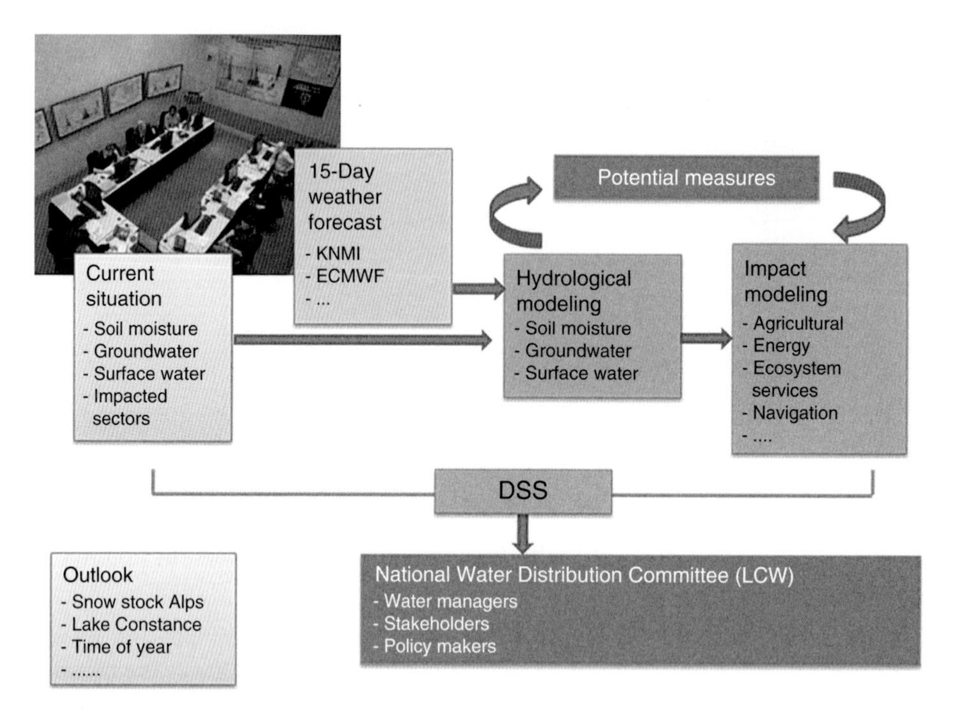

Figure 3.9.6 Procedure for drought management in the Netherlands. *Source*: Andreu et al. (2015).

3.9.2.3 Stakeholder involvement in drought management

In the Netherlands, national coordination committees come into action when 'extreme conditions' occur. There are three national coordination committees: for floods, for drought, and for pollution. These committees coordinate the information from WMCN (in Lelystad) to be shared with all stakeholders. The national committees produce reliable and useful information about the expected conditions with respect to floods, drought, and pollution. In coordination with KNMI, the committees mobilise all reliable and useful information on the condition of waters to attain a national overview of the water conditions. Based on this national overview, the committees provide advice to water managers about measures that need to be taken – such as, for example, closing regulators in case of flooding risk, increasing water levels in water storage reservoirs when water shortage is imminent, or taking measures to control pollution in order to reduce impact on environment or drinking water intakes.

WMCN consists of three national coordination committees:

- *National Coordination Committee for Flooding Threat.* This committee has a crucial role in providing timely warnings regarding increased flooding risk and informing the threatened areas.
- *National Water Distribution Committee* (LCW). This committee comes into action when there is a prolonged period of drought and low river flows, and when the demand for water is higher than supply.
- *National Coordination Committee for Environment Pollution.* This committee functions when there are reports of nuclear, biological, or chemical pollution events, and whenever there is a water-related risk or danger.

WMCN also provides the 'Helpdesk Water' service, which is primarily designed to answer questions from people who are (professionally) involved in water policy, water management, and water safety issues in the Netherlands. This water helpdesk was created by the Dutch government, provinces, municipalities, and the Dutch Association of Regional Water Authorities (previously referred to as 'Association of Water Boards').

For accident and disaster management, the Netherlands is divided into 'safety regions' (currently 25). In such safety regions, many public parties collaborate on being prepared for crises and disasters, also cooperating in actual crisis conditions. This can deal with any disaster – for example, flooding, animal diseases, terrorism, or forest fires.

In safety regions, municipalities and services cooperate with 'crisis partners', for example:

- Water boards (local water managers)
- Regional departments of 'Rijkswaterstaat' that are responsible for the management of the main water system
- Regional military commands

In addition, other (private) partners too are involved, including:

- Hospitals
- Red Cross
- Dutch rail system
- Energy companies
- Chemical industry

For drought, the National Water Distribution Committee (LCW) is important. Its goals are to distribute the available water as equitably as possible and to manage water shortage crises when they occur.

The LCW consists of representatives from:

- Ministry of Infrastructure and Environment
- Dutch Association of Regional Water Authorities (Association of Water Boards)
- Inter-provincial Consultation
- Rijkswaterstaat
- KNMI

When there is an 'upcoming' drought, preventive measures (e.g. increasing the levels of freshwater reservoirs) are considered, and can be implemented whenever feasible. LCW issues a (fortnightly) 'drought message' to keep the stakeholders and the public informed. These messages (and more) are published on a publicly accessible website.

3.9.2.4 Example of the 2015 drought monitoring and management

Whenever LCW convenes a meeting, one of the first topics discussed is the 'precipitation deficit', which is taken as a measure of drought, and is the difference between evaporation and precipitation. Figure 3.9.7 shows both the development of the precipitation deficit throughout the year 2015 as well as its spatial distribution over the country. The precipitation deficit was at its maximum by the end of July, and it continued more or less until mid-August. Afterwards, there was a quick recovery. By early September, the precipitation deficit was around the median.

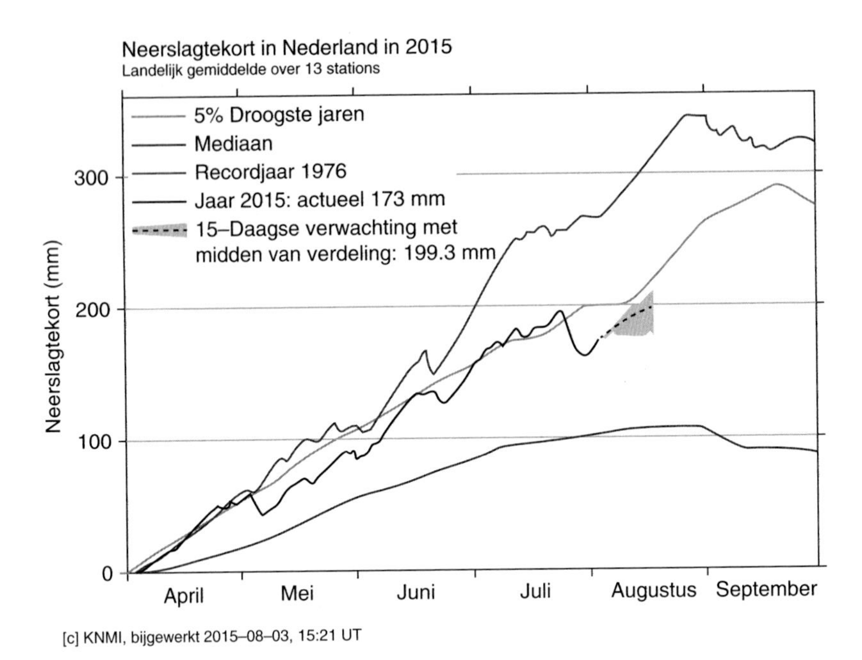

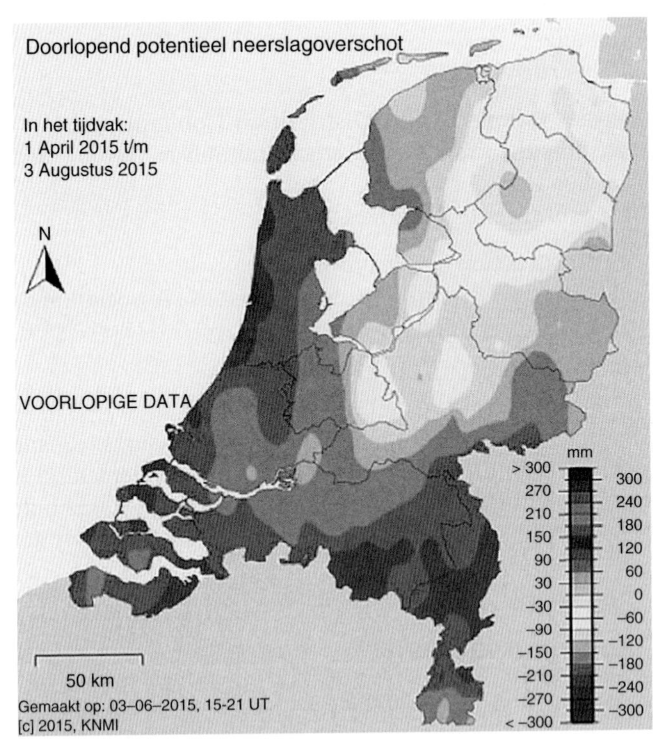

Figure 3.9.7 The precipitation deficit throughout the year 2015 until August. Top: Precipitation deficit in The Netherlands, August 2015; Bottom: Cumulative Potential Rainfall Surplus (1 April 2015–3 August 2015). *Source*: KNMI (2015b). (*See colour plate section for the colour representation of this figure.*)

This data is published in the weekly 'Drought Monitor', which is publicly available on the Internet. The message of the Monitor (#2015-10, dated 4 August 2015; KNMI, 2015b) also includes other information, such as:

- The rainfall during last week reduced the precipitation deficit somewhat in the Netherlands. Its value is around 200 mm. Meanwhile, the precipitation deficit is increasing again.
- The discharge of the Rhine decreased further, but it is still sufficient to meet water demand and to control salt intrusion from the sea. The situation of the river was slightly improved by the rains. This means that the overall picture in the next week remains the same: a relatively high rainfall shortage in the west and south of the Netherlands, with sufficient water supply still from the major rivers.
- Because of the drought situation, the water boards and Rijkswaterstaat remain alert, taking measures to supply or retain water. In western Netherlands, there are still inspections of peat dikes, although the situation has improved owing to the recent rainfall.
- In parts of southern Netherlands, the restrictions on withdrawal from surface water remain in place.
- The water temperatures at the beginning of the week were relatively low but have now risen again to 22°C and 23°C, respectively, at Eijsden (Meuse) and Lobith (Rhine). These temperatures are now falling again.
- Because of the low water levels in the Waal and the Ijssel rivers, the depth for navigation is less than normal. At the locks at Doesburg and Eefde, there are some restrictions on the passage of ships (with further information available on the Internet).

3.9.2.5 Comparison of the 2015 drought with 2003

In comparison with the summer of 2003, the precipitation deficit, a determining factor for the agriculture sector, was 230 mm in 2015, which is not extreme.

The flow of the Rhine was also not extremely low in 2003 (about 780 m^3/sec), although the period of low flow was long. The drought was not limited to the Netherlands, and the upriver countries also suffered from it – that is, the drought had a European dimension.

The water temperature in 2003 was recorded as extremely warm. In fact, the water temperatures had never been measured so high (28°C in early August) in the preceding 100-year period. This condition almost led to the 'step-plan' to cut power generation (colour codes with measures to reduce power generation, with serious consequences for industries) being activated; the intake of cooling water is restricted in such a step-plan, to prevent damage to the ecosystem.

In 2003, the affected sectors included: safety (two peat dikes collapsed due to drying out); energy production (cooling water); agriculture (less yield, not necessarily less income for farmers, as prices often increase due to low availability of products); navigation (restricted passage through locks and one-way traffic on rivers); and nature (e.g. death of birds due to algal growth resulting from high water temperature).

In summary, the summer of 2003 was not an extreme drought period. It nevertheless yielded valuable lessons, including increasing awareness that drought will become a structural problem in the Netherlands due to climate change, and that water policy and planning in general needs investigation. Before 2003, 'drought' had somewhat lost attention (after 1976), also because there were some near-disasters concerning flooding. Nevertheless, since 2003, drought has been on the agenda again; water management requires continuous care, which is necessary to create a sustainable habitat for people, plants, and animals.

3.9.2.6 Responses to past drought events and assessment of their effects on drought impact mitigation

There are significant differences in the regulatory and policy frameworks between the two most severe recent droughts in the Netherlands – the one of 1976 and the one of 2003. The drought in 1976 was a crisis that was addressed with emergency interventions. This led to the bestowing of 'priority ranking' for water use, and the establishing of a drought monitoring system and the LCW. The 'drought season' starts on 1 April, with the calculation of the evaporation surplus, that is, precipitation minus evaporation. There is also continuous monitoring of the temperature and discharge of the rivers (Rhine, Scheldt, Meuse) entering the country. This implies that, from 1 April every year, there is an 'early warning' system in place. When there is an 'upcoming' drought, preventive measures (such as, for example, increasing the levels of freshwater reservoirs) are considered that can be implemented whenever feasible. NCW issues a fortnightly 'drought message' to inform both the stakeholders and the public. These messages (and more) are published on a freely accessible website.

The 'priority ranking' system introduced after the drought in 1976 was not used until 2003. A well-learned lesson from the 2003 drought is that communication with partners and stakeholders is vital to prevent economic losses. Moreover, in 2003, environmental needs were not adequately addressed. This led to a slight adjustment of the 'priority ranking' to address 'irrecoverable damage to nature' within the first priority. This applies especially to valuable peat areas, which experience total destruction once deprived of moisture.

The 2003 drought demonstrated that both the operation and policy of the current water management system were nearing their limits. While the short-term focus is on flexibility in the system (creation of water storage through higher water levels and alternative water supply routes), a fundamentally different approach is required for the long term. Table 3.9.1 shows how the policy and regulatory framework for drought in the Netherlands has evolved over the years. This new approach is in development through the Delta Programme, a national programme in which the national government, provinces, municipalities, and water boards cooperate, with inputs from civil society organisations and the private sector. It has nine subdivisions, of which one is on freshwater. Other subdivisions include safety, construction and restructuring, Lake Ijssel area, rivers, and coastal areas.

In 2007, the Dutch government established a 'water vision' for sustainable and climate-proof water management, which also covered drought issues. Drought studies concluded that large-scale infrastructural measures in the main system to redirect water, as well as the construction of large-scale reservoirs, were neither feasible nor affordable. It was found that, at the local scale, the balance of cost and benefit might be different, implying that solving the drought challenge would be especially a question of 'tailor-made' solutions at a local scale, also taking the conditions of water surplus and water quality into account. The recent work in the Delta Programme built on those findings, leading to the 'service levels' approach.

Table 3.9.1 Key elements of policy and regulatory framework for drought in the Netherlands (De Stefano et al., 2012).

Instrument	Key elements
Drought policy	• *Description*: Formulated as part of the National Water Policy (4th Note) following the 1976 drought • *Objective*: To deal with drought both nationally and regionally; establishes an order of priority for water use • *Year of approval*: Valid since 1985 (latest update in 2009); latest version in the National Water Plan 2010–2015 • *Geographic scope*: Country level (Rhine and Meuse river systems) • *Responsible organisations and involved actors*: Ministry of Infrastructure and Environment, the national government, together with provinces, municipalities, and water boards
LCW	• *Description*: Committee that proposes decisions for changes in national and regional water management during droughts • *Year of approval*: Developed after the 1976 drought, continuous operation since then • *Geographic scope*: Country level (Rhine and Meuse river systems) • *Responsible organisations and involved actors*: Ministry of Infrastructure and Environment for the national system, water boards for regional systems, Association of Water Boards, and representatives of provinces
Legal framework: Law on Safety Regions	• *Description*: Specifies who will have executive powers during a crisis • *Objective*: Aims to regulate responsibility • *Year of approval*: 2010, to date • *Geographic scope*: Country level • *Responsible organisations and involved actors*: Ministry of Home Affairs, Queen's Commissioner, all involved authorities, mayors, etc.
Legal framework: Delta Law	• *Description*: Legal basis for Delta Programme funds, Delta Commissioner, and related programmes • *Objective*: Aims to protect against floods and provide freshwater • *Main policy instruments*: Allocating responsibility and funds • *Year of approval*: 2011, to date • *Geographic scope*: Country level • *Responsible organisations and involved actors*: Ministry of Infrastructure and Environment, and the national and regional governments

3.9.3 Conclusions – Future needs

3.9.3.1 Conclusions

The two last major droughts in the Netherlands occurred in 1976 and 2003. Lessons learnt from their management have led to a drought monitoring and drought management system that fulfils its purpose. There is a priority-based ranking of water allocation during droughts. Also, the water system management and water users meet bi-weekly during droughts and discuss all drought issues and measures that need to be taken. This has led to drought in the Netherlands being managed in a way accepted by the water users and the public.

There are significant differences in the regulatory and policy frameworks between the two most severe recent droughts in the Netherlands. The drought in 1976 was a crisis that was addressed with emergency interventions. This led to the 'priority ranking' of water use and the establishment of a drought monitoring system (LCW).

A major drought can happen any time again. It should be noted that the Netherlands is more vulnerable to drought now than it was in 1976 or 2003, owing to the socioeconomic development that took place over time. Time will tell whether the drought management system, as it is in place at this point in time, will be able to cope with future droughts of greater intensity or frequency than what happened so far. The efforts invested in public campaigns and the facilities that provide publicly available information whenever a drought is building up are expected to improve drought preparedness and management.

3.9.3.2 Preparing for the future through establishing 'service levels'

Overall, the water system management and users are coordinating to face future droughts through preparedness based on 'service levels', which are defined as 'the availability of freshwater in normal and drought conditions in an area'. Governments and users jointly identify these levels, in the form of region-specific agreements, specifying the responsibilities and obligations that the government has in terms of ensuring water availability, and establishing what the responsibilities and remaining risks are for the users. This concerns surface and groundwater, as well as quality and quantity. The planned timeline for detailing the service levels is as follows:

- *2015*: Plan of action made – for the main system – for defining the 'service levels' (i.e. process, starting point, methodology, and relation to other processes)
- *End of 2015*: Plan of approach made – by regions (provinces and water boards) – for the 'service levels' (which areas first, etc.). Roles: who will lead what, etc.
- *End of 2017*: Method available to establish 'service levels' for the first group of areas
- *2018*: Evaluation of applicability and experiences from reality
- *Until end of 2021*: Continued establishment of 'service levels'

3.9.3.3 International dimension

The national government is working towards putting freshwater issues on the international agendas in river committees (Meuse and Rhine). It is deemed important to create support for joint exploration of the freshwater issues faced by the Netherlands. In the Rhine Ministers Conference in 2013, it was agreed to investigate drought as part of a climate adaptation strategy. In the Meuse Committee, some model calculations have been started (Delta Programma, 2013). The Netherlands also actively follows the developments in relevant European guidelines (such as the Blueprint). The approach of the Netherlands for this is twofold – on the one hand, exploiting opportunities for a joint EU approach, while, on the other hand, reducing risks of European developments that would not be in agreement with the national preference strategies.

References

Alderlieste, M.A.A. and Van Lanen, H.A.J. (2012). Trends in low flow and drought in selected European areas derived from WATCH forcing dataset and simulated multi-model mean runoff. *DROUGHT–R&SPI Technical Report No. 1*, 110 pp. Available from: http://www.eu-drought.org/technicalreports.

Alderlieste, M.A.A. and Van Lanen, H.A.J. (2013). Change in future low flow and drought in selected European areas derived from WATCH GCM forcing data and simulated multi-model runoff. *DROUGHT–R&SPI Technical Report No. 5*, 316 pp. Available from: http://www.eu-drought.org/technicalreports.

Alderlieste, M.A.A., Van Lanen, H.A.J., and Wanders, N. (2014). Future low flows and hydrological drought: how certain are these for Europe? In: *Hydrology in a Changing World: Environmental and Human Dimensions* (ed. T.M. Daniell, H.A.J. Van Lanen, S. Demuth, G. Laaha, E. Servat, G. Mahe, J.-F. Boyer, J.-E. Paturel, A. Dezetter, D. Ruelland), 60–65. Montpellier, France: IAHS Publ. No. 363.

Andreu, J., Haro, D., Solera, A., Paredes, J., Assimacopoulos, D., Wolters, W., Van Lanen, H.A.J., Kampragou, E., Bifulco, C., De Carli, A., Dias, S., González-Tánago, I., Massarutto, A., Musolino, D., Rego, F., Seidl, I., and Urquijo Reguera, J. (2015). Drought indicators: monitoring, forecasting and early warning at the case study scale. *DROUGHT-R&SPI Technical Report no. 33*. Available from: http://www.eu-drought.org/technicalreports.

De Stefano, L., Urquijo, J., Acácio, V., Andreu, J., Assimacopoulos, D., Bifulco, C., De Carli, A., De Paoli, L., Dias, S., Gad, F., Haro Monteagudo, D., Kampragou, E., Keller, C., Lekkas, D., Manoli, E., Massarutto, A., Miguel Ayala, L., Musolino, D., Paredes, J., Rego, F., Seidl, I., Senn, L., Solera, A., Stathatou, P., and Wolters, W. (2012). Policy and Drought Responses – Case Study Scale. DROUGHT–R&SPI *Technical Report No. 4*. Available from: http://www.eu-drought.org/technicalreports.

De Wit, M., Buiteveld, H., Van Deursen, W., Keller, F., and Bessembinder, J. (2008). Klimaatverandering en de afvoer van Rijn en Maas (Climate change and discharge of Rhine and Meuse). *Stromingen* 14 (1): 13–24.

Delta Programme (2011). Synthese van de landelijke en regionale knelpuntenanalyses. Delta Programme, deelprogramma Zoetwater (Available in Dutch only). Title in English: Synthesis of the national and regional bottleneck analyses.

Delta Programma (2013). Water voor economie en leefbaarheid, ook in de toekomst. Kansrijke strategieën voor zoet water. Bestuurlijke rapportage fase 3. (available in Dutch only). Title in English: Water for economy and quality of life in the future. Promising strategies for freshwater. Administrative reporting phase 3.

Janssen, L.H.J.M., Okker, V.R., and Schuur, J. (2006). Welvaart en Leefomgeving, een scenariostudie voor Nederland in 2040. Centraal Planbureau, Milieu- en Natuurplanbureau en Ruimtelijk Planbureau (Available in Dutch only). Title in English: Prosperity and living environment, a scenario-study for The Netherlands in 2040.

Klijn, F., Kwadijk, J., De Bruijn, K., and Hunink, J. (2010). Overstromingsrisico's en droogterisico's in een veranderend klimaat, verkenning van wegen naar een klimaatveranderingsbestendig Nederland (Available in Dutch only). Title in English: Flood and drought risks in a changing climate, exploring ways to a climate-proof Netherlands. Deltares, Delft.

KNMI (2015a). Climate Atlas. Royal Meteorological Institute. De Bilt, The Netherlands: KNMI (www.klimaatatlas.nl, accessed: 15 September 2015).

KNMI (2015b). Web-based information portal on precipitation deficit throughout the years, freely available information for the public. https://www.knmi.nl/nederland-nu/klimatologie/geografische-overzichten/historisch-neerslagtekort

KNMI & PBL (2015). Klimaatverandering. Samenvatting van het vijfde IPCC assessment en een vertaling naar Nederland (Climate Change. Summary of the 5th IPCC Assessment and an Interpretation for The Netherlands). Den Haag/De Bilt: PBL/KNMI.

Kwadijk, J., Jeuken, A., and Van Waveren, H. (2008). De klimaatbestendigheid van Nederland Waterland. Verkenning van knikpunten in beheer en beleid voor het hoofdwatersysteem (Available in Dutch only). Title in English: 'Climate proofing of The Netherlands. Exploring tipping points in management and policy for the main water system'. Deltares, Delft.

Van Den Hurk, B., Siegmund, P., and Tank, A.K. (Eds) (2014). KNMI'14: Climate Change scenarios for the 21st Century – A Netherlands perspective; by Jisk Attema, Alexander Bakker, Jules Beersma, Janette Bessembinder, Reinout Boers, Theo Brandsma, Henk van den Brink, Sybren Drijfhout, Henk Eskes, Rein Haarsma, Wilco Hazeleger, Rudmer Jilderda, Caroline Katsman, Geert Lenderink, Jessica Loriaux, Erik van Meijgaard, Twan van Noije, Geert Jan van Oldenborgh, Frank Selten, Pier Siebesma, Andreas Sterl, Hylke de Vries, Michiel van Weele, Renske de Winter, and Gerd-Jan van Zadelhoff. *Scientific Report WR2014-01, KNMI*, De Bilt, The Netherlands. www.climatescenarios.nl.

3.10
Improving Drought Preparedness in Portugal

Susana Dias, Vanda Acácio, Carlo Bifulco, and Francisco Rego
Centro de Ecologia Aplicada "Prof. Baeta Neves" (CEABN), Instituto Superior de Agronomia, Universidade de Lisboa, Lisboa, Portugal

3.10.1 Local context

3.10.1.1 Climate and land use

Located in the Iberian Peninsula (southwestern Europe), mainland Portugal has a mild Mediterranean climate, with mean annual temperature varying between 7°C in the inner central highlands and 18°C in the southern coastal region (Figure 3.10.1). On average, about 42% of the annual precipitation occurs during the 3-month winter season, while only 6% of annual precipitation falls between June and August (de Lima et al., 2013). The highest values (above 3000 mm) occur in the highlands of the Atlantic northwest region, and the lowest values (around 500 mm) in the southern coast and in the Continental eastern part of the territory (bellow or around 500 mm) (Miranda et al., 2002). The geographic location of mainland Portugal is favourable to the occurrence of regular drought episodes, especially in the south (Pires et al., 2010; Santos et al., 2010). A major forcing factor for precipitation in Portugal, and therefore an important indicator of drought magnitude and duration, is the North Atlantic Oscillation (NAO), with a clear spatial gradient of increasing influence from north to south (Trigo et al., 2004).

Forests and agricultural land dominate as land cover in Portugal (Figure 3.10.1). Trends from the last three decades show an increase of forest area (3.1%), uncultivated soils (0.9%), artificial land (2%), and water bodies (0.3%) (Vale et al., 2014). The latter have registered the largest growth: a total of 18000 ha between 2000 and 2006, mainly due to the construction of the large Alqueva Dam (Guadiana River, southern Portugal). Artificial areas increased – due to urban sprawl in coastal areas, illegal dwellings, and tourism pressure – at the expense of agriculture and forest land (Vale et al., 2014). On the other hand, agriculture abandonment took place in marginal areas with scarce water resources, favouring forest and shrub encroachment (EEA, 2006). Intensification

Drought: Science and Policy, First Edition.
Edited by Ana Iglesias, Dionysis Assimacopoulos, and Henny A.J. Van Lanen.

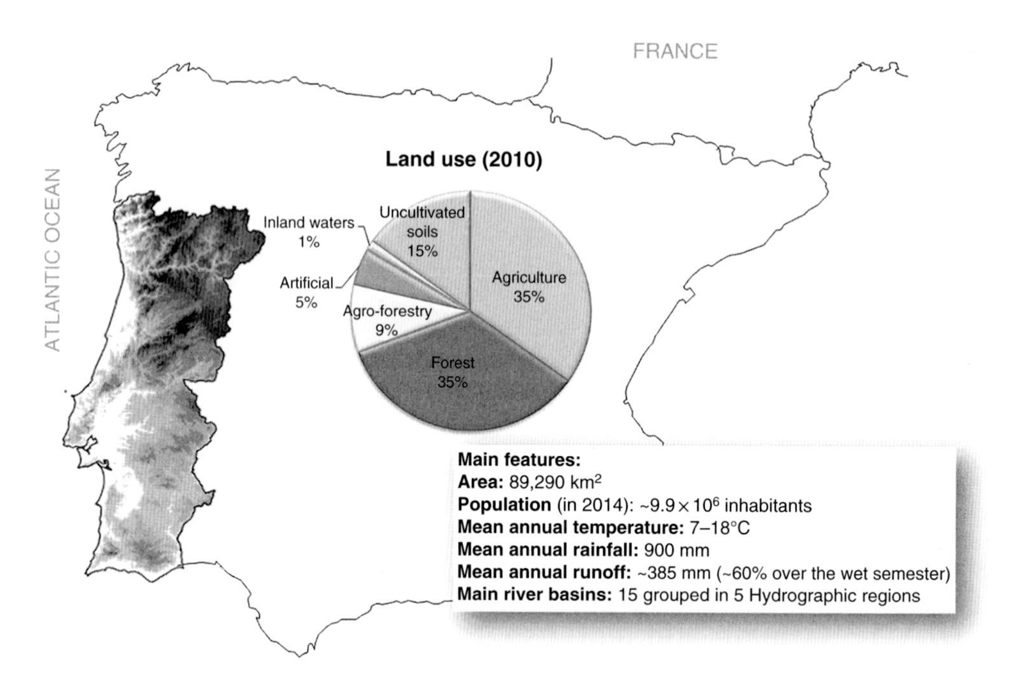

Figure 3.10.1 Elevation map of mainland Portugal; main features and land use composition. *Source*: Pie chart adapted from Vale et al. (2014). (*See colour plate section for the colour representation of this figure.*)

of agriculture was also a significant process in areas with water availability, as a response to EU policies and market incentives (Diogo and Koomen, 2010). Currently, olive orchards and vineyards represent 66% of permanent crops, and about 35% of the total arable land is irrigated (MAOTDR, 2007).

In the next decades, urban sprawl and soil sealing are expected to increase in Portugal (Rounsevell et al., 2006), particularly in the northern regions and around Lisbon (10–30%). Conversely, the land occupied by agriculture is expected to continue on the downwards trajectory throughout the country (EC, 2007), whereas a reduction in forested land is only forecast for the Lisbon region (Vale et al., 2014).

3.10.1.2 Water resources – use and consumption

Portuguese freshwater resources fulfil the national demand, and about 80% of water resources remain unexploited. Of the remaining 20% that are exploited, 54% is groundwater, and the rest is surface water (PCM, 2005).

Water use in Portugal equals 167 litres per person per day. Agriculture is the sector with the highest level of water use (estimated to be 75% of total demand, ~6.55 × 10^9 m^3/year), but it is also responsible for 44% of the water returns to the system (INAG, 2001). About half of the Portuguese farms are irrigated (54%, equivalent to 92 000 farms), with a total irrigated surface of 464 627 ha. Water for irrigation comes mostly from underground aquifers (82%), and irrigation systems are predominantly privately owned. The energy sector is also responsible for a significant share of water use (14%, mainly derived and thus not consumed), corresponding to 27% of the water returns to the system (INAG, 2001). The remaining 10% of the exploited water is used for urban supply (6%) and industry (4%). Installed hydropower capacity is about 4406 MW, and annual production is 4533804 MWh (2005 data).

The share of distributed water for consumption by main sectors varies between northern and southern river basins (Figure 3.10.2). In the northern basins, values are basically shared between urban supply and industry, given that other sectors extract water directly from the origin (river, wells, etc.). In the southern basins (mainly in the Sado, Mira, and Guadiana rivers in Alentejo), agriculture, supported by extensive irrigation infrastructures, is the sector with the highest consumption (APA, 2012).

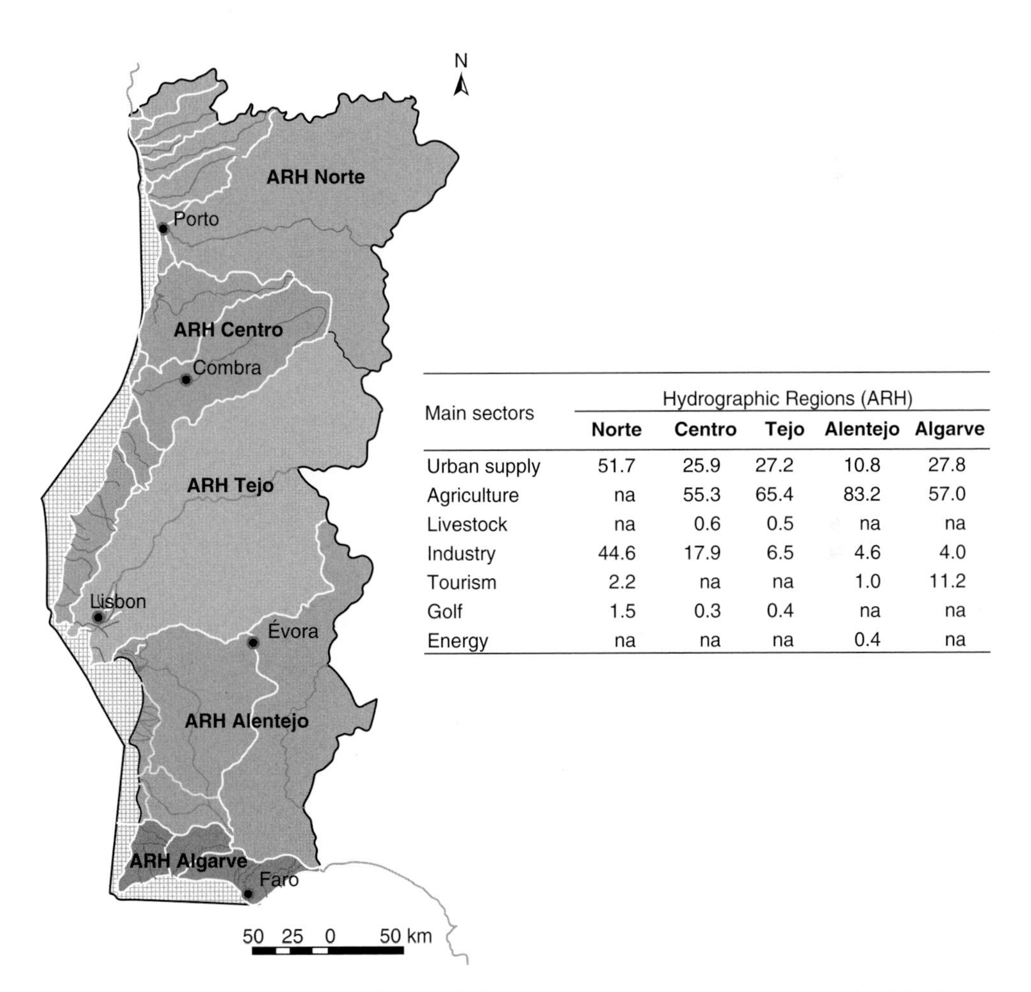

Main sectors	Hydrographic Regions (ARH)				
	Norte	Centro	Tejo	Alentejo	Algarve
Urban supply	51.7	25.9	27.2	10.8	27.8
Agriculture	na	55.3	65.4	83.2	57.0
Livestock	na	0.6	0.5	na	na
Industry	44.6	17.9	6.5	4.6	4.0
Tourism	2.2	na	na	1.0	11.2
Golf	1.5	0.3	0.4	na	na
Energy	na	na	na	0.4	na

Figure 3.10.2 Share of distributed water (%) consumed by the main sectors in each of the hydrographic regions (ARH) in Portugal (map on the left). Negligible amounts (na) for livestock were included in the agriculture sector, for tourism included in urban supply, and golf included in tourism. *Source*: Adapted from APA (2012).

In 2000, the market value of water supply was of €2 billion/year, which is equivalent to about 2% of the Portuguese economy. Water supply derives from freshwater abstraction (837×10^6 m^3, 69% from surface water and 31% from groundwater), and freshwater treatment (756×10^6 m^3, 74% from water treatment plants and 26% from chlorine (bleaching) stations). The water supply system is divided into municipalities, which serve the population (36%), municipal utilities (29%), concessionaires (24%), and local companies (11%).

Diffuse pollution and industrial contamination occur in all river basins, although with differing intensities (INAG, 2001; APA, 2009; Pereira et al., 2009), and some aquifers are overexploited, particularly in the south. Furthermore, water use inefficiency in Portugal was about 41% of total withdrawals in 2000 (INAG, 2001). Nevertheless, water consumption efficiency gains in the agricultural sector have reached 46% (15×10^3 m^3/ha/year) in the last decades (Nuncio and Arranja, 2011), and, according to the Portuguese National Program for the Efficient Use of Water (PNUEA), a more ambitious target has been set for attaining short-term efficiency (80% improvement) in water use for human consumption (PCM, 2005).

Portugal intends to increase hydroelectricity production from 46% to 70% (~7000 MW) of the hydrological potential by 2020 (Cortes, 2013). According to projections made under the river basin management plans (APA, 2012), the level of water consumption for the two most water-demanding sectors (agriculture and urban supply) will be maintained or decreased in the future, with the exception of the southern basins. Tourism and urban sprawl will put extra pressure on these southern basins, particularly where overexploitation and contamination of groundwater resources is already a reality (Stigter et al., 2013). Moreover, irrigated agriculture surface will continue expanding, profiting mainly from the Alqueva Dam reservoir and its subsidiaries' infrastructures, at the expense of extensive agriculture and more environmentally friendly and drought-resistant land use systems (such as the Iberian 'montado') (Máñez Costa et al., 2011; Jongen et al., 2013), which will likely increase conflicts.

3.10.1.3 Past drought events, impacts, and forecasted trends

In the last seven decades, there were 12 major drought events experienced in mainland Portugal (Figure 3.10.3); the last one occurred in 2017. Severe or extreme droughts exhibited a 10–15-year return period, lasted for 1–3 years, and affected more than half of the territory (Pires et al., 2010; Sousa et al., 2011). In addition, analysis of events from the last century shows an increase in drought frequency and intensity since the 1980s, particularly from February to April (Silva et al., 2014; Vicente-Serrano et al., 2014).

The drought episode of 2004–2006 can be considered the most severe of all drought events recorded in Portugal in the last 75 years, in terms of meteorological data, extent of the area affected (100% of mainland Portugal was affected), consecutive months with severe drought index (Palmer drought severity index, or PDSI), and impacts on different socioeconomic and environmental sectors (MAMAOT, 2013a). By the end of hydrological year 2005 (30 September), 97% of the territory was still in severe or

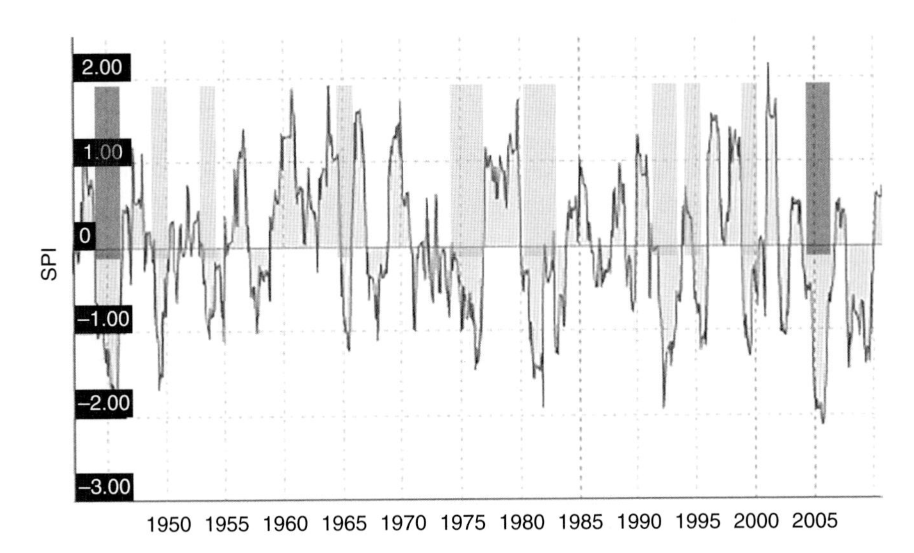

Figure 3.10.3 Six decades of annual precipitation anomalies estimated by the standardised precipitation index (SPI) for mainland Portugal, regarding the reference period of 1970–2000 (IPMA, 2015). Major drought events are highlighted: red – extreme; yellow – moderate or severe. (*See colour plate section for the colour representation of this figure.*)

extreme drought (CPS, 2006). The severity and extent of the 2004–2006 drought episode resulted in the decrease of water storage for domestic and industrial consumption, energy, and irrigation of agricultural crops, as well as other impacts of different nature. The long period without soil humidity during 2005 also affected non-irrigated crops, causing severe reduction in wheat yield (decrease of more than 50% when compared with the previous 15 years) (Gouveia et al., 2009). The water levels of important aquifers decreased in comparison with average values (1987–2005), and, in some aquifers (e.g. Querença-Silves, the most important aquifer in Algarve), the water level reached the lowest ever recorded values. Also, in Algarve, two reservoirs were completely depleted of water. In August 2005, about 100 000 people were using water from alternative systems (CPS, 2006). In some rivers, fish populations were severely depleted due to very low water levels, and a recent study indicates that the 2004–2006 drought was the main driver of a significant crash in southern steppe bird populations (Moreira et al., 2012). Regardless of the efforts, the summer of 2005 had been the second worst in recorded history until then, with 325 000 ha burned in forest fires (Pereira et al., 2013). Very dry forests and difficulties in water availability for firefighting resulted in much more intense fire behaviour (CPS, 2006; Gouveia et al., 2012). The total direct costs associated with the 2004–2006 drought amounted to about €286 million (CPS, 2006), the largest portion being associated with reduced energy production (€182 million, without considering CO_2 emissions).

Climate change precipitation models predict a drier climate in Portugal, with a shorter and wetter rainy season, followed by a long dry summer (Santos et al., 2002; MAMAOT, 2013b). The increasing trend in temperature (Ramos et al., 2011) and the decreasing trend in precipitation will significantly influence the frequency and severity of drought events in Portugal (Vicente-Serrano et al., 2014; van Lanen et al.,

2013). When in drought, seawater intrusion is likely to occur mainly in the south, as a consequence of increased pumping rates near the coast and lower recharge (Ribeiro, 2013). A future significant reduction in precipitation and groundwater resources, foreseen mainly for the southern river basins, coupled with an increase in water demands, will boost regional and seasonal asymmetry in water availability, and thus result in higher exposure and sensitivity to droughts (Pereira et al., 2009; MAMAOT, 2013b).

3.10.1.4 Lessons learned from the 2004–2006 drought event

The 2004–2006 drought was an opportunity for evaluating and improving the efficacy of measures to avoid, mitigate, and compensate drought impacts. The importance of this extreme event is evident from the creation, by governmental initiative in March 2005, of the Drought Commission, in charge of monitoring the progress of the drought, and assisting in mitigating its effects. With a multidisciplinary and pluri-sectoral composition, this commission improved circulation of data and information, and facilitated agreements between different sectors regarding water use (Do Ó and Roxo, 2009). Also, during the summer of 2005, Portugal and Spain came together to create a commission to manage the reservoirs located at river basins shared by the two countries, and to promote relevant exchange of information. The drought of 2004–2006 was therefore very illustrative of the importance of cooperation among national entities under the coordination of the National Water Authority (currently the Portuguese Environment Agency, or APA, of the Ministry of Agriculture and Sea), and between countries. Moreover, since the drought episode of 2004–2006, many water supply entities started implementing contingency plans, and information exchange between Portugal and Spain on shared water resources has been consolidated. Table 3.10.1 shows some of the measures and processes that contributed most to minimising drought impacts during the extreme drought event of 2004–2006.

Table 3.10.1 Measures and processes that contributed most to minimising drought impacts during the 2004–2006 drought. *Source*: Adapted from De Stefano et al. (2012).

Measure or process
Creation of a Drought Commission (concerted group of stakeholders for decision-making)
Law changes (exceptional and transitional arrangements)
Sensitisation campaigns for responsible water use in urban areas
Increased water pricing in some irrigation perimeters (e.g. Algarve)
Investments made to improve efficiency and reduce losses of urban water supply
Use of treated wastewater for garden irrigation and to reinforce irrigation
Good understanding between Portuguese and Spanish governments
Regulation of all public water supply entities to improve efficiency, and water loss reduction of distribution networks

3.10.2 Current approach to drought monitoring and management

3.10.2.1 Drought monitoring systems

Currently, there are two drought monitoring systems in Portugal managed by two different entities: (i) the *Drought Observatory* ('Observatório de Secas'); and (ii) the *National Information System for Water Resources* ('Sistema Nacional de Informação de Recursos Hídricos', or SNIRH), both of which are discussed in the following text.

The Drought Observatory was established in 2009, and is coordinated by the Portuguese Sea and Atmosphere Institute (IPMA, Ministry of Agriculture and Sea). Data (precipitation and temperature) is collected by a national network of meteorological stations. The occurrence of meteorological droughts is based on calculations of SPI and PDSI, averaged for main river basins and for mainland Portugal. The occurrence of agricultural droughts are calculated based on soil water content (%). Drought is forecast with the PDSI for the following month based on three scenarios of precipitation occurrence: (1) lower than average, precipitation values only attained on 20% of the years (decil 2); (2) precipitation values occurring on 50% of the years (decil 5); (3) values above average, only attained in 20% of the years (decil 8). Since 2002, IPMA also calculates the fire weather index (FWI) daily for Portugal. The Drought Observatory can be consulted online at http://www.ipma.pt/pt/oclima/observatoriosecas, and by using their monthly climate reports.

The National Information System for Water Resources (SNIRH) is coordinated and managed by the APA since 1995. SNIRH monitors water resources with four national networks of stations, including meteorological (742 stations), hydrometric (419 stations), superficial water quality (275 stations), and groundwater (1487 stations) networks. Collected data can be freely consulted online and downloaded from http://snirh.pt. A subsystem for Drought Monitoring and Early Warning ('Sistema de Vigilânca e Alerta de Secas', or SVAS) was developed by the APA within SNIRH. Meteorological and hydrological droughts are monitored based on precipitation, streamflow, reservoir storage, groundwater storage, and water quality, and the relevant data is presented at http://snirh.pt. This information (including maps, graphs, and tables) is published in monthly reports on hydric resources edited by the APA.

3.10.2.2 Existing framework for drought management

There is no officially approved drought declaration process in Portugal, where droughts have been managed as a crisis event. From the 2004–2006 drought onwards, the process adopted has been as follows: when a meteorological drought is detected by IPMA, this entity informs the Drought Commission for Reservoir Management, which analyses the situation, based essentially on the water storage levels in some reservoirs, particularly multipurpose reservoirs. The Drought Commission for Reservoir Management is a national permanent commission that follows drought/flood situations through periodical meetings and includes important stakeholders such as reservoir main user entities, the National Authority for Civil Protection (ANPC), the General Directorate of Agriculture and Rural Development (DGADR), the Regulatory Entity for Water and Waste (ERSAR), and the Institute for Nature Conservation and Forests (ICNF).

If a drought is detected, this commission makes a recommendation to the government to declare a state of drought and draw up a Drought Monitoring and Impact Mitigating Program. After governmental approval, an institutional solution for managing the drought is established, which comprises two action levels: (i) the Drought Commission for political and strategic issues; and (ii) a working group for technical and operational issues. The organisational model is focused on the permanent availability of information to all authorities, economic agents, and the public in general, using information and communications technology. A website is set up on the Internet and managed for this purpose. All the available information from monitoring is used to support planning and decision-making.

The drought evolution is evaluated with real-time data based on the quantification of water availability in rivers, reservoirs, and groundwater, as well as the water requested by the different users. The water storage in reservoirs and aquifers is subjected to detailed monitoring during the drought, and appropriate measures are taken according to three levels of severity: A1 – Pre-alert, corresponding to moderate drought; A2 – Alert, when severe drought is detected and A3 – Emergency, for extreme drought. Measures vary from information campaigns on water saving and animal fodder storage, to careful analysis of new water abstraction applications, irrigation restrictions, namely for permanent crops, or fish removal in reservoirs. At the highest level, exceptional measures, such as water transfer between reservoirs, or building of emergency infrastructures and new wells, can be implemented according to current needs. When the results of drought monitoring show a normal situation based on precipitation and reservoir storage levels, the Drought Commission proposes a declaration of the end of the drought, which also needs to be approved by the government. The Drought Commission ceases its activity with the release of a drought balance report that contains the main results and lessons obtained.

3.10.3 Improving drought preparedness and drought management

3.10.3.1 Stakeholder involvement in drought management

Stakeholder involvement in the water management process has so far been limited to consultation in various levels of water management initiatives. This situation is, however, gradually changing, owing to commitments from different sectors, both public and private, to reduce costs and enhance resource use in the coming decades. When a drought is declared, representatives of different economic sectors cooperate with state institutions and local administration in technical working groups and drought commissions. Information transfer is facilitated by these representatives to other local end-users. Meanwhile, most of the end-users (e.g. public/private companies for energy and water supply, and farmers' associations) have been developing strategies to cope with climate change and related events such as drought. Some are engaged in R&TD projects aimed at improvements in water conservation and water reuse (e.g. Jacinto et al., 2011; Nuncio and Arranja, 2011; Jongen et al., 2013), or building drought coping capacity, particularly in southern river basins (Jacobs et al., 2008; Máñez Costa et al., 2011; Stigter et al., 2013; Ribeiro, 2013). Under the FP7 DROUGHT–R&SPI project, about

20 of those stakeholders were involved in the process of evaluating past impacts and responses to drought, identifying major past and future vulnerabilities, and in selecting best measures to cope with future droughts (Kampragou et al., 2015). The outcomes of this approach allowed a commonly accepted framework for improving drought preparedness in Portugal.

Socioeconomic factors, more than climate change, will likely be the main drivers of drought vulnerability in the next decades, particularly in the southern basins, owing to their dependence on the agriculture and tourism sectors. The biggest challenge is thus to be able to provide water where and when it is needed. Therefore, stakeholder involvement in water and drought management is and will be of major importance.

3.10.3.2 Vulnerability to drought: Analysis of SPIs as indicators of future drought impacts

According to stakeholder perceptions, most vulnerability factors in societal, technical/economic, and institutional/political dimensions will eventually decrease in importance in the future, whereas environmental factors are perceived to sustain or increase their importance in the coming decades (Kampragou et al., 2015). The accrued relevance of environmental drivers was seen as the result of an external evolution (such as climatic change), not under the control of national or international institutions, coupled with the low impact or sensitivity of the policy implemented, such as control of diffuse pollution or of water abstraction.

The most important drought impacts reported for past events in mainland Portugal have been agricultural losses (particularly a reduction in rain-fed crop production), interruptions in public water supply, and wildfires. Although water supply systems for human use and agriculture are foreseen to improve, both agricultural losses and wildfire extent are expected to increase in magnitude in the future (Bedia et al., 2014; Sousa et al., 2015). A way of increasing drought preparedness is to identify specific periods where water shortage is crucial for maximising impacts. Exploring the links between meteorological drought indicators (such as SPI) and variables expressing drought impacts was considered to be a step towards the improvement of an early warning system for seasonal drought impacts in Portugal, and other countries sharing similar climatic conditions.

The approach used for Portugal, relating long-term data series of crop production and wildfire extent (extracted from EUROSTAT and European Fire Database, respectively) with SPIs (from WATCH forcing data methodology applied to ERA-interim data), revealed that different drought accumulation periods (SPI-1–6) over various time scales could play an important role in explaining changes in annual burned area or crop production (Bifulco et al., 2014; Rego et al., 2015). For example, winter rain was positively associated with larger burned areas, since water surplus by the end of winter benefits the accumulation of biomass, which burns easily during the Mediterranean summer (Pellizzaro et al., 2007; Gouveia et al., 2012). Thus, the assessment of SPI-1 for February can provide important warning information to decide whether or not to enact measures, in the subsequent 4 months, to reduce herbaceous biomass, in order to decrease wildfire risk (Bifulco et al., 2014). Furthermore, an accumulated deficit of precipitation from May to July (SPI-3) is a good predictor of wildfire extent for Portugal and other European regions (Stagge et al., 2014). It can thus be used in real time, yet giving some forecast lead time, to evaluate possible major risks of forest fires and to engage in field measures to implement in the forthcoming months.

Significant correlations between SPIs and crop yield anomalies were also found, although they varied markedly according to crop type. Rainfed winter cereals were negatively affected by winter precipitation and positively by spring rain (Table 3.10.2). According to the modelling results, wheat and olives will experience the largest changes in production due to drought. These crops are mainly concentrated in the south of the country, and are currently subjected to external drivers such as CAP subsidies (Rego et al., 2015).

Understanding drought consequences on Portuguese producers and consumers of these crops has to include a study of the relationships between crop yields and prices. As expected from economic theory, the identified changes in crop yield (Table 3.10.2) were positively related to changes in price (on any year in relation to the previous year). This elasticity of supply was especially important for maize, wheat, and, to a lesser extent, for rice. In addition, the relative changes in price tend to show weak negative correlations with the relative changes in quantity, and this correlation was more marked (higher elasticity of the price) for maize than for wheat and rice. In any case, the increase in price in the years when crop yield was low (in those affected by drought episodes) only partly compensates the producer for the yield losses (4.5% for rice, 11.5% for wheat, and 29.1% for maize), and these price increases were borne by the consumer. The study carried out for major crops in Portugal led to the conclusion that, in general, producers react rapidly and tend to adjust their crop yields to the changes in the previous prices of the crops. Furthermore, the sign and magnitude of drought impact would differ between crops that have a global market and those crops associated with small local markets (Rego et al., 2015).

Table 3.10.2 Changes in crop production for SPI = −1 (severe droughts) and SPI = −2 (extreme droughts), predicted with regression modelling (GLM). When the most parsimonious model selected two variables, the effect of each is presented separately. *Source*: Adapted from Rego et al. (2015).

Crop	Selected predictor	Anomaly production (%) when	
		SPI = −1	SPI = −2
Wheat	SPI-3 December (with SPI-3 May = 0)	18	36
	SPI-3 May (with SPI-3 December = 0)	−16	−32
Rye	SPI-3 March	8	15
Grain maize	SPI-2 May	−3	−6
Potato	SPI-1 August	−8	−16
Rice	SPI-3 August	−7	−14
Olive trees	SPI-6 $\text{November}_{\text{previous}}$ (with SPI-1 August = 0)	22	45
	SPI-1 August (with SPI-6 $\text{November}_{\text{previous}}$ = 0)	−14	−29
Vineyards	SPI-2 August (with SPI-2 June = 0)	−6	−11
	SPI-2 June (with SPI-2 August = 0)	6	12

3.10.3.3 Strengthening national drought information systems

Current outputs from the monitoring systems are of good quality and have been efficiently used to monitor the last severe drought events (2004–2005 and 2012). However, the operability of drought monitoring is limited by the lack of information regarding several exploitation systems; there is also an incomplete record of groundwater abstractions and volumes, particularly for private irrigated systems. After 2008, the regular maintenance of monitoring stations decreased sharply due to lack of funds, and about 70% of the APA's monitoring network stopped functioning, particularly hydrological and water quality stations. Furthermore, SNIRH is more oriented towards flood evaluation, thus not tailored to provide detailed data allowing drought evaluation with a single and common classification system across the country. In addition, current indicators do not include socioeconomic impact assessments.

A system for early warning and drought management based on evaluation indicators is currently being developed by the APA, but has not been applied yet (Do Ó, 2011; Vivas and Maia, 2011). A study that aims to create and calibrate a global indicator to identify droughts is also underway, using not only meteorological and hydrological data but also drought impact indicators. However, this global indicator has not yet been officially published and applied.

Specific measures to strengthen national drought information systems should include:

1. Improvement of monitoring programs on available water resources using reliable and well-structured networks, particularly for groundwater and private irrigated areas
2. Establishment of monitoring programs with well-structured information networks for transboundary waters
3. Development of a permanent system for drought forecasting, early warning, monitoring, and for providing real-time information on water availability and water requirement for each sector, regarding both quantity and quality
4. Definition of thresholds indicators for drought onset and ending, and of the institutions responsible for their implementation
5. Linking drought indices with drought impacts to develop a global drought indicator (e.g. a composite socioeconomic index) that is useful to evaluate risks related to drought
6. Accounting for regional and local specificities when developing drought indicators

3.10.3.4 Policy gaps and measures to improve drought preparedness and management

In Portugal, current drought management is still based on a crisis management approach. Until 2017[1] there was no policy or plan developed specifically for droughts, legislation was outdated or absent, and river basin management plans did not

[1] On June 2017, the Council of Ministers, through Resolution nr 80/2017, created a permanent commission for prevention and monitoring of drought and drought impacts. This inter-ministerial commission is technically assisted by a working group that remains active even after and between drought events.

include drought management plans. Drought planning was addressed through a set of policy and management instruments. In addition, drought management–related topics were dispersed in several regulations for the different affected sectors (such as urban supply and irrigated agriculture), and across numerous municipal contingency plans (Table 3.10.3) (Do Ó, 2011). Furthermore, measures established for the protection and sustainable use of Portuguese–Spanish river basins (transboundary waters) are very generic. Also, they lack a harmonisation of methodologies – based on constant exchange of information, establishment of common objectives, monitoring networks, and socioeconomic indicators – for water resources monitoring and management.

Nevertheless, a proactive approach based on drought preparedness and long-term risk reduction has received increasing attention, particularly after the 2004–2006 drought (Afonso, 2007). Several national programs attest the need for an action plan to enhance drought preparedness and list adaptation measures focusing on increasing water reservoir capacity, ensuring planning and legal measures, and improving water use efficiency (PCM, 2005; MAMAOT, 2013a, 2013b). Following the 2012 drought, a permanent working group was established to create a drought action plan[2] observing three operational levels: prevention; monitoring and emergency; and definition of entities responsible for their implementation. According to the outcomes of the DROUGHT–R&SPI project, the action plan should elaborate on the following long-term options to reduce drought sensitivity, build coping capacity, and improve operational drought management in Portugal:

1. Contingency plans for each public water management entity (with a harmonised approach), supervised by the Commission for Reservoir Management to ensure a coordinated approach and common criteria

Table 3.10.3 Main policy and management instruments that cover drought planning in Portugal. *Source*: Adapted from Do Ó (2011).

Instrument	Scale
Convention on cooperation for the protection and sustainable use of Portuguese–Spanish river basins (1998)	Portuguese–Spanish river basins
National Water Plan (2002)	Country
National Water Law (2005)	Country
Drought Commissions (2005 and 2012)	Country
National Program for Efficient Water Use (2006)	Country
River Basin Plans (2001) and River Basin Management Plans (2011–2012)	River basins
Municipal contingency plans	Municipality
Regulations for irrigated areas	Local
Water-specific management plans	Local
National Strategy for Climate Change adaptation – ENAAC (2010)	National
Strategic Plan for water supply and treatment of wastewater (PEAASAR 2200/06 and 2007/12)	Country

[2] A plan for prevention, monitoring and contingency for drought situations was finally approved in August 2017, after the completion of the DROUGHT-R&SPI Project.

2. A legal definition of exploitation regimes for reservoirs with quantification of water uses to ensure ecological flows at international basins and improve mitigation/management of conflicts during droughts
3. Establishment of a central organisation to elaborate adaptive management plans under increased drought frequency and intensity; such plans should clearly define levels of action, articulation, and responsibility of public administration, ensuring the pre-operationalisation of measures that need to be implemented when in drought
4. Improvement of water-saving best practices for all sectors, including reuse of treated wastewater for irrigation and recycling of industrial waste waters, to complement water conservation measures
5. Improvement of water efficiency and water-saving programs – including: (i) artificial recharge of overexploited aquifers in rainy years; (ii) reducing water leaks in urban distribution systems; and (iii) using water in agriculture more efficiently (e.g. through new farming practices and selection of less-water-demanding crops) – and knowledge dissemination on these subjects, especially in the regions more affected by drought
6. Implementation of a national investment plan to build 'green' and 'blue' infrastructures, taking advantage of investment priorities established in programmes using the European Cohesion Fund and European Regional Development Fund.

Both water-saving and water conservation options should be supported by an effective monitoring program during the different stages of implementation, allowing its continuous evaluation and improvement.

3.10.4 Conclusions

Drought in Portugal has been managed as a crisis, and the measures adopted to overcome past drought events have been mostly oriented to agricultural and urban supply sectors, usually with high economic and environmental costs (such as improvement and construction of water reservoirs). Increasing water supply during drought has to be addressed as a long-term strategy, particularly in water-scarce regions. Such a strategy also needs to take into consideration specific environmentally oriented measures, promoting best farming practices and clean technologies within water-saving and water efficiency programs. Investment in overcoming institutional and policy gaps in the frame of water management instruments is, therefore, crucial. Furthermore, policies currently fomented by the European Commission to protect and preserve 'green and blue infrastructures' (EC, 2012, 2013) should be evaluated and explored for drought mitigation, at least in a water conservation framework strategy.

Acknowledgements

This research was undertaken as part of the European Union (FP7)–funded integrated project DROUGHT–R&SPI ('Fostering European Drought Research and Science–Policy Interfacing'), contract no. 282 769. The authors are particularly thankful to all the stakeholders involved in the project, representing six national and regional public authorities, four water-user entities and federations, two risk expert institutions, and two environmental NGOs.

References

Afonso, J.R. (2007). Water scarcity and droughts: main issues at European level and the Portuguese experience. In: *Water Scarcity and Drought – A Priority of the Portuguese Presidency* (ed. MAOTDR), 93–103.

APA (2009). *Relatório do Estado do Ambiente 2009. Agência Portuguesa do Ambiente.* [Online] Available from http://sniamb.apambiente.pt/infos/geoportaldocs/REA/rea2009.pdf (accessed 10 December 2010).

APA (2012). Regional hydrographic management plans. *Final Technical Report*. Part 4 – Prospective scenarios MAMAOT. [Online] Available from http://www.apambiente.pt/?ref=16&subref=7&sub2ref=9&sub3ref=834 (accessed 10 December 2012).

Bedia, J., Herrera, S., Camia, A., Moreno, J.M., and Gutiérrez, J.M. (2014). Forest fire danger projections in the Mediterranean using ENSEMBLES regional climate change scenarios. *Climatic Change* 122 (1–2): 185–199.

CPS (2006). *Relatório de Balanço. Seca 2005*. Comissão para a seca. [Online] Available from http://www.drapc.min-agricultura.pt/base/geral/files/relatorio_seca_2005.pdf (accessed 10 December 2010).

Bifulco, C., Rego, F.C., Dias, S., and Stagge, J.H. (2014). Assessing the association of drought indicators to impacts. The results for areas burned by wildfires in Portugal. In *Advances in Forest Fire Research* (ed. D.X. Viegas), 1054–1060. Coimbra, Portugal: Imprensa da Universidade de Coimbra.

Cortes, R. (2013). *River Regulation and Climate Change in Portugal: New Challenges to Preserve Biodiversity*. Paper presented to 3rd Workshop DROUGHT–R&SPI 11 October 2013 ISA, Lisbon.

de Lima, M.I.L.P., Espírito Santo, F., Ramos, A.M., de Lima, J.L.M.P. (2013). Recent changes in daily precipitation and surface air temperature extremes in mainland Portugal, in the period 1941–2007. *Atmospheric Research* 127: 195–209.

De Stefano, L., Reguera, J.U., Acácio, V., Andreu, J., Assimacopoulos, D., Bifulco, C., De Carli, A., De Paoli, L., Dias, S., Gad, F., Monteagudo, D.H., Kampragou, E., Keller, Lekkas, D., Manoli, E., Massarutto, A., Ayala, L.M., Musolino, D., Arquiola, J.P., Rego, F., Seidl, I., Senn, L., Solera, A.S., Stathatou, P., and Wolters, W. (2012). Policy and drought responses – case study scale. *DROUGHT–R&SPI Technical Report No. 4.*

Diogo, V. and Koomen, E. (2010). Explaining land-use change in Portugal 1990–2000. *13th AGILE International Conference on Geographic Information Science*, 1–11. Guimarães, Portugal.

Do Ó, A. (2011). Gestão transfronteiriça do risco de seca na bacia do Guadiana: análise comparativa das estruturas nacionais de planeamento. *VII Congreso Ibérico sobre Gestión y Planificación del Agua 'Ríos Ibéricos +10. Mirando al futuro tras 10 años de DMA'*, Talavera de la Reina. [Online] Available from http://www.fnca.eu/images/documentos/VII%20C.IBERICO/Comunicaciones/A8/02-DoO.pdf (accessed 10 December 2010).

Do Ó, A. and Roxo, M.J. (2009). Drought response and mitigation in Mediterranean irrigation agriculture. In: *Water Resources Management*, 515–524. WIT Press, Southampton.

EC (2007). *Scenar 2020* – Scenario study on agriculture and the rural world. *European Communities Report*, p. 232.

EC (2012). *Blue Infrastructure (BI) Blue Growth Opportunities for Marine and Maritime Sustainable Growth.* EU Commission Communication. Brussels, COM (2012) 494 Final, 13.9.2012.

EC (2013). Green Infrastructure (GI) – Enhancing Europe's Natural Capital. EU Commission Communication. Brussels, COM (2013) 249 Final, 6.5.2013.

Gouveia, C.M., Trigo, R.M., and DaCamara, C.C. (2009). Drought and vegetation stress monitoring in Portugal using satellite data. *Natural Hazards and Earth System Sciences* 9 (1): 185–195.

Gouveia, C.M., Bastos, A., Trigo, R.M., and DaCamara, C.C. (2012) 'Drought impacts on vegetation in the pre- and post-fire events over Iberian Peninsula'. *Natural Hazards and Earth System Sciences* 12: 3123–3137.

INAG (2001). *Plano Nacional da Água.* Instituto da Água Ministério do Ambiente e Ordenamento do Território. [Online] Available from http:// www.cm-sever.pt/ambiria/Download.aspx?ent=ambi_anexo&id=43 (accessed 10 December 2010).

Jacinto, R., Cruz, M.J., and Santos, F.D. (2011). *Adaptation of a Portuguese Water Supply Company (EPAL) to Climate Change: Producing Socio-Economic and Water Use Scenarios for the XXI Century* (31 Oct–4 Nov). YSW, International Water Week Amsterdam 2011.

Jacobs, C., Wolters, W., Todorovic, M., and Scardigno, A. (2008). *Mitigation of Water Stress Through New Approaches to Integrating Management, Technical, Economic and Institutional Instruments.* Aquastress Integrated Project. Report on Water saving in Agriculture, Industry and Economic Instruments. Part A – Agriculture. 58 pp.

Jongen, M., Unger, S., Fangueiro, D., Cerasoli, S., Silva, J.M.N., and Pereira, J.S. (2013). Resilience of montado understorey to experimental precipitation variability fails under severe natural drought. *Agriculture, Ecosystems & Environment* 178: 18–30.

Kampragou, E., Assimacopoulos, D., Andreu, J., Bifulco, C., de Carli, A., Dias, S., González Tánago, I., Haro Monteagudo, D., Massarutto, A., Musolino, D., Paredes, J., Rego, F., Seidl, I., Solera, A., Urqujo Reguera, J., and Wolters, W. (2015). Systematic classification of drought vulnerability and relevant strategies – case study scale. *DROUGHT–R&SPI Technical Report No. 24*, Athens, Greece, p. 43.

MAMAOT (2013a). *Seca 2012. Relatório de balanço. Ministério da Agricultura, do Mar, do Ambiente e do Ordenamento do Território.* 132 pp. [Online] Available from http://www.portugal.gov.pt/media/916024/Relatorio_Balanco_GTSeca2012_v1.pdf (accessed 10 December 2013).

MAMAOT (2013b). *Estratégia de adaptação da agricultura e das florestas às alterações climáticas. Portugal Continental.* Ministério da Agricultura, do Mar, do Ambiente e do Ordenamento do Território. 88 pp. [Online] Available from http://www.apambiente.pt/_zdata/Politicas/AlteracoesClimaticas/Adaptacao/ENAAC/RelatDetalhados/Relat_Setor_ENAAC_Agricultura.pdf (accessed 10 December 2013).

Máñez Costa, M.A., Moors, E.J., and Fraser, E. (2011). Socioeconomics, policy, or climate change: what is driving vulnerability in southern Portugal? *Ecology and Society* 16 (1): 28. [Online] Available from http://www.ecologyandsociety.org/vol16/iss1/art28/ (accessed 10 December 2011).

MAOTDR (2007). *Programa Nacional da Política de Ordenamento do Território.* Ministério do Ambiente, do Ordenamento do Território e do Desenvolvimento Regional, Lisboa.

Miranda, P., Coelho, F.E.S., Tomé, A.R., and Valente, M.A. (2002). 20th century Portuguese climate and climate scenarios. In: *Climate Change in Portugal. Scenarios, Impacts and Adaptation Measures – SIAM Project* (ed. F.D. Santos, K. Forbes, and R. Moita), 23–83. Gradiva, Lisboa.

Moreira, F., Leitão, P., Synes, N., Alcazar, R., Catry, I., Carrapato, C., Delgado, A., Estanque, B., Ferreira, R., Geraldes, P., Gomes, M., Guilherme, J., Henriques, I., Lecoq, M., Leitão, D., Marques, A.T., Morgado, R., Pedroso, R., Prego, I., Reino, L., Rocha, P., Tomé, R., Zina, H., and Osborne, P.E. (2012). Population trends in the steppe birds of Castro Verde for the period 2006–2011. Consequences of a drought event or land use changes? *Airo* 22: 79–89.

Nuncio, J. and Arranja, C. (2011). Water management in collective irrigation districts: water use quantification and irrigation systems efficiency. In: Leão and Ribeiro (eds.) *O uso da água em Agricultura*, 43–54. INE, I.P., Lisbon, Portugal.

PCM (Presidency of the Council of Ministers) (2005). Programa Nacional para o Uso Eficiente da Água – Bases e Linhas Orientadoras (PNUEA). Resolução do Conselho de Ministros n.

113/2005, 30 June. [Online] Available from http://www.iapmei.pt/iapmei-leg-03.php?lei=3550 (accessed 10 December 2010).

Pellizzaro, G., Cesaraccio, C., Duce, P., Ventura, A., and Zara, P. (2007). Relationships between seasonal patterns of live fuel moisture and meteorological drought indices for Mediterranean shrubland species. *International Journal of Wildland Fire* 16 (2): 232–241.

Pereira, H.M., Domingues, T., Vicente, L., and Proença, V. (Eds.) (2009). *Ecossistemas e bem estar humano. Avaliação para Portugal do Millennium Ecosystem Assessment*, 734 pp. Escolar Editora. Lisboa.

Pereira, M.G., Calado, T.J., Dacamara, C.C., and Calheiros, T. (2013). Effects of regional climate change on rural fires in Portugal. *Climate Research* 57: 187–200.

Pires, V., Silva, A., and Mendes, L. (2010). Risco de secas em Portugal Continental. *Territorium* 17: 27–34.

Ramos, A.M., Trigo, R.M., and Santo, F.E. (2011). Evolution of extreme temperatures over Portugal: recent changes and future scenarios. *Climate Research* 48: 177–192.

Rego, F., Bifulco, C., Dias, S., Massarutto, A., Musolino, D., and Carli, A. (2015). Crop yields and prices as affected by drought. In: *International Conference on Drought, Research and Science–Policy Interfacing. Book of Abstracts and Program* (ed. D. Monteagudo and J. Andreu), 123–125. Valencia, Spain: University of Valencia. March 2015.

Ribeiro, L. (2013). Groundwater: availability and prospects for an integrated and sustainable use of water resources. In: Leão and Ribeiro (eds.) *O uso da água em Agricultura*, 55–56. INE, I.P., Lisbon, Portugal.

Rounsevell, M.D.A., Reginster, I., Araújo, M.B., Carter, T.R., Dendoncker, N., Ewert, F., House, J.I., Kankaanpa, S., Leemans, R., Metzger, M.J., Schmit, C., Smith, P., and Tuck, G. (2006). A coherent set of future land use change scenarios for Europe. *Agriculture, Ecosystems and Environment* 114: 57–68.

Santos, F.D., Forbes, K., and Moita, R. (Eds.) (2002). *Climate Change in Portugal. Scenarios, Impacts and Adaptation Measures – SIAM Project*, 456 pp. Lisboa: Gradiva.

Santos, J.F., Pulido-Calvo, I., and Portela, M.M. (2010). Spatial and temporal variability of droughts in Portugal. *Water Resources Research* 46: W03503. doi:10.1029/2009WR008071.

Silva, Á., de Lima, M.I.P., Espírito Santo, F., and Pires, V. (2014). Assessing changes in drought and wetness episodes in drainage basins using the standardized precipitation index. *Die Bodenkultur* 65 (3–4): 31–34.

Sousa, P.M., Trigo, R.M., Aizpurua, P., Nieto, R., Gimeno, L., and Garcia-Herrera, R. (2011). Trends and extremes of drought indices throughout the 20th century in the Mediterranean. *Natural Hazards and Earth System Science* 11: 33–51. doi:10.5194/nhess-11-33-2011.

Sousa, P.M., Trigo, R.M., Pereira, M.G., Bedia, J., and Gutiérrez, J.M. (2015). Different approaches to model future burnt area in the Iberian Peninsula. *Agricultural and Forest Meteorology* 202: 11–25.

Stagge, J.H., Dias, S., Rego, F., and Tallaksen, L.M. (2014). Modeling the effect of climatological drought on European wildfire extent. *Geophysical Research Abstracts* 16: EGU2014-15745.

Stigter, T., Bento, S., Varanda, M., Nunes, J.P., and Hugman, R. (2013). Combined assessment of climate change and socio-economic development as drivers of freshwater availability in the south of Portugal. Transboundary water management across borders and interfaces: present and future challenges. In: *Proceedings of the TWAM2013 International Conference & Workshops*, pp. 1–6. [Online] Available from http://ibtwm.web.ua.pt/congress/Proceedings/papers/Stigter_Tibor.pdf (accessed 10 December 2013).

Trigo, R.M., Pozo-Vazquez, D., Osborn, T.J., Castro-Diez, Y., Gamiz-Fortis, S., and Esteban-Parra, M.J. (2004). North Atlantic oscillation influence on precipitation, river flow and water resources in the Iberian Peninsula. *International Journal of Climatology* 24: 925–944.

Vale, M.J. (Coord.) (2014). *Uso e ocupação do solo em Portugal continental: Avaliação e Cenários Futuros. Projeto LANDYN.* Direção-Geral do Território (DGT), Lisboa. [Online] Available from http://www.dgterritorio.pt/a_dgt/investigacao/landyn/ (accessed 10 December 2014).

van Lanen, H., Alderlieste, M., van der Heijden, A., Assimacopoulos, D., Dias, S., Gudmundsson, L., Haro Monteagudo, D., Andreu, J., Bifulco, C., Rego, F., Parede, J., and Solera, A. (2013). Likelihood of future European drought: selected European case studies. *DROUGHT–R&SPI Technical Report No. 11*. [Online] Available from http://www.eu-drought.org/ (accessed 10 December 2013).

Vicente-Serrano, S.M., Lopez-Moreno, J.-I., Beguería, S., Lorenzo-Lacruz, J., Sanchez-Lorenzo, A., García-Ruiz, J.M., Azorin-Molina, C., Morán-Tejeda, E., Revuelto, J., Trigo, R., Coelho, F., and Espejo, F. (2014). Evidence of increasing drought severity caused by temperature rise in southern Europe. *Environmental Research Letters* 9(4): 044001 (9 pp.). doi:10.1088/1748-9326/9/4/044001.

Vivas, E. and Maia, R. (2011). A gestão de escassez de água e secas enquadrando as alterações climáticas. The water scarcity and droughts' management framing climate change issues. *Recursos Hidricos* 31 (1): 25–37.

3.11

Drought Management in the Po River Basin, Italy

Dario Musolino[1], Claudia Vezzani[2], and Antonio Massarutto[3]

[1]*Università Bocconi – Centro di Economia Regionale, dei Trasporti e del Turismo, Milan, Italy*
[2]*Autorità di Bacino Distrettuale del Fiume Po, Parma, Italy*
[3]*Dipartimento di Scienze Economiche e Statistiche, Università di Udine, Udine, Italy*

3.11.1 General context

3.11.1.1 Physical and socioeconomic system

The Po River Basin area spans 74 700 km^2 (4000 km^2 Delta area), and has a total population of about 17 million. The average demographic density is 225 inhabitants/km^2, higher than the average density in Italy (180 inhabitants/km^2).

The Po River Basin is the biggest hydrographic basin in Italy. It spans seven Italian regions,[1] Canton Ticino (Switzerland), and some areas in France (Figure 3.11.1). About 3200 municipalities, some big urban agglomerations (such as Milan and Turin), and several medium-sized urban centres are part of this territory.

Po River is 652 km long and has 141 tributary rivers. The average annual precipitation in the Po River Basin is 1200 mm, with the maximum rainfall values reached in the spring. The total annual water volume supplied by precipitation amounts to 78 billion m^3. The average annual flow (1923–2006) of the Po River is 1500 m^3/sec. The total water available in the region from Alpine lakes is approximately 1.04 billion m^3. The average annual temperature in the basin is around 5°C in the high Alps, 5–10°C in the medium-height mountains, and 5–10°C in the other zones.

The Po River Basin is a highly developed area in economic terms: 34% of the value added created in Italy comes from the Po River Basin, owing to the remarkable concentration of a wide range of agricultural, manufacturing, and services activities (29% of the Italian industrial and services firms are located there). In addition, there are several sectoral specialisations, such as mechanics, textile and clothing, and food. The value

[1] Piedmont, Aosta Valley, Liguria, Lombardy, Veneto, Emilia-Romagna, Tuscany, and Provincia di Trento.

Drought: Science and Policy, First Edition.
Edited by Ana Iglesias, Dionysis Assimacopoulos, and Henny A.J. Van Lanen.

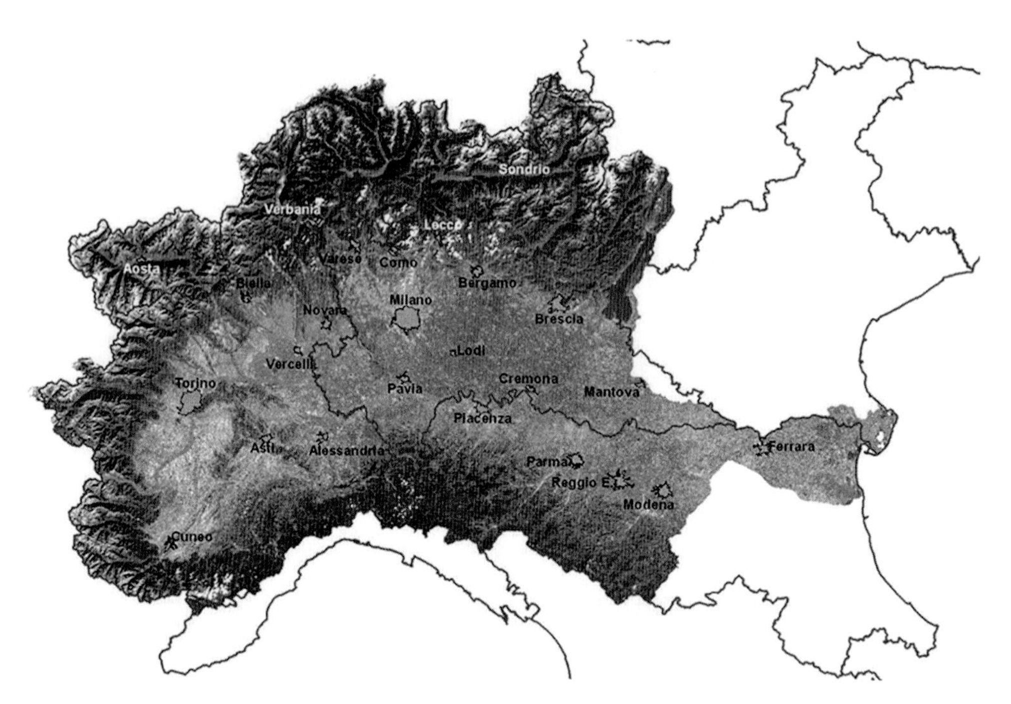

Figure 3.11.1 Po River Basin: regional and provincial administrative boundaries. *Source*: Po River Basin Authority. (*See colour plate section for the colour representation of this figure.*)

added produced by food in the Po basin accounts for 41% of the sectoral value added in Italy.[2]

As far as agriculture is concerned, several figures show its importance in the Po basin. About 35% of the national agricultural production comes from the basin, and 55% of Italian livestock comes from only five provinces of the basin. About 2 700 000 ha in the Po basin are classified as 'utilised agricultural area' (about 40% of the total basin area), of which 59% is irrigated. Agriculture is composed of permanent forage crops, covering about 85% of the irrigated utilised agricultural area (grain corn 32.5%; rice 14.5%), and arboreal crops (fruit orchards 4.5%; horticulture 3.6%) (Autorità di bacino del fiume Po, 2009).

Another important sector in the Po basin, significant in drought management, is the power industry, particularly hydropower and thermal power, which are highly hydro-demanding activities. About 890 hydropower plants are located in the basin. The total hydropower installed capacity is slightly over 8 GW, equal to 48% of the total power production in Italy. Most of the installed capacity is concentrated in a few plants. About 46% (19 TWh) of the hydropower generated in Italy comes from the Po basin.[3] Similarly

[2] Figures calculated using Istat data about value added at current prices, and about number of active firms (2013; dati.istat.it). The area taken into consideration for these calculations includes Lombardy, Piedmont, Aosta Valley, and part of Emilia-Romagna (the provinces of Piacenza, Reggio Emilia, Parma, Modena, and Ferrara).

[3] A total of 174 reservoirs, which regulate 1.86 billion m^3 of water, are located there; these are predominantly used for power generation (143 are used exclusively for hydropower).

important is thermal power generation. In fact, the Po basin hosts about 400 thermal power plants, whose installed capacity totals 19 GW (with 45% of the installed capacity concentrated in eight plants). The amount of thermal power coming from the basin corresponds to 32% of the total Italian power production (76 TWh) (Autorità di bacino del fiume Po, 2009).

3.11.1.2 Drought characteristics and water availability

The average summer quantity of rain considerably diminished in the Po River Basin in the last decades (Figure 3.11.2). In the meantime, an increase in rainfall intensity (and, consequently, in the number of floods) was observed. The number of rainy days decreased, particularly in spring and summer, resulting in the decrease of river flow during the dry season (the worst reduction occurred in the years between 1980 and 2010). Moreover, since 1960, the average yearly temperature increased by about 2°C, raising the water needs for agriculture and power generation (due mainly to air conditioning).

Analysing the distribution of the water volume in the basin throughout the year, it appears that the whole storage capacity, distributed in natural regulated alpine lakes and in artificial reservoirs, is enough to cover the water demand (agriculture, industry, power production, domestic uses) in absence of rainfall until the end of June. A dry period in spring, from March to April, is usually sufficient to trigger the activation of extraordinary water management measures for the next irrigation season, unless there

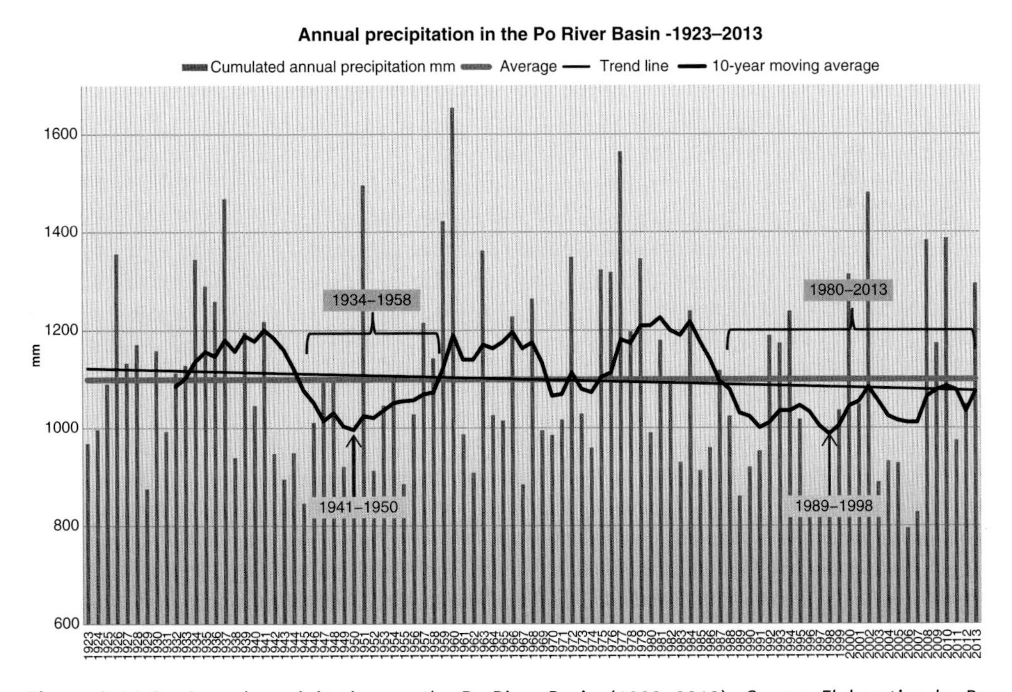

Figure 3.11.2 Annual precipitation on the Po River Basin (1923–2013). *Source*: Elaboration by Po River Basin Authority. (*See colour plate section for the colour representation of this figure.*)

is a successive occurrence of rainfalls. In the same way, a dry period during the start of irrigation season (April, May, and June) causes early release of all the spare accumulated water resources. The major drought events experienced in the last decades in the Po River Basin occurred in 2003, 2005/2007, 2012, and 2015.

The 2003 and 2005/2007 drought events In 2003, after a scarcely snowy winter, precipitations were very infrequent in spring, particularly in May and June. As a consequence, in the plain areas of Lombardy and Emilia-Romagna and in the Apennine region, the reduction of water flows ranged about 50–75%, and in Piedmont about 60–65%. The scarcity of water was further aggravated by the increase in temperatures over the seasonal average. Intense rainfalls on 24 and 25 July, and then precipitations from mid-August on, reduced the gravity of the situation and brought it to normalcy.

In 2006, again in May, June, and July, because of low precipitation levels and limited snow melting (caused by the very low temperatures in mountain areas), low water flow levels were registered, although not as dramatically low as in 2003. Water deficiency was registered again shortly after, between February and May 2007, since, in the last months of 2006 and in January 2007, rain and snow precipitations were mostly under the seasonal average (20–40%). From February to May 2007 water flows in Po river were under the level registered in the same period in 2003 and 2006. In June 2007, thanks to frequent and intense rainfalls, the normal situation was restored.

The 2012 and 2015 drought events The most recent drought/water scarcity events occurred in 2012 and 2015. In 2012, a severe drought, lasting from autumn to spring, created some concern for the following irrigation season, which however faded as a result of abundant rainfall in spring and early summer. The discharge measured at Pontelagoscuro station (located in the province of Ferrara, in Emilia-Romagna) went below 400 m^3/sec only at the end of the irrigation season, without creating problems to farmers or other users. In general, winter droughts in the Po river do not have major impacts, since low flow rates in January are normal owing to both poor rain or frost.

The 2015 event, in particular, shows some interesting features as water scarcity occurred under normal meteorological conditions registered until the end of June. This fact is clearly highlighted by the performance of SPI indices at 3 and 6 months (Figures 3.11.3 and 3.11.4).

Only when looking at the value of SPI in 1 month (Figure 3.11.5) are values observed that correspond to a moderately dry situation for the month of July.

Although the rainfall scenario does not indicate the occurrence of a water shortage, the persistence of high temperatures and their exceptional durations, both for daily minimum and maximum levels, has led to a surge in the water consumption of all sectors (civil, irrigation, hydropower) associated with the considerable rate of natural evapotranspiration. Finally, snow accumulation was also affected by the heat wave, which drove the height of the freezing level to its historic maximum, over 5200 m above sea level. From a hydrological point of view, all these factors, along with a moderate lack of precipitation, resulted in significant decreases in river flows throughout the whole basin.

In the Po valley, a decreasing hydrometric trend was observed from the middle of June until the first week of August, reaching significantly low levels. Figure 3.11.6 shows

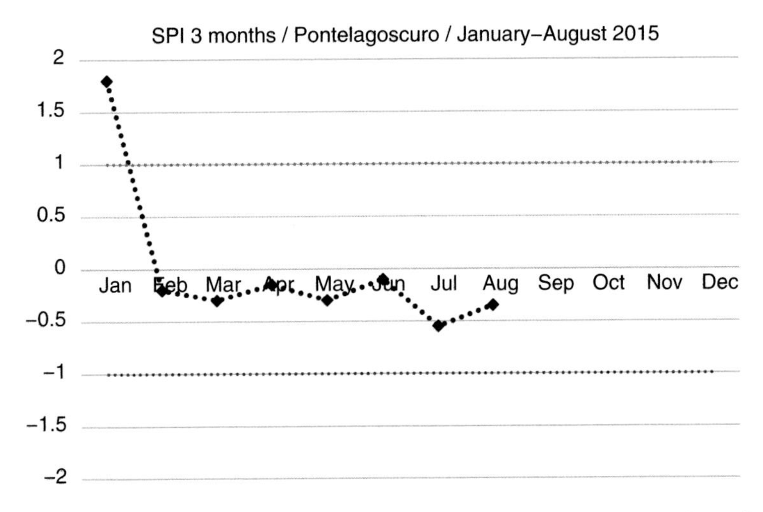

Figure 3.11.3 SPI-3 values at Pontelagoscuro. *Source*: Elaboration by Po River Basin Authority – Data from Po Drought Early Warning System Data (2016).

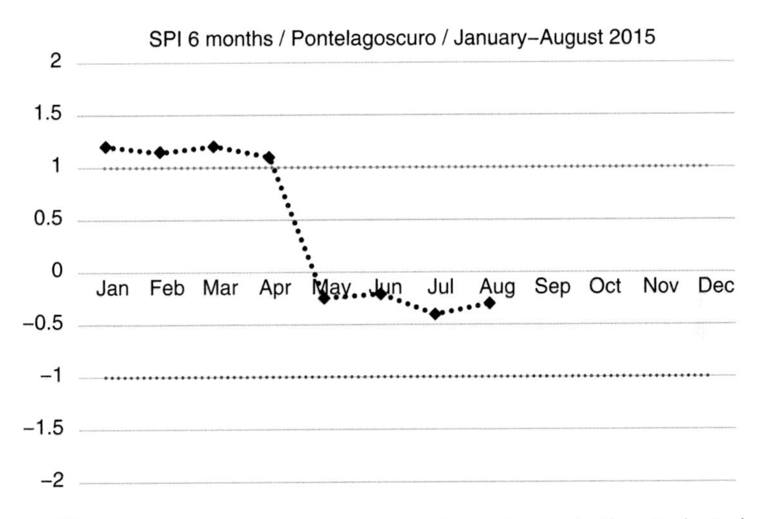

Figure 3.11.4 SPI-6 values at Pontelagoscuro. *Source*: Elaboration by Po River Basin Authority – Data from Po Drought Early Warning System Data (2016).

the values of average monthly flows of the Po river at Pontelagoscuro, calculated using the decade 2002–2011 and the 50 years 1921–1970 as reference periods, and then compared with values observed monthly from January to August 2015.

Data indicate a normal trend of monthly outflows from January to May, while the observed values are below the long-term average in June. In July, the most critical conditions occurred, with monthly average flow much lower than normal. The occurrence of precipitation, and the simultaneous decrease in water demand, led to less critical values in August, even if remaining below the long-term average.

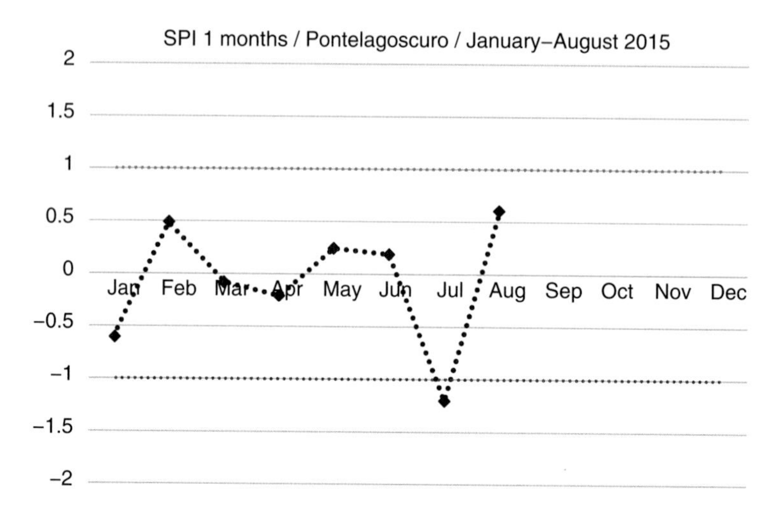

Figure 3.11.5 SPI-1 values at Pontelagoscuro. *Source*: Elaboration by Po River Basin Authority – Data from Po Drought Early Warning System data (2016).

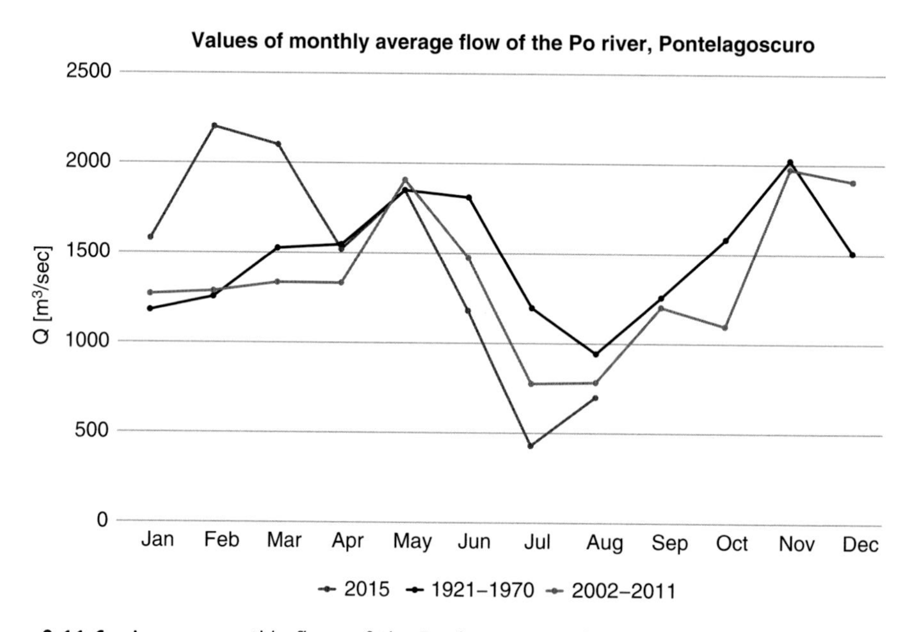

Figure 3.11.6 Average monthly flows of the Po river at Pontelagoscuro. *Source*: Elaboration by Po River Basin Authority. (*See colour plate section for the colour representation of this figure.*)

3.11.2 Drought risk and mitigation

3.11.2.1 Vulnerability to drought

The analysis of the drought events in the last decade shows that the Po River Basin was vulnerable to drought, although the area is characterised by considerable systemic resilience, due to the large storage capacity and regulation of water flows. The lack of a drought management plan is a critical vulnerability factor. This lack creates the so-called 'institutional vulnerability'. The Italian legal system provides several planning tools for regional governments to address drought management, such as water protection plans ('Piani di Tutela delle Acque'), but these are absent from most of the water plans implemented by the regional governments, except in the cases of Emilia-Romagna and Veneto. Emilia-Romagna, in particular, within its water protection plan (Regione Emilia Romagna, 2005), has identified the areas threatened by drought risks, and is currently drafting a drought management plan, which includes the creation of a monitoring system; the analysis of economic, social, and territorial impacts and vulnerabilities; and the definition of responses to drought crises.

On top of the deficiencies of the water management framework, vulnerability to drought depends on the socioeconomic characteristics of the area. The economic sector in the Po basin that consumes the highest quantity of water is agriculture (Table 3.11.1), which, given the 1 355 258 ha of irrigated area (Zucaro, 2013), accounts for an annual water demand volume of more than 16×10^6 m^3/year – more than 80% of the total annual water demand volume used in the Po basin. This being such an important area for agriculture in Italy, and, broadly speaking, such an important sector in the economic system of northern Italy, as discussed in the first paragraph, the importance of drought management in the Po basin is apparent.

Water supply for agriculture is predominantly ensured by Irrigation and Drainage Consortia, which manage water schemes made of a dense network of canals. Surface irrigation is the most common irrigation system (almost 50% of irrigated areas), although this is considered inefficient and inflexible, since it works in turns. In fact, agriculture in the northern part of the basin benefits from the proximity to the Alps, as well as to lakes, glaciers, and Alpine reservoirs. However, this geographical advantage in terms of water availability also serves as a handicap to the northern part of the basin, since it receives much less investment in more efficient technologies compared to the southern part of the basin.

Moreover, the insufficient distribution of water storage infrastructures and the limited extension of the irrigation network are additional critical aspects of irrigated agriculture in the Po basin, and even experts and stakeholders frequently complain about these (see Assimacopoulos et al., 2014, chapter 6.3). However, the high level of crop diversification, also related to the existence of several quality agri-food value chains, is another important characteristic that might make agriculture in the Po basin less vulnerable.

Table 3.11.1 Annual water demand volume by use (Autorità di bacino del fiume Po, 2009).

Type of use	Volume (10^6 m^3/year)	Surface water (%)	Groundwater (%)
Drinking	2500	20	80
Industrial*	1537	20	80
Irrigation	16 500	83	17
Total	20 537	63	37

**Energy production excluded.*

As far as the other sectors are concerned, the industrial sector has been progressively becoming less hydro-demanding, mostly because of the increasing use of less-hydro-demanding technologies in manufacturing processes, owing to the introduction of specific environmental regulations. On the other hand, the power industry, and particularly hydropower, remains highly hydro-demanding. Its potential vulnerability is increased by the old age of power plants (65% of them began operating before 1950), and by the lack of planning in their spatial development, particularly for small plants. Hydropower is anticipated to grow, as it represents one of the best energy production alternatives for reducing greenhouse effects.

3.11.2.2 The framework for drought management: Current situation and on-going changes

The drought management plan for the Po River Basin (Po-DMP) and the development of regional/sub-basin-level DMPs are being carried out within the Po Water Balance Plan (Autorità di bacino del fiume Po, 2015a), adopted in 2017.[4] Po-DMP is structured to be consistent with the guidelines provided by the European Environmental Agency (2009). Given the absence of any drought management plan at the basin or at the regional/local levels, the dominant response/reaction in previous drought events (2003, 2005/2007, 2012, and 2015) was set on a double approach, allowing the introduction of tools that have already been run in, and largely accepted by water managers and stakeholders.

On the one hand, the response was based on voluntary agreements among all relevant stakeholders aimed at maintaining minimum water flow levels for irrigation and thermal power plants, employing water from alpine hydroelectric reservoirs and fostering provident and efficient water use for irrigation.

A technical board, named *Cabina di Regia*, was established during the 2003 drought event, and met at regular intervals to analyse the real-time situation and to bring different needs to the table, in order to coordinate a water management effort aimed at reducing major impacts at the local and basin levels. Since then, the *Cabina di Regia* meets at least once per year at the beginning of the irrigation season, and, based on observation data and forecasting, sets the outline for water management for the current year. In case of water shortage, the *Cabina di Regia* continues to meet at intervals to re-evaluate the situation and make decisions using updated information. The *Cabina di Regia* benefits from a much-enhanced early warning and drought modelling system, named 'Drought Early Warning System for the Po River Basin' (DEWS-Po), used as a major tool in the decision-making process (Autorità di Bacino del fiume Po, 2015b). These actions led to the signing of a memorandum of understanding *Protocollo d'Intesa* (Autorità di bacino del fiume Po, 2005), mainly promoted by the Po River Basin Authority, which is the key institutional actor in drought management. On the heels of the activity of the *Cabina di Regia* which has been recognized as a good practice at a national level, the Italian Ministry of Environment and the National Department of Civil Protection promoted the creation of similar boards within each district authority

[4]The Po Water Balance Plan was recently adopted with DPCM (Decreto del Presidente del Consiglio del Ministri Italiano) on 11 December 2017.

in Italy, enlarging the area of competence to the permanent control of water balance; this brought about a new memorandum of understanding, involving also the national administrative level, which came into effect on 13 July, 2016 (Autorità di Bacino del Fiume Po, 2016).

On the other hand, the declaration of a state of emergency in case of drought is the responsibility of the national government, and a commissioner is appointed in order to carry out such extraordinary and urgent activities, coordinating several temporary government bodies at national and regional levels.

Although the approach described in the preceding text has been shown to be effective, it has some limits, mainly resulting from the preponderance of reactive rather than proactive actions.

The main gaps that have been detected, and which are addressed in Po-DMP as urgent actions, are: the lack of a comprehensive and shared knowledge of drought vulnerability and impacts across the basin, which necessarily causes an approximate management of water resources in real time; and the lack of an 'a priori' definition and timing of actions that each actor can put forward, and the consequent effects. Moreover, an economic assessment of the impacts, critical to the implementation of cost–benefit approaches, can be implemented only after the completion of the vulnerability and impact assessments. Another main point relates to the lack of a quantitative definition of the conditions of drought and prolonged drought as requested by the WFD 2000/60 to access to the exemption to the WFD objectives under article 4.6.[5]

To overcome these gaps, the new Po-DMP proposes simple but effective tools that have already been presented and shared in the public participation process conducted for drafting the Po Water Balance Plan.

The first measure aims at strengthening the monitoring system DEWS-Po, enabling its use also to regional nodes, made possible by the distributed architecture that characterises the DEWS-Po system, which allows users access through client interfaces. The action consists of training specialised personnel in the use of local nodes of the system (already existing), in order to ensure continuous updating of the data and reasoned evaluation of local results. This action should lead to effective network management across the basin, which has in fact already begun with an agreement between Emilia Romagna and Lombardy, and with actions carried on with the involvement of the Po River Basin Authority.

Another important measure is impact and vulnerability assessment, carried out by the Po River Basin Authority, and by local actors for regional and lower districts, with the participation of universities and local administration. The main tool, called '*Sicc*-Idrometro' ('Drought-hydrometer'), is a shared visualisation tool of the impacts of low flow periods on river discharge, developed for the entire basin. It consists of a thematic map of a whole river in which, at every reference cross-section, the major impacts are represented versus discharge value, in order to make the effects of water management (effects of upstream withdrawals or release, etc.) clear to all the upstream and downstream users. Based on *Sicc*-Idrometro, draft sets of actions to be carried out at each local node during real-time management will be identified, to be submitted for discussion during the *Cabina di Regia* meetings.

[5]Terms as referred in "Guidance document on exemptions to the environmental objectives" – Common Implementation Strategy for the Water Framework Directive (2000/60/EC) – Technical Report 2009-027 – Guidance document. n. 20.

In addition, a specific study is expected to set the thresholds between drought and prolonged drought.

Other measures include the definition of mitigating actions, the constitution of a comprehensive network for the real-time monitoring of water withdrawals (linked to the DEWS-Po system), and the performing of economic evaluations of drought impacts and water use.

Beyond the completeness of the package of measures, and the consistency with European directives, it is assumed that the value of these tools lies in sharing the goals, use, and design among the whole system of water users and administrators, in a strongly inclusive path involving all the stakeholders and the different levels of governance, which should lead to an increase of awareness about drought and water scarcity in the Po basin, and to the choice of well-accepted measures. Among others, the measure involving the net reduction of abstraction during the irrigation season, to be implemented in the next 5 years with the agreement of all the Po water users, should be highlighted.

3.11.2.3 Policy responses to the 2003 drought event: A qualitative assessment

The water shortage caused by the 2003 drought event had negative consequences both on farming (farmers had to start irrigation earlier in the season) and on public water supply (in some areas close to the Apennines). High temperatures also entailed high energy consumption, forcing electricity companies to produce at their peak. Many thermal power plants, which use surface water for their processes, were affected so much by the water deficiency that some of them could not work in June, leading to several service interruptions (in addition, other plants were closed for renovations at the time). *Cabina di Regia* decided to reduce the quota of water set aside for irrigation by 10% for a certain period, and at the same time to release a certain daily amount of water from Alpine reservoirs, until intense rainfalls at the end of July re-established the normal situation. It was found that the main economic consequences were suffered by the final consumers of agricultural products, because of price increase effects. For farmers, on the contrary, the gains fetched by the price increase was much bigger in absolute value than the loss caused by the reduction of crop volumes (Massarutto and De Carli, 2009; Musolino et al., 2015). With regard to power generation, producers did not suffer any loss, as there were no decreases in hydropower generation (thanks to precipitation in August, Alpine reservoirs filled up again); instead, consumers suffered some costs associated with service interruptions (Massarutto et al., 2013).

The responses to the 2003 drought event were evaluated by means of some field qualitative surveys (questionnaire survey and direct interviews) targeted at experts and stakeholders in the Po River Basin. The broad picture that emerges from these surveys is mostly positive, as the interviewees pointed out several good aspects of the drought management efforts. Nevertheless, some points of criticism were raised as well, quite often related to the lack of a definite and formal institutional framework for the drought management endeavours in the Po basin (again, the 'institutional vulnerability').

The assessment of drought management of the 2003 and 2005/2007 drought events was focused on three themes: *participation and coordination*; *information and communication*; *implementation and effects of policy measures*.

Overall, there was a positive evaluation of the efforts made concerning the first theme (*participation and coordination*). The participation of the stakeholders in the management of the drought event was assumed to be wide enough in terms of types and numbers of actors (Figure 3.11.7). Both users and managers, and several potentially conflicting interests, were widely represented around the table, particularly during the emergency phase. Almost all actors were involved in the process, according to the interviewees: the only actors not adequately involved were the environmental organisations, the farmers' associations, and some regional and local governments (provinces) and local bodies. The benefit of this wide participation was that it enabled the creation of a network among actors, who until then were isolated, and helped them to better exchange information.

The creation of the *Cabina di Regia*, the informal body responsible for the coordination efforts headed by the Po Authority, was also appreciated, as it represents a unique body that is responsible for drought management, which improves the decision-making process. In so doing, it can be thought as a 'prototype' of a possible future structure of drought management in the Po basin. However, some points of criticism were raised, mostly related to the informal character of this body. These included the lack of clarity on the roles played by each actor; the overlapping of institutional, political, and technical competences; the lack of scheduling of the meetings; and the difficulty in identifying the priorities.

As far as the issues of information and communication are concerned, stakeholders highlighted their satisfaction in the availability of data and information about water

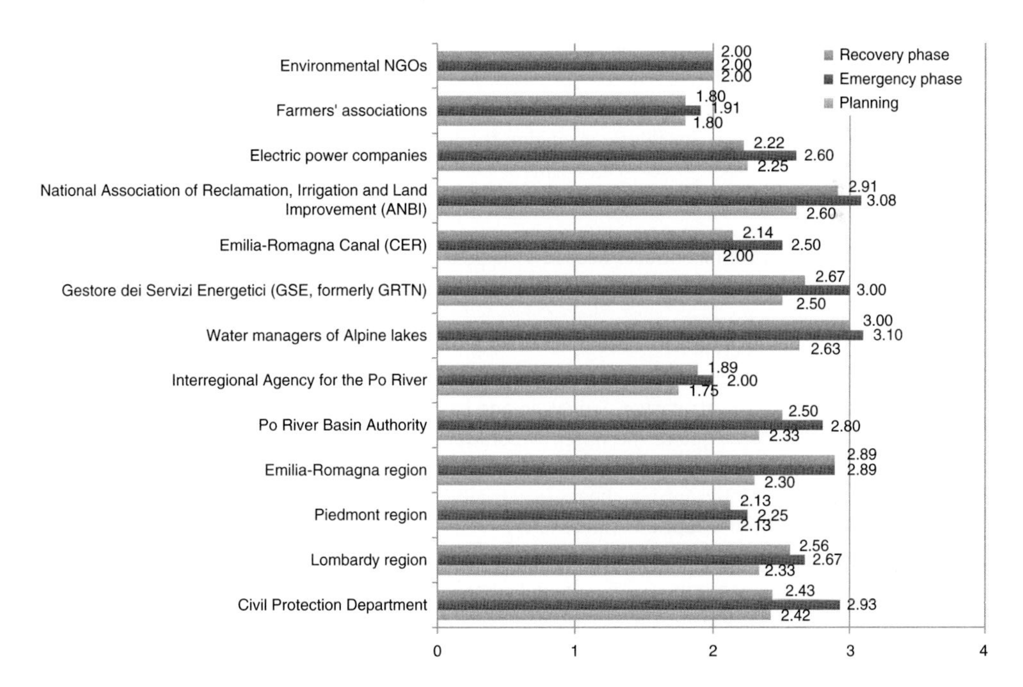

Figure 3.11.7 Average evaluation of the stakeholders' participation (rating: 4 = very adequate; 3 = adequate; 2 = not so adequate; 1 = not adequate) – survey conducted for the FP7 DROUGHT-R&SPI project. (*See colour plate section for the colour representation of this figure.*)

resources (for the stakeholders and for the public) during the drought event. The same applies for information concerning the measures that were planned and adopted. However, they did not evaluate positively the information available regarding environmental effects and the social and economic impacts (of the drought and of the responses); hence, it was not clear, according to some experts, which were the actual conflicting interests raised by the drought events.

Regarding the effects of policy measures and the evaluation of their implementation, the field qualitative surveys revealed that most of the measures designed and implemented had a positive influence (Table 3.11.2). Concerning the strategic measures, again, the creation of *Cabina di Regia* and the *Protocollo d'Intesa*, together with the creation of a daily information exchange system among stakeholders, were the most appreciated measures. On the contrary, the data collection activities, together with the implementation of the strategic measures by the regional governments, were aspects not particularly appreciated by the stakeholders. With regard to the operational measures, all demand management measures were assumed to be satisfactorily planned and implemented (the design of the measure and its implementation are the criteria against

Table 3.11.2 Evaluation of the quality of strategic measures and operational demand–supply management measures in the Po River Basin (De Stefano et al., 2012).

Strategic measures	Total score*	Ranking
Creation of a committee including the stakeholders by signing the voluntary agreement (*Protocollo d'intesa*)	2.71	1
Creation of a daily information exchange system on average water volumes and streamflows	2.61	2
Definition and signing of the voluntary agreement (*Protocollo d'Intesa*) by the relevant stakeholders for jointly managing the responses to a drought event	2.55	3
Survey, gathering, and provision of all relevant data for situation monitoring	2.29	4
Issuing of the regional government acts needed to implement the *Protocollo d'intesa*	2.15	5
Demand management measures	Total score*	Ranking
Control of electricity production	2.63	1
Reduction of 10% in water withdrawal for irrigation, thanks to the National Association for Reclamation, Irrigation and Land Improvement, which can regulate the water withdrawal from their derivations	2.46	2
Planned disconnection of interruptible clients (electricity)	2.44	3
Planned rotation of retail interruptions (electricity)	2.30	4
Supply management measures		
Increase of the water volumes released by alpine hydroelectric reservoirs, and direct transfer of the additional quantities downstream of the lakes	2.62	1
Controls on water withdrawals by Interregional Agency for the Po River, with regards to the rivers under its authority	1.89	2

**Average of the mean scores received for each of the evaluated aspects; results based on 19 questionnaire replies.*

which they are best evaluated, while activation time and contribution to impact reduction are the ones for which they receive a lower mark). As to supply management measures, the releases of water from the lakes and from the Alpine hydroelectric reservoirs received the highest appreciation, while the controls on water withdrawal received the lowest evaluation.

An additional point raised by the interviewees concerns the fact that, during the drought episode, there were few financial resources available for drought management. Only a few farmers in Emilia-Romagna could use Law 225/92 (*Legge contro le calamità naturali*) during natural disasters to get subsidised loans and exemptions on social security payments. However, they also underlined that the human resources involved in drought management were prepared and skilled enough, and that they could cooperate and get external support from some research centres and specialised entities (studies and analyses on drought management could benefit from the cooperation of experts in hydraulics and hydrology, but not from input by economists and social scientists).

Finally, the majority of the interviewees said that the 2003 drought event played the role of a training round for all actors involved in drought management. In fact, the 2005/2007 drought episode that followed was better managed – for several reasons, mostly related to the experience gained in 2003, and by learning from the successes and failures experienced then. In 2005/2007, the available knowledge and information were more complete and detailed; the priority uses were already clear according to the experience of 2003; there was an authority available to make decisions; a stronger and wider involvement of relevant stakeholders, as well as more coordination; and higher attention paid by the public.

3.11.3 Conclusions

The latest drought events provided important lessons concerning drought management in the Po basin. For example, they demonstrated the importance of gathering all relevant decision-makers, managers, and users; sharing information and ideas; and taking decisions jointly coordinated by a unique public body. Another critical lesson was the need to systematically collect data and information on hydrometeorological variables, in order to assess, monitor, and forecast drought (having done this, the Po River Basin Authority is now able to develop and use a forecasting model). Moreover, the importance of keeping the planning process updated in relation to the rapid evolution of the external environment/context (technological and economic, such as changes of crops, irrigation periods, electricity prices, etc.) was highlighted.

However, the system is still characterised by a reactive approach. The next step, therefore, is to shift from the reactive to a proactive approach, designing proactive and structural measures in order to prevent, in advance, the negative effects of droughts (Kampragou et al., 2015). This means that, in addition to reactive measures (such as the water releases from Alpine reservoirs and lakes), demand-side and supply management measures too should be taken into account, aimed at reducing the vulnerability of the socioeconomic and environmental systems of the Po basin. For example, control on illegal withdrawals, improvement of irrigation technologies, investment in infrastructures and technologies for water saving (e.g. storage of groundwater), and evaluation of possible new alternative measures (such as barriers against salty water in the Po delta, and barriers or dams at different points along the river) are some of the key policy options that stakeholders and experts frequently point out and highlight. Part of such a proactive approach is also the promotion of an integrative and comprehensive evaluation of drought impacts, including social and environmental impact assessment and economic impacts analysis.

All these actions are considered necessary parts of the strategy for increasing the preparedness of the Po basin; they should be planned and implemented for better tackling the risks deriving from drought events. The new drought management plan will be fundamental for designing this future strategy, and for addressing the issue of institutional vulnerability highlighted in this document.

References

Assimacopoulos, D., Kampragou, E., Andreu, J., Bifulco, C., de Carli, A., De Stefano, L., Dias, S., Gudmundsson, L., Haro-Monteagudo, D., Musolino, D., Paredes-Arquiola, J., Castro Rego, F., Seidl, I., Solera, A., Urquijo, J., van Lanen, H., and Wolters, W. (2014). Future drought impact and vulnerability – case study scale. *DROUGHT–R&SPI Technical Report No. 20.* [Online] Available from www.eu-drought.org/technicalreports (accessed 7 April 2018).

Autorità di bacino del fiume Po (2005). Protocollo d'intesa "Attivita' unitaria conoscitiva e di controllo del Bilancio idrico volta alla prevenzione degli Eventi di magra eccezionale nel bacino Idrografico del fiume Po, 8 June 2005.

Autorità di bacino del fiume Po (2009). *Il territorio del fiume Po. L'evoluzione della pianificazione, lo stato delle risorse e gli scenari di riferimento*, Edizioni Diabasis, Reggio Emilia.

Autorità di bacino del fiume Po (2015a). *Piano di Bilancio Idrico del distretto idrografico del fiume Po.Allegato 3 alla Relazione Generale. Piano per la gestione delle siccità e Direttiva Magre.* Parma, 8 July 2015. Available from: http://www.adbpo.gov.it/it/piani-di-bacino/piano-bilancio-idrico (accessed 29 February 2016).

Autorità di bacino del fiume Po (2015b). *Piano di Bilancio Idrico del distretto idrografico del fiume Po. Allegato 4 alla Relazione Generale. Drought Early Warning System Po – Sistema di modellistica di distretto.* Parma, 8 July 2015. Available from: http://www.adbpo.gov.it/it/piani-di-bacino/piano-bilancio-idrico (accessed 29 February 2016).

Autorità di bacino del fiume Po (2016). "Istituzione dell'Osservatorio Permanente sugli utilizzi idrici nel distretto idrografico del Fiume Po". Protocollo d'Intesa, ROMA, 13 luglio 2016.

De Stefano, L., Urquijo, J., Acacio, V., Andreu, J., Assimacopoulos, D., Bifulco, C., De Carli, A., De Paoli, L., Dias, S., Gad, F., Haro Monteagudo, D., Kampragou, E., Keller, C., Lekkas, D., Manoli, E., Massarutto, A., Miguel Ayala, L., Musolino, D., Paredes Arquiola, J., Rego, F., Seidl, I., Senn, L., Solera, A., Stathatou, P., and Wolters, W. (2012). Policy and drought responses – case study scale. *DROUGHT–R&SPI Technical Report No. 4.* [Online] Available from www.eu-drought.org/technicalreports (accessed 26 February 2016).

European Environmental Agency (2009). Water resources across Europe – confronting water scarcity and drought. *EEA Report No. 2/2009.*

Kampragou, E., Assimacopoulos, D., De Stefano, L., Andreu, J., Musolino, D., Wolters, W., Van Lanen, H., Rego, F., and Seidl, I. (2015). Towards policy recommendations for future drought risk reduction. In: *Drought: Research and Science–Policy Interfacing* (ed. J. Andreu, A. Solera, J. Paredes-Arquiola, D. Haro-Monteagudo, and H. van Lanen). Leiden, the Netherlands: CRC Press, Taylor & Francis Group, A Balkema Book.

Massarutto, A. and De Carli, A. (2009). I costi della siccità: il caso del Po, Economia delle fonti di Energia e dell'Ambiente, 2. Pp. 123–143.

Massarutto, A., Musolino, D., Pontoni, F., De Carli, A., Senn, L., De Paoli, L., Rego, F.C., Dias, S., Bifulco, C., Acácio, V., Andreu, J., Assimacopoulos, D., Ayala, L.M., Gad, F., Monteagudo, D.H., Kampragkou, E., Kartalides, A., Paredes, J., Solera, A., Seidl, I., and Wouters, W. (2013). Analysis of historic events in terms of socio-economic and environmental impacts. *DROUGHT–R&SPI Technical Report No. 9*. [Online] Available from www.eu-drought.org/technicalreports (accessed 9 March 2016).

Musolino, D., Massarutto, A., and De Carli, A. (2015). Ex-post evaluation of the socio-economic impacts of drought in some areas in Europe. In: *Drought: Research and Science–Policy Interfacing* (ed. J. Andreu, A. Solera, J. Paredes-Arquiola, D. Haro-Monteagudo, and H. van Lanen). Leiden, the Netherlands: CRC Press, Taylor & Francis Group, A Balkema Book.

Regione Emilia Romagna (2005). *Piano di Tutela delle Acque*. [Online] Available from http://ambiente.regione.emilia-romagna.it/acque/informazioni/documenti/piano-di-tutela-delle-acque (accessed 11 March 2016).

Zucaro, R. (eds) (2013). Analisi territoriale delle criticità: strumenti e metodi per l'integrazione delle politiche per le risorse idriche. Volume I – Applicazione nel Nord e Sud Italia, INEA (accessed 7 April 2018).

3.12
Experiences in Proactive and Participatory Drought Planning and Management in the Jucar River Basin, Spain

Joaquin Andreu[1], David Haro[2], Abel Solera[1], and Javier Paredes[1]

[1]*Instituto Universitario de Investigación de Ingenieria del Agua y Medio Ambiente, Universitat Politècnica de València, València, Spain*

[2]*School of Water, Energy and Environment, Cranfield University, Bedfordshire, United Kingdom*

3.12.1 Introduction

The Jucar River Basin District (JRBD), with an area of 42 989 km^2, is located in the eastern part of Spain (Figure 3.12.1). It is composed of several adjacent basins flowing into the Mediterranean Sea, from the northern Cenia River Basin, to the Southern Vinalopo and Alacanti River Basins. The average precipitation is 493 mm/year, although it is highly variable in time and space. Many areas can be classified as semi-arid, and some as arid. The two main river basins are Jucar (22 378 km^2), and its northern neighbour Turia (6913 km^2). The average total runoff is 3250 hm^3/year (Jucar 1300 hm^3; Turia 285 hm^3), with an irregular hydrology very characteristic of Mediterranean basins. Therefore, episodes of droughts and floods are very common, sometimes even both taking place at the same time.

The permanent population of the JRBD is 4.3 million inhabitants, with the main cities being Valencia (1.5 million in the metropolitan area), Alicante (330 000), Castellón (180 000), and Albacete (170 000).

The main economic activities are tourism (2.6 million overnights/year), irrigated agriculture (400 000 hectares, mainly citrus fruits and vegetables), hydropower (1417 MW of installed capacity), shipping and commerce, and several industrial sectors

Drought: Science and Policy, First Edition.
Edited by Ana Iglesias, Dionysis Assimacopoulos, and Henny A.J. Van Lanen.

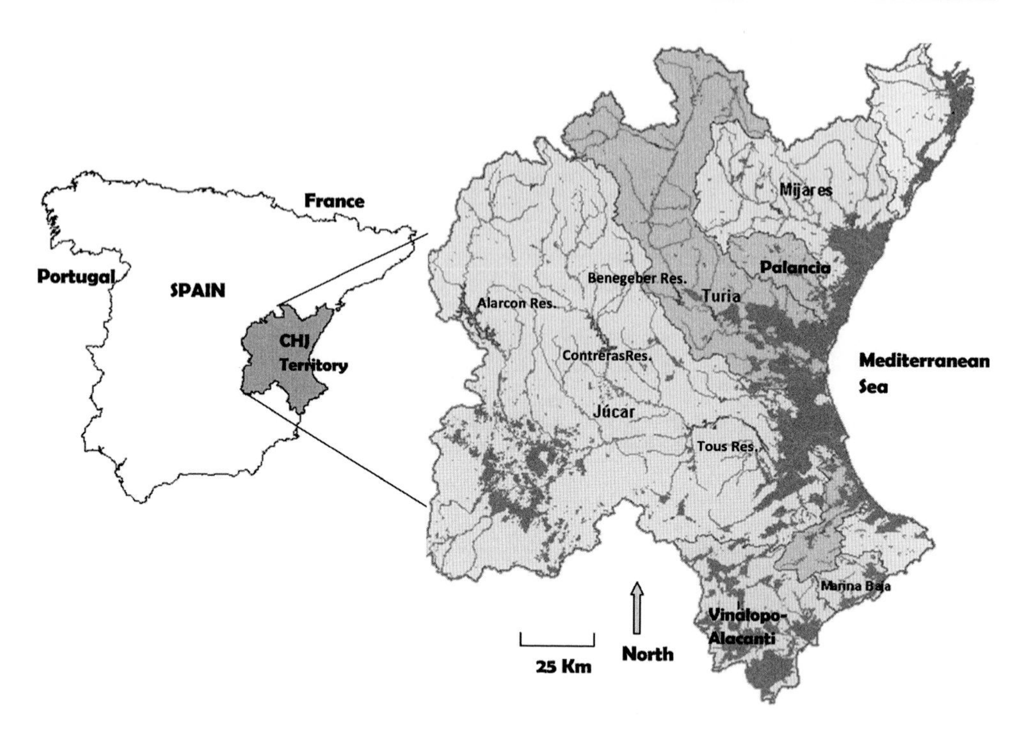

Figure 3.12.1 Location of the Jucar River Basin District, in Spain. *Source*: Self elaboration from Confederacion Hidrografica del Jucar originals, with permission. (*See colour plate section for the colour representation of this figure.*)

(automotive, furniture, tiles, etc.). Irrigated agriculture accounts for nearly 80% of water demand. In general, agricultural demand appears to be established or is decreasing, whereas urban/industrial demand is forecasted to rise.

In the Valencia coastal plain, where rivers Jucar and Turia have their final parts, and between both river mouths, there is a shallow lake called 'Albufera', with an associated wetlands area. Both the lake and the wetlands depend on return flows from irrigation areas belonging to both basins, and also on groundwater flows from the coastal aquifer beneath the plain.

Water demand from the population and economic activities is high as compared to available natural resources, resulting in very high water exploitation indexes in most basins within the district, and therefore water stress and water scarcity, and also high environmental stress and water quality deterioration. These problems are exacerbated during drought episodes.

For the purpose of the present chapter, the analysis of drought characterisation and of the most recent extreme episode refers mainly to the Jucar River Basin instead of the entire JRBD. The reason is that, in the last extreme episode (2004–2008), the most affected basin was the Jucar River Basin, and the main experience gained was in the management of the drought in this basin.

More detailed information about the JRBD can be found in a study by the Confederacion Hidrografica del Jucar (CHJ 2004).

3.12.2 Droughts characterisation

3.12.2.1 Past droughts

The characterisation of drought events in the JRBD is done from both meteorological and hydrological perspectives. The meteorological approach studies the evolution of precipitation, and the hydrological approach takes into account the flows in rivers and the stored volume in reservoirs. More information on this can be found in Van Lanen et al. (2013a), and a very detailed characterisation of the different types of droughts for the JRBD using multiple indices and establishing new proposals is available (in Spanish) in Villalobos (2007), the more important results of which are extracted and reflected in Andreu et al. (2007).

Due to the climatic characteristics of the JRBD, and the high intra-annual variability, the study of drought is done on an annual basis, since a lower-scale study would derive a high frequency of droughts. Another reason for the annual study of drought occurrence is the size of the reservoirs existing in the basin, since they can store water for several years and release it during dryer periods. Moreover, the main concern at CHJ is operative drought – that is, those situations when water demands cannot be fully met due to the lack of available water, either in reservoirs, aquifers, or streams.

Meteorological characterisation of past droughts in the JRBD The characterisation of meteorological droughts has been made with regard to the precipitation over the whole territory of the Jucar River Basin Authority using annual precipitation data for the period from 1940/1941 to 2008/2009. Figure 3.12.2 shows the evolution of the annual precipitation (in mm) in the territory of the Jucar River Basin Authority. The average annual precipitation is 493 mm/year, and annual precipitation was above average on 33 occasions (48%) and below average on 36 occasions (52%). It must be stressed that the probability of having 2 or more continuous below average precipitation years is high.

Figure 3.12.3 shows the evolution of the 12-month *standardised precipitation index* (SPI12) (McKee et al., 1993) for the period 1940/1941–2008/2009. Similar to the previous

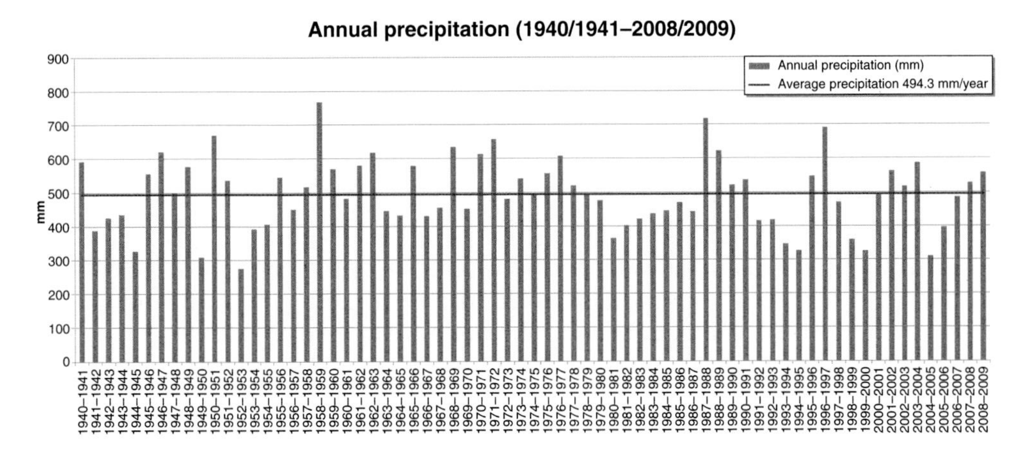

Figure 3.12.2 Annual precipitation over the territory of the Jucar River Basin Authority.

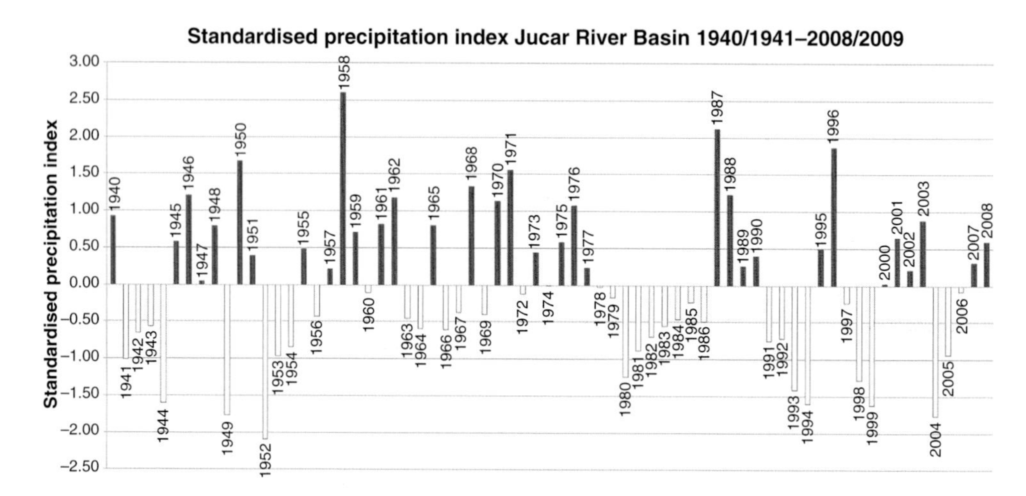

Figure 3.12.3 Evolution of the 12-month-lag SPI in the territory of the Jucar River Basin Authority.

figure, it is possible to observe the existence of long dry periods followed by relatively wet ones. It is also possible to appreciate a difference between the period before 1980 when dry years alternated with wet years almost on an annual basis, and the period after 1980 with longer drought periods. The lack of longer data records makes it very difficult to determine whether this is an effect of global climate change or a punctual climate anomaly, although recent specific climate studies (Lavorel et al., 1998; Vicente-Serrano et al., 2004) predict an increase in the variability and severity of droughts in the region.

Hydrological characterisation of droughts in the Jucar river basin Hydrological drought characterisation is restricted to the Jucar River Basin, instead of the entire JRBD. Figure 3.12.4 shows the total annual flow of the Jucar river in the location corresponding to the last reservoir in the river (Tous reservoir) for the period 1940/1941–2008/2009.

It should be noted that a strong sudden decrease in the annual inflow can be observed from 1980 onwards. The average annual inflow before 1980 is 1734 mm^3/year, and after 1980 it is 1196 mm^3/year. At present, the decrease in the average inflows from 1980 onwards is accepted as a permanent fact, and calculations regarding drought are made considering the two separate periods. This is in line with the meteorological aspects considered in the previous section.

To determine the intensity and duration of the cycles of minimum inflow, comparing them with the cycles of minimum precipitation and mean annual stored volume in reservoirs, a 'standardised inflow index' was used. The calculation process is identical to that used for calculating the SPI, but using the highest average for the period 1940/1941–1979/1980 and the lowest for the period 1980/1981–2008/2009. The results for the Jucar River Basin are shown in Figure 3.12.5.

As can be seen, hydrological droughts are very frequent, and they are mostly multi-year episodes. Very extreme drought episodes were recorded in 1940–1942, 1952–1957 (6-year episode), 1979–1983, and 1992–1995. The last extreme drought started in hydrological year 2004/2005 and lasted for 4 years, and was the second-most intense

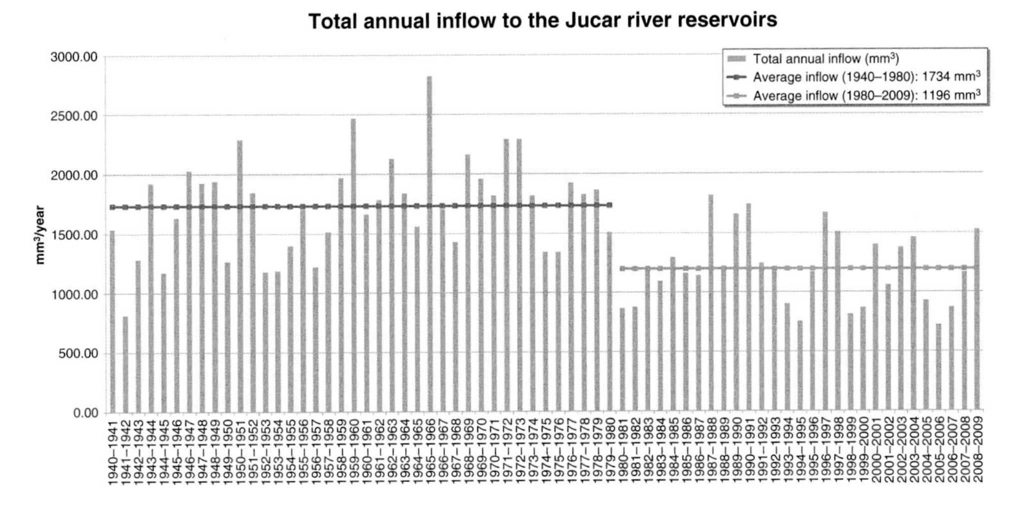

Figure 3.12.4 Total annual inflow to the reservoirs of the Jucar river.

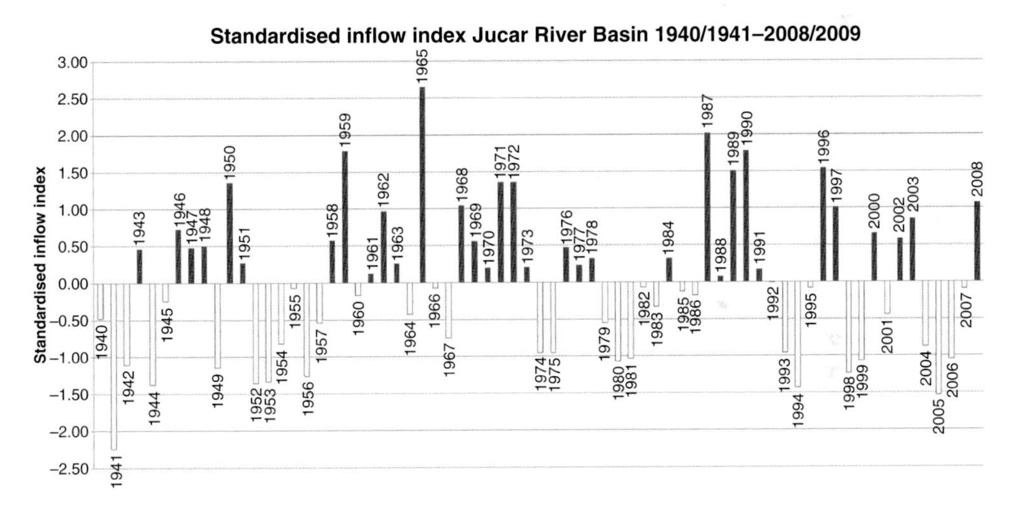

Figure 3.12.5 Standardised inflow index for the Jucar River Basin.

drought on the record and the worst in modern times, with much higher water demands than the ones existing in the 1940s.

In the Jucar River Basin, hydrological droughts have also been characterised by other methodologies, such as, for instance, time-runs analysis (Yevjevich, 1967). This characterisation is very convenient for estimating the statistical figures, such as the duration, intensity, and magnitude (total volume deficit) of hydrological droughts, which are very meaningful for the analysis of operational droughts and the design of measures to decrease vulnerability to them (Sanchez-Quispe et al., 2000; Villalobos, 2007). Threshold analysis has also been used in the Jucar River Basin in order to verify that these statistical figures related to drought are preserved in synthetic hydrological scenarios generated either by classical autoregressive–moving-average (ARMA) modelling (Fernández et al., 1998), or by artificial neural network approaches (Ochoa-Rivera et al., 2007), as well as to characterise the expected future droughts under climate change scenarios.

Drought impacts As already mentioned, the most recent extreme droughts happened in 1992–1995 and 2004–2007. Since the hydrological year in the JRBD goes from October to September of the next year, it is usual to compute most variables in terms of hydrological years. Hence, the values of SPI in Figure 3.12.3 are for hydrological years, and, therefore, the last two extreme droughts cover hydrological years 1992/1993–1994/1995 and 2004/2005–2007/2008.

It should also be clarified that this chapter is focusing mainly on hydrological droughts, and on the so-called 'operational drought', or water resources drought. Therefore, the impacts and management of droughts on unirrigated land and forests are not included.

The most important impacts in the event of 1992–1995 were on agriculture (restrictions in water supply, reduced yields, loss of annual crops, and permanent damages to permanent crops), on water quality (deterioration of water quality in natural surface waters, reservoirs, and groundwater; increased algal bloom – toxic species; eutrophication of surface waters; increased temperature and decreased oxygen saturation levels in surface waters; increased pollution loads in surface waters; increased salinity of surface waters and groundwater), on hydroelectricity (reduced production), and on the environment (increased mortality of aquatic species, including endangered/protected species; migration and concentration of wildlife; increased population of invasive – exotic – aquatic species; observation of adverse impacts on populations of rare/endangered – protected – riparian and wetland species and loss of biodiversity; deterioration of wetlands). It should also be highlighted that, during 2 months in the summers of 1994 and 1995, a stretch of almost 30 km of the Jucar river in Albacete plain dried up due to reversal of the hydraulic connection with the Mancha Oriental aquifer, resulting in net loss of water from the river to the aquifer. Of course, under these circumstances, water management was complicated by regional/local user conflicts.

In the second drought event from 2004/2005 to 2007/2008, most of these impacts had only marginal importance (low-ranked impacts), showing that drought preparedness and management had greatly improved by then, as will be discussed in the following text. In irrigated agriculture, restrictions were experienced, but resulted in only minor reductions of yields, even though, in many areas, irrigation was more complicated (e.g. turns), and in some, water was more expensive due to the use of alternative resources. Hydropower experienced a decrease in production. The drought resulted in the deterioration of water quality in natural surface waters in the lower basin, reservoirs, and groundwater; increased algal bloom (toxic species); eutrophication of surface waters; increased temperature and decreased oxygen saturation levels in surface waters; and increased pollution loads in surface waters. In environmental systems, impacts were less severe, even though there was some mortality of aquatic species (including endangered/protected species) and some increased species concentration near the water surface. The Albufera wetlands did not experience any additional deterioration in status; on the contrary, a slight improvement was perceptible at the end of the drought episode. It should also be highlighted that management of the drought made it possible to maintain the stretch of the Jucar River Basin in the Albacete plain with water flowing, mainly thanks to measures of temporary water rights acquisition by the Jucar River

Basin Authority from groundwater users in the Mancha Oriental aquifer, and owing to Alarcon reservoir management, as described later in text.

Even though economic quantification of drought impacts for the Jucar River Basin has not been officially carried out, an exercise can be found in Massarutto et al. (2013), where the economic impacts on agriculture and power sector are estimated for the 1992/1993–1995/1996 and 2004/2005–2007/2008 drought episodes. As an interesting conclusion, economic losses are mainly experienced by consumers, rather than by producers, even though some indirect costs are not quantified in the analysis, and could increase the total costs and change the proportions between producers and consumers.

3.12.2.2 Future droughts

The assessment of the impact of climate change and the exploration of future droughts in Spain have been made jointly by the State Agency for Meteorology (AEMET) and the Centre of Hydrographical Studies (CEH) from CEDEX (Spanish Governmental Research Institute). AEMET (2008) used four well-known global circulation models (ECHAM4, CGCM2, HadCM3, and HadAM3) to calculate future precipitation and temperature projections for emission scenarios A2 and B2. The GCM results obtained were downscaled using four different regionalisation approaches (FIC, SDSM, PROMES, and RCAO). The projections reveal consistent results with a gradual increase of maximum temperatures, especially in the inland areas, and a lower increase of minimum temperatures, increasing daily temperature oscillations during the twenty-first century. However, precipitation projections show a major dispersion, although there seems to be a tendency towards reduction of precipitation in the south of Spain with a north–south gradient. Superimposed to this gradient, there is also a west–east gradient. This results in a major reduction of precipitation in the southeastern basins in the Iberian Peninsula. The study concludes that precipitation projections are not very robust, owing to the uncertainty introduced by the regionalisation methods. One of the main reasons for the uncertainty is the location of the Iberian Peninsula in the transition zone between high latitudes, where precipitation tends to increase, and the low ones, where precipitation tends to decrease. More detailed information can be found in Van Lanen et al. (2013b).

CEDEX (2010) used these climate projections and a precipitation-runoff model to assess the impact of climate change on natural streamflows in Spanish river basins. They concluded that hydrological droughts tend to become more frequent in all periods.

3.12.3 Methods for drought vulnerability and risk assessment

In the JRBD, it was realised early enough that, while the classical indices related to hydrological variables may tell us the natural situation of drought, they cannot communicate the possible impacts or the efficiency of the measures addressed at minimising them. Therefore, for practical purposes, and in order to provide a better perception of the reliability, risk, vulnerability, and resilience relevant to drought, it is very convenient to define operative indices, and to use the outcomes of models (desirably integrated in decision support systems – DSSs). In fact, this is done in the Jucar case at two time horizons. For the long term (basin and drought planning), specific criteria (vulnerability) are defined and used to assess and classify the vulnerability of different water uses to drought (acceptable or not acceptable), and, therefore, to propose long-term proactive measures to lower vulnerability and increase resilience. And for the short term (real-time drought management), DSSs are used to predict the short-term (12–24 months) impacts of the ongoing drought, in deterministic and probabilistic fashion, using flow forecasts. This provides an excellent indicator of risk, which is used in participatory meetings in order to define and refine the degree of application of the measures contemplated in the plan for each scenario, testing their effectiveness by means of the DSS, as will be discussed later in text.

3.12.3.1 Assessment of vulnerability during the planning phase

In the making of river basin plans, the Jucar River Basin stakeholders and managers have agreed to consider that a particular demand has a *low* (acceptable) or *high* (unacceptable) vulnerability to drought if it meets some specified reliability/vulnerability criteria.

- In the case of urban demands, the criterion defines the vulnerability as *high* if:
  - The expected maximum deficit in 1 month exceeds 10% of the monthly demand
  - The expected maximum deficit accumulated in 10 years exceeds 8% of the annual demand
- On the other hand, for agricultural demands, the criterion to determine their vulnerability is defined as high if:
  - The expected maximum deficit in 1 year cannot exceed 50% of the annual demand
  - The expected maximum deficit in 2 consecutive years cannot exceed 75% of the annual demand
  - The expected maximum deficit of 10 consecutive years cannot exceed 100% of the annual demand

The evaluation of the vulnerability criteria is aided by the simulation models of system management developed with Aquatool DSS Shell (Andreu et al., 1996). The normal procedure involves feeding the model with a set or sets of streamflow scenarios and water demand scenarios. The resulting flows along the system will be related to its management rules, and the failures of supply represented as water shortages in the different demands will determine whether the system can comply with the vulnerability criteria (Figure 3.12.6).

Finally, the allocation of water as well as the programme of measures proposed in the river basin plans are such that the vulnerability of all water uses is *low*, while meeting environmental objectives, including the attainment of good ecological status, and the ecological flows of surface water bodies.

3.12.3.2 Assessment of vulnerability during the management phase (Real time)

In the management phase, vulnerability to drought is estimated in two complementary ways:

- Using the standardised operative drought monitoring indicators (SODMIs), as stated in the special drought plan (SDP). This is performed systematically every

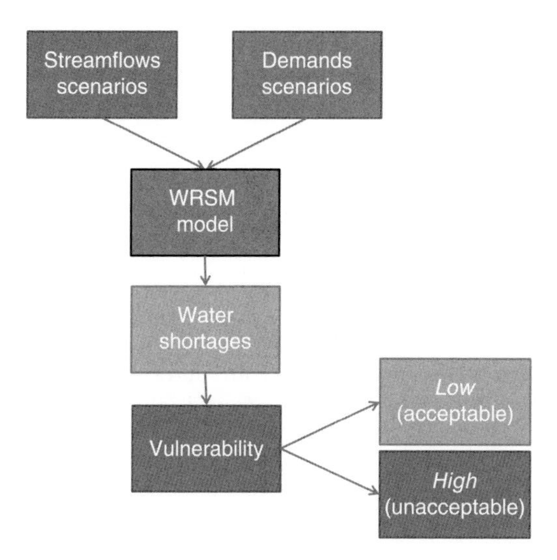

Figure 3.12.6 Drought vulnerability calculation process used in the Jucar River Basin.

month, by means of the procedure explained in the following text. These indicators are displayed in maps for the JRBD, constituting an early warning system device for a proactive approach to drought management, by means of drought stages definition, tied to the measures to be applied in each of the drought stages.

- Using DSSs to make more precise deterministic and probabilistic assessments of risk and vulnerability, used in turn to get a finer estimation of the degree of the measures to be applied.

Use of SODMIs The SODMI system was developed by CHJ (Estrela et al., 2004). In essence, SODMIs use real-time information provided by CHJ's automatic data acquisition system on the state of the reservoirs, aquifers, rivers, and precipitation to produce standardised indices for some selected elements in the basin. These indices are then combined into a single standardised index for each basin.

The SODMI system essentially has a hydrological and water resources system character, since its practical application depends on its ability to serve as a decision-making instrument regarding water resources management in the basin.

The value of the state index (I_e) for each selected variable has the following expression:

$$If\ V_i \geq V_{av} \rightarrow I_e = \frac{1}{2} \cdot \left[1 + \frac{V_i - V_{av}}{V_{\max} - V_{av}} \right] \qquad (1)$$

$$If\ V_i < V_{av} \rightarrow I_e = \frac{1}{2} \cdot 1 + \frac{V_i - V_{\min}}{V_{av} - V_{\min}} \qquad (2)$$

where V_i is the value of the variable in month i; V_{av} the average value of the variable in the historic series considered; and V_{max} and V_{min} the maximum and minimum values, respectively, of the variable in the historic series considered.

The compounded index for the basins is defined as the weighted sum of the I_e values of each of the selected variables. Then, using these compounded indices, four different levels of drought, or scenarios, are defined:

- Normality (green): $I_e \geq 0.5$
- Pre-alert (yellow): $0.5 > I_e \geq 0.3$
- Alert (orange): $0.3 > I_e \geq 0.1$
- Emergency (red): $0.1 > I_e$

For the Jucar River Basin, the drought index is calculated from 12 different variables selected for their relevance to the drought status. Figure 3.12.7 depicts the evolution of the drought index in the Jucar River Basin from October 2001 to October 2009. More information on this indicator system can be found in Acacio et al. (2013) and in the Jucar River Basin Drought Plan (CHJ, 2007).

SODMIs are computed in CHJ not only for the Jucar River Basin itself, but also for the other six neighbouring basins that form the territory of CHJ, and are displayed in maps as a monitoring and early warning system, updated every month, as illustrated in Figure 3.12.8, and as reported in Ortega et al. (2015).

In turn, this information is used as input to the National Drought Monitoring System for Basin Water Resources Systems of Spain. Hence, in a 'bottom-up' approach, drought indicators and scenarios are provided from the river basin partnerships to the General Directorate for Water in order to compile and publish the monthly *National Monitoring Report* and drought maps.

3.12.3.3 Use of DSSs for drought management in real time

The SODMI system provides useful information for early warning and action against drought, as well as for risk perception by the public. Yet, in order to manage droughts,

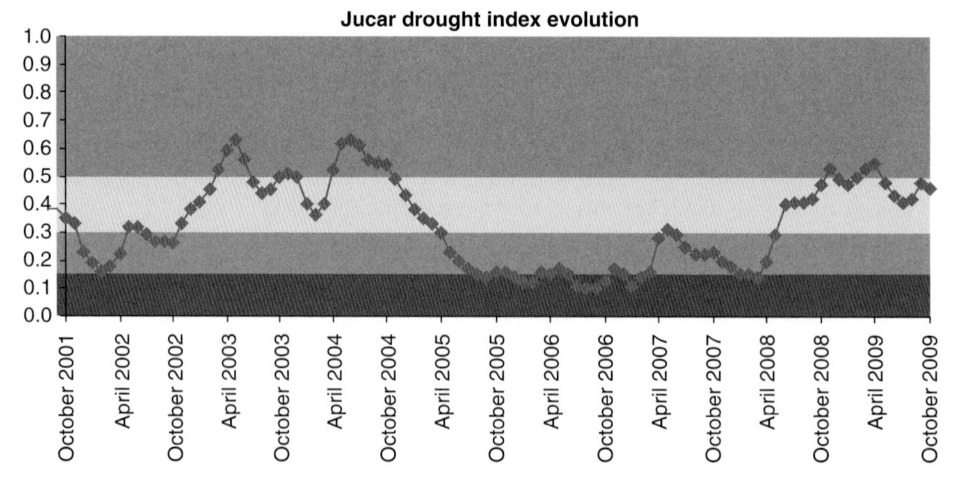

Figure 3.12.7 Evolution of the state index of the Jucar River Basin from October 2001 to October 2009. *Source*: Self-elaboration with data provided by CHJ. (*See colour plate section for the colour representation of this figure.*)

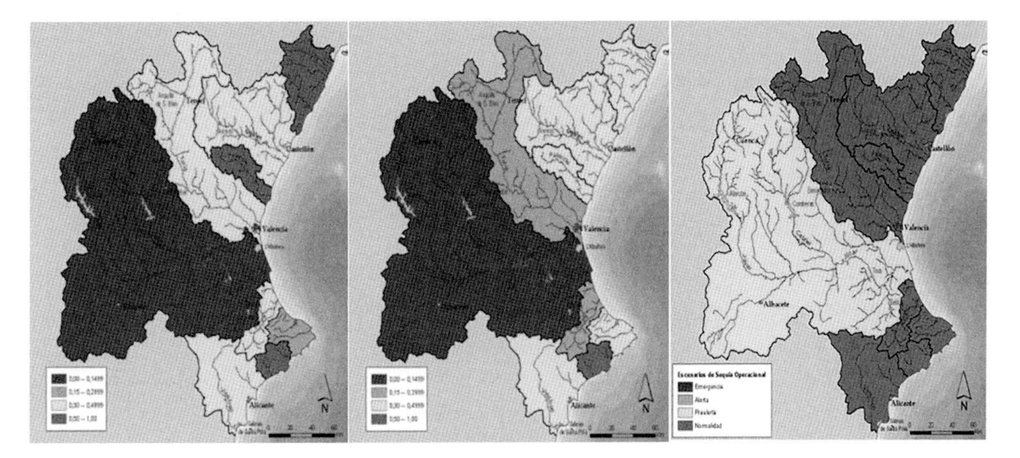

Figure 3.12.8 SODMIs for the CHJ basins corresponding to March 2006, January 2007, and March 2009 (from left to right). *Source*: Self elaboration from public domain information. (*See colour plate section for the colour representation of this figure.*)

a more elaborate and detailed information system is needed to better assess the risks, estimate the effectiveness of measures that can be used to modify the risks, and to mitigate the effects of drought on both the established uses and on the environment. Therefore, in addition to the use of SODMIs to monitor operative drought, CHJ has gone one step further, demonstrating that DSS can be very useful for the real-time management of basins, especially during drought episodes and their associated conflict situations. As mentioned in the preceding text, Aquatool DSS allows for the development and use of real-time management models that are able to assess, for instance, the risk of drought and the efficacy of proactive and reactive measures (Capilla et al., 1998). Indeed, it is applied on a regular basis for the management of the Jucar River Basin.

In order to decide real-time water allocation, each basin has a water allocation committee (WAC) that meets every month. Depending on the situation, the committee decides how much water will be delivered from each source, and how much water will remain in reservoirs. These are participatory committees in which users are represented and information for making decisions is provided by CHJ technicians, including the results of the risk assessment models (Figure 3.12.9). Following the methodology of drought risk and assessment depicted in Andreu and Solera (2006), and using the SIMRISK module of Aquatool, the probability distributions for all variables of interest are obtained (e.g. deficits in water demands, volumes in reservoir storage, deficits in ecological flows) for every month over the assumed time horizon (e.g. 12, 24, or 36 months). The hydrological inputs to this risk assessment methodology are obtained through stochastic modelling of the observed variables, generating multiple equiprobable and conditioned streamflow series that are later fed to the model. For this kind of modelling, the advances presented in Ochoa-Rivera (2002) are used to generate a series that respect not only the basic statistics of the historic streamflow values but also the historical drought characteristics.

The DSS can show these results in tabular or graphical form, highlighting the evolution of probabilities and percentiles for water demand and reservoir storage (see Figure 3.12.10). Cumulative distribution functions of any state or quality variable at

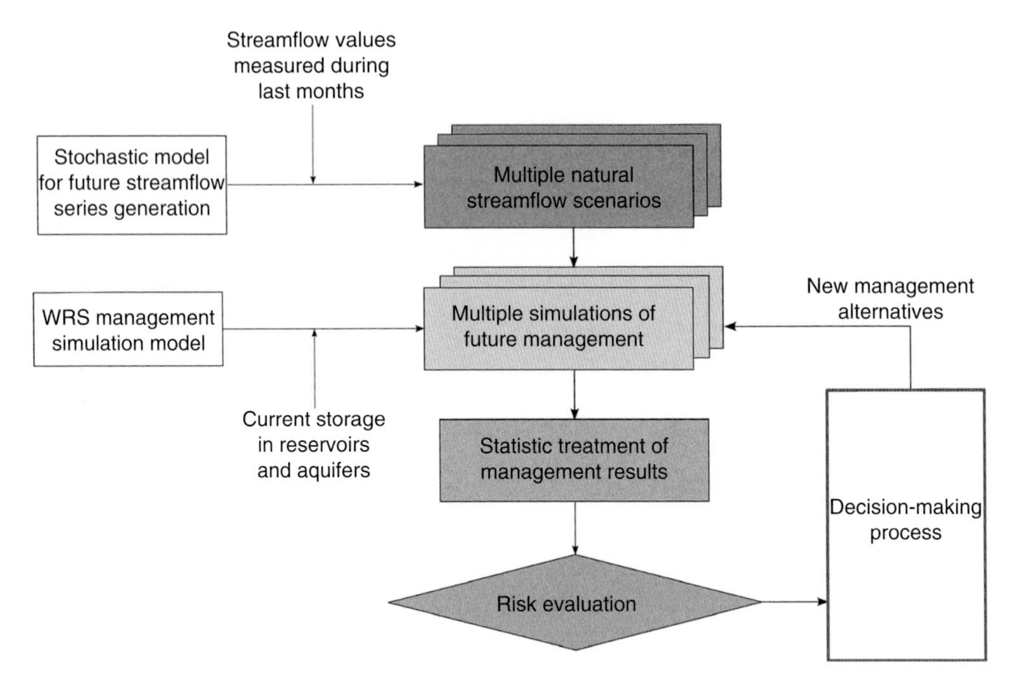

Figure 3.12.9 Scheme of the risk assessment methodology used in the Jucar River Basin. *Source*: SIMRISK.

any time can be obtained. If the estimated risks are acceptably low, then there is no need to undertake measures. However, if the estimated risks are seen as unacceptably high, then some measures must be applied. In that case, alternatives with sets of measures are formulated, and the modification of risks and the efficiency of measures are assessed. This iterative process can be continued until, eventually, an acceptable value of risk is reached, and the process ends. The approach can be applied directly to any complex water resources system, thanks to the models and data management modules included in Aquatool. In fact, this is also implemented in other Spanish basins, such as Tajo and Segura. Without the development of the DSSs, it would be very difficult, if not impossible, to estimate risks with such a complete vision of the consequences of decisions, concerning both management and infrastructure (Andreu et al., 2013).

3.12.4 Proactive and participatory drought management

3.12.4.1 Culture of adaptation to droughts in the JRBD

For more than two millennia, there has been a continuous adaptation to water scarcity, and also towards becoming less vulnerable and more resilient in face of droughts. The adaptation has been done by means of infrastructures, management, legal development, development of institutions and partnerships, and planning. Infrastructures (e.g. irrigation systems, ditches, and reservoirs) were developed since ancient times, but a very big growth occurred in the 1900s (large reservoirs, wells, water transfers, etc.). An

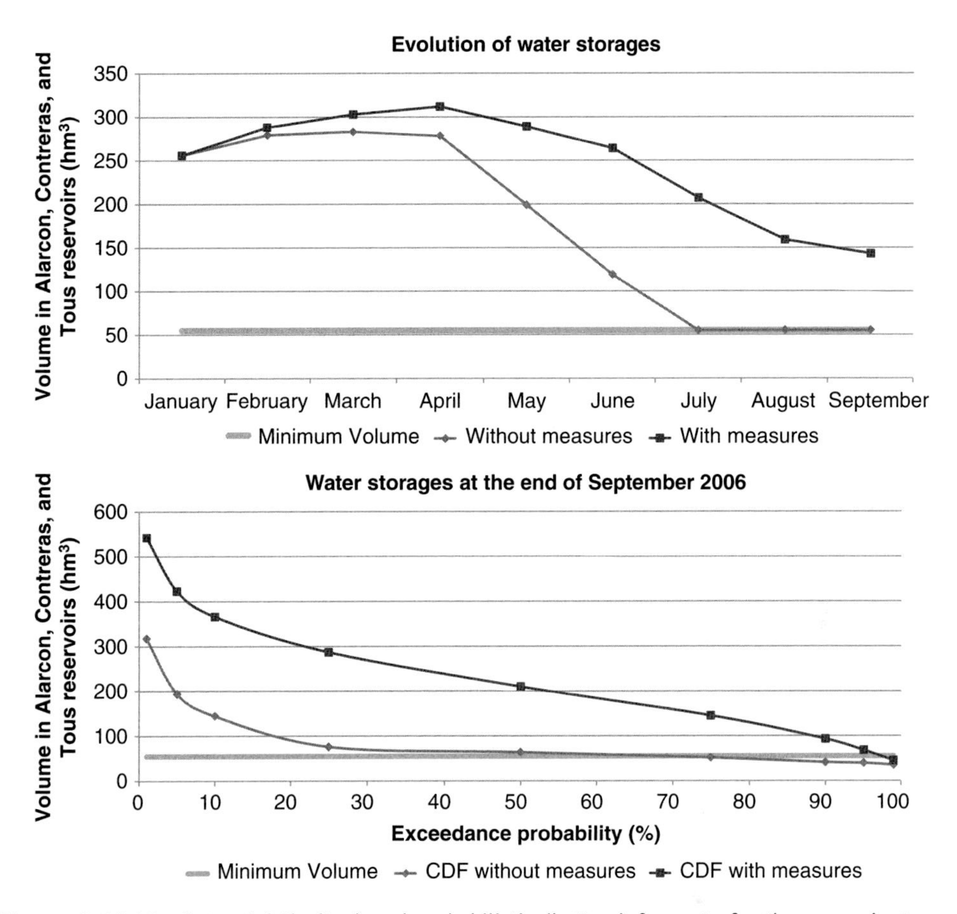

Figure 3.12.10 Deterministic (top) and probabilistic (bottom) forecasts for the reservoir storage evolution in 2006 campaign – without measures, and with the agreed measures. *Source*: Self elaboration.

example of sophisticated management is the conjunctive use of surface and groundwater performed in several basins, some since the late nineteenth century. Organisation of farmers was historically significant, as reflected by the existence of water tribunals since around the year 1000 AD (in fact, the Valencia Water Tribunal still works with complete legal authority). However, a big advance was made in 1936, when the Jucar River Basin Partnership (Confederacion Hidrografica del Jucar, or CHJ) was created jointly by the state, the agricultural users, the urban users, and the hydroelectric users. Since then, water has been managed on a basinwide basis (hydrological division) rather than along political divisions (regions). CHJ has evolved with time, and currently includes representatives from all sectors of society, as well as from regional and local administrations, NGOs, and other public interest representatives. Legal development was linked to Spanish Water Laws (1879 and 1985) and to the European Water Framework Directive (2000).

An important role in adaptation to drought has been played by the planning activity that took place in Spain from 1902 (only infrastructure planning at that time), continuing in 1933 and 1940, evolving into a more modern and holistic version since the 1980s, which culminated in the JRBD Plan of 1998, and more recently in the JRBD Plan of 2014. As explained in the preceding text, these plans assess vulnerability to droughts by means of indicators, and adopt measures in order to obtain an 'acceptable' (low) vulnerability for all water uses present in the basin district, which is already the beginning of a more proactive approach to drought management.

Finally, and triggered by the extreme drought of 1992–1995, which affected most river basin districts in Spain, the start of a fully proactive approach to drought management happened in 2001, when the Spanish water law enforced the development of the SDP, which was finalised in 2007.

3.12.4.2 Institutional, legal, and normative framework for drought planning and management

As mentioned in the preceding text, the Jucar River Basin Partnership (CHJ) is the basin organisation responsible for river basin planning, and some of its main objectives are the (planned) allocation of water for economic uses and the environment, and the selection / definition of necessary measures in infrastructures and management to provide acceptable (low) levels of vulnerability during droughts. Each basin within the CHJ territory is managed by a WAC, a participative commission that decides the real amount of water that will be delivered to each user. The WAC for the Jucar River Basin meets almost monthly, and, depending on the situation (hydrology, water storage, etc.), the water delivered corresponds to normal (planned) values, or to restricted values.

An important institutional development in order to manage droughts in emergency situations was the setting up of the permanent drought committee (PDC) by a Royal Decree in 1981 (during another extreme drought), which included representatives of the state, the governors of the affected provinces, the technical director of CHJ, and the water commissary (at that time not a part of CHJ). Later on, PDCs were also set up during the extreme situations in 1983, 1994, and 2005. Over time, the composition of the committee has evolved, and nowadays the PDC includes representatives of CHJ; of regional governments (Castilla la Mancha, Valencia, and Catalonia regions); of agricultural, industrial, and urban water consumption sectors; of the Ministry of Agriculture; of the Spanish Geological Institute; of nongovernmental environmental organisations; and of labour unions. The PDC is clearly a participatory committee in which most stakeholders are represented, and its main missions are to:

- Make decisions on water management during drought in order to achieve equilibrium between the interests of different sectors, different groups of users in the same sector, and environmental needs, as well as to to mitigate the impacts of the drought
- Perform continuous monitoring in order to control the implementation of decisions, and to follow the evolution of droughts and their impacts on users, on water quality, and on the environment (water quality in the lower Jucar river and Albufera wetlands were critical issues, as well as low flows in the middle Jucar river and low inflows to the Albufera wetlands)
- Authorise emergency works in order to improve control and efficiency of water use, connectivity, and development of additional sources (e.g. drought wells, direct treated wastewater reuse, etc.) in order to improve the reliability of supply

The Royal Decree that sets up a PDC every time it is needed gives temporary special powers to the PDC, such as to override the water rights status, if necessary, in order to find an equilibrium between economic uses and environment, and for equitable allocation of water. This is very helpful for governance in these emergency situations.

However, as mentioned in the preceding text, the real push for a proactive approach regarding droughts happened in the year 2000 (owing to the extreme drought of 1992–1995), when Spanish law made the stipulation that all river basin partnerships should design SDPs. The SDP for the JRBD (CHJ, 2007) includes monitoring for early drought detection, drought stages definition, and the measures to be applied in each of the stages, which will be explained later in this chapter.

Another important factor in this framework is the science–policy interface (S&PI) that was developed in CHJ over time, starting in the 1980s. Since then, researchers and technicians have developed different types of models that help in decision-making in CHJ (and other Spanish river basin partnerships) by responding to real questions related to water planning, and thereby contributing to the improvement of knowledge about basins and water resource systems, to the assessments of the efficiency of programs of measures, and to the estimation of risks (CHJ, 1998, 2014; Andreu et al., 2007; Ochoa-Rivera et al., 2007; Paredes-Arquiola et al., 2010; Ferrer et al., 2012; Haro et al., 2012, 2014; Lerma et al., 2013; Pulido-Velazquez et al., 2013; Pérez-Martín et al., 2014). Since the 1990s, all these have been incorporated into friendly DSSs that can be used systematically by the decision-makers, both for regular planning as well as for regular real-time management and drought management (Andreu et al., 1996, 2013).

Moreover, within CHJ, significant knowledge brokering has taken place, a major achievement being the joint development of a DSS for the Jucar River Basin in a participatory committee set up to analyse the viability of a project, as reported by Andreu et al. (2009). The committee devoted 4 months of regular meetings for reviewing all the components and models included in the DSS, resulting in an improved version for the Jucar River Basin that was considered by all parties to be a common shared vision for the water systems, and as a reliable tool to assess the alternatives. This updated version of the DSS was developed using the Aquatool DSS (Andreu et al., 2009) shell, and facilitated its use for each party in the committee. This significantly contributed to facilitating transparency, participation, and negotiation, which are all essential factors in conflict resolution. It also contributed to creating an atmosphere of more confidence and understanding from the non-technical parties regarding the techniques, models, and tools. Moreover, using the DSS as a driving force, more rationality is introduced into the debate, as opposed to debates that are based fundamentally on opinions, or on political positions. This experience paved the road for the management of the 2005–2008 drought episode in the Jucar River Basin. This participatory approach to problem and conflict resolution was very good for the development of the drought management and mitigation process.

Stakeholder involvement As discussed in the preceding text, all stakeholders are represented both in the CHJ and in the PDC. Therefore, the involvement of their representatives in the activities related to planning for droughts and real-time drought management is high. Moreover, all planning activities have a public information phase prior to the final debate and approval session in the corresponding board. Nevertheless, in the public information phase, more intense and active participation by individuals would be desirable.

3.12.4.3 Measures included in the SDP

As explained in more detail by Estrela and Vargas (2012), the objectives of the SDP are to guarantee water availability to urban demands, to avoid or minimise negative effects on the status of water bodies, and to minimise negative effects on economic activities. To reach these objectives, the SDP defines the already described SODMI, and the corresponding drought scenario definition, to be used not only as a monitoring system, but also as an early warning system, since it establishes a link between drought scenarios and a nearly complete set of measures for anticipation and mitigation, adapted to the different scenarios.

Under the normal scenario, the measures derive from regular management practices. As the drought progresses and approaches a more critical situation, measures go from control and information to conservation and restriction. These measures are developed and implemented by CHJ through the WAC until the alert scenario is triggered. When the emergency scenario is reached, however, the PDC is set up, and measures are adopted in the body of the PDC, with special powers, as already described, to improve governance in case of water conflicts.

The main measures included in the SDP can be grouped into different categories: structural measures (new pumping wells, new pipes, use of new desalination plants, etc.) and non-structural measures (changing the priority of users, water savings and demand reductions, intensification in the practice of conjunctive use of surface water and groundwater, etc.). Some of the measures will be discussed in the following text when describing the responses to the most recent extreme drought in the Jucar River Basin.

Finally, the SDPs also include a management and follow-up system that allows analysing the implementation of measures, using corrective measures in case the established objectives are not met. It also includes the obligation to develop follow-up reports and analyses of each drought episode as it occurs.

3.12.4.4 Responses to past drought events and assessment of their effects on drought impact mitigation

Andreu et al. (2013) report the case of the 2004/2005–2007/2008 drought episode in the Jucar River Basin in detail. As described and shown in Figure 3.12.5, this hydrological drought episode has been one of the most intense historical episodes on record for the Jucar River Basin. If we take into account that water demands are higher than before, the imbalance due to the drought has probably been the worst in history. The hydrological year 2004/2005 was ranked third in terms of lowest total inflows to the ensemble of Alarcon, Contreras, and Tous reservoirs, while 2005/2006 was ranked the lowest.

The drought monitoring system of CHJ was already functioning at that time, and an early warning was issued in October 2004 when the indicator entered the pre-alert scenario. After a dry autumn and winter, in February 2005, probabilistic forecasts performed with the DSS gave an early warning signal, with probabilities greater than 50% that the hydrological year would end with low storage (i.e. 10–20% of reservoir capacity). In response, water savings in agricultural uses was encouraged, 'drought wells' in the Plana de Valencia Coastal aquifers were prepared for service, and new drought wells were drilled. Besides reducing application rates, the agricultural use of surface water was reduced further by engaging in conjunctive use with groundwater. It is interesting to note that energy consumption in these wells in the lower basin is not paid for by the traditional users, but by the junior rights users and by urban users, who benefit from the surface water in exchange. At the end of the 2004/2005 hydrological year, the hydrological situation did not improve, the monitor system entered into the emergency scenario, and a PDC was set up. In March 2006, the SODMI provided the image shown in Figure 3.12.8 (left), suggesting a maximum emergency situation in the Jucar River Basin. The deterministic and stochastic forecasts for the evolution of reservoir storage obtained with the DSS were as depicted in Figure 3.12.10. As the figure shows, if no additional measures were undertaken, with respect to the ones taken in the previous year, then the total storage in the three main reservoirs of the Jucar River Basin would reach values below 55 hm^3 (minimum environmentally and technically admissible value), and the probability of ending the campaign above 193 hm^3 would be less than 5%. Hence, surface water allocated to irrigation was reduced to 43% of normal supply to traditional users, and to 30% of normal supply to junior water rights. Supplementary supplies from groundwater of about 40 hm^3 were mobilised, as well as 62 hm^3 from recirculation of water in the rice fields in the wetland area (under very strict water quality control to avoid salinity build up over tolerable limits). Another measure was to increase the amount of water supplied to the metropolitan area of Valencia from the Turia basin to 49 hm^3/year, thereby reducing the supply from the Jucar river. In order to achieve this without harming the reserves in the Turia basin, it was agreed that approximately 36 hm^3 of adequately treated wastewater should be directly supplied to traditional agricultural users in the lower Turia basin as a partial substitute for surface water. Furthermore, CHJ temporarily purchased 50 hm^3 of water rights from agricultural users of the Mancha Oriental aquifer to avoid extracting this groundwater, which resulted in a decrease of water seepage to the aquifer from the middle course of the Jucar river and an improvement in environmental flows. The decision to adopt these measures was based on the results of the DSS. Figure 3.12.10 presents the deterministic and stochastic forecasts for the evolution of reservoir storage, as calculated in April 2006, from the application of these measures. It can be seen in the figure that final storage was significantly improved (i.e. from 55 to 143 hm^3 in the deterministic forecast, and a 50% probability of ending with more than 210 hm^3), and the plan, including the measures, was approved and implemented.

Additional measures included increase of efficiency in water use (a main pipeline and two distribution pipelines that were under construction were speeded up and adapted as a substitute for the old main canal and ditches for traditional farmers in the lower Jucar); and improvement in control devices for surface water diversion. Also, special monitoring programs were included in the measures adopted for the environmental protection of critical spots, with a focus on the middle reach of the Jucar river, where groundwater abstraction in Mancha Oriental aquifer can cause river depletion up to the point of drying up the river bed; the lower reach of the Jucar river, where low flows and high pollution loads from urban waste water can cause severe problems; and the Albufera lagoon, which depends on irrigation returns for its inflows (additional inflows were provided by 26 hm^3 of treated waste water with nutrient removal and green filtering). Water quality improvement measures (e.g. removal of algae and artificial aeration) were applied in the lower Jucar river. Although not a specific measure of drought management, a waste water treatment plant that was under construction was put in service, which reduced the above-mentioned pollution loads considerably. Finally, a special groundwater monitoring program was also included among the measures in order to control the volume of water extracted, water levels, and the quality of the water.

In the hydrological years 2006/2007 and 2007/2008, total hydrological inflows were higher than in the year 2005/2006 in the Jucar river, but still under the average, and the distribution was irregular. As a result, the reservoirs in the upper part of the basin received less water, aggravating the low-flow problem in the middle Jucar river. Hence, the plans for the 2007 and 2008 campaigns approved by the PDC adopted very similar measures to those included in the 2006 campaign, but with greater intensity in order to achieve the same degree of drought mitigation and environmental protection as in 2006/2007. In the Turia basin, the situation passed from pre-alert in March 2006 to alert in January 2007, due to low hydrological inflows (Figure 3.12.10). On 30 September 2008, as a consequence of the continued measures and a slightly better hydrology, the total reservoir storage in the Jucar River Basin was 260 hm^3 – the best end to a campaign since 2004. In the hydrological year 2008/2009, furthermore, abundant rainfall brought the basins near normality, as shown in Figure 3.12.10 for SODMI in March 2009. Normality was attained in the Jucar River Basin by March 2010, after a very wet winter.

3.12.5 Conclusions

A case study of the JRBD has been presented in this chapter, where aridity, water scarcity, and recurrent multi-year meteorological and hydrological drought episodes have continuously stimulated adaptation over several centuries. Historically, one of the main objectives of the development of infrastructures, institutions, legal frameworks, and planning exercises for water resources has been to decrease vulnerability and increase resilience. The extreme drought of 1992–1995 triggered the shift to a proactive approach, leading to the creation of an SDP, as well as the development of a compounded drought operative index, used to define drought scenarios from normal to emergency, and serving as an early warning system to put into action predefined measures attached to each scenario in order to decrease vulnerability and increase resilience. The set of measures includes improvements in the efficiency of water uses; water-saving practices; conjunctive use of surface and groundwater, with economic compensation from users benefiting from the measures to users giving up surface water; public offers of water rights purchase for environmental protection; irrigation sluice water recirculation; reclaimed wastewater direct reuse; improvements in the control of water uses, water quality, and ecological status of water bodies; and a review of actions and post-analysis after a drought episode.

The participation of stakeholders is enhanced by the fact that the Jucar River Basin Partnership is in charge of all related water planning activities, and all sectors of society are included in the partnership. Moreover, a participatory PDC with special powers is set up in emergency scenarios in order to improve governance aspects.

An important reason for the better attainment of the objectives is that, for many decades now, a science–policy interface has been promoted, and models and DSSs have been developed in collaboration between scientists and practitioners at CHJ, including knowledge brokering directed at stakeholders in order to have a common shared vision of the basin through DSS, contributing to building an atmosphere of transparency and trust. It has been shown that such DSSs constitute a complementary and more precise early warning system with probabilistic assessments of risks and efficacy of measures in risk reduction, facilitating the refinement of measures to be applied and the achievement of a consensus in water conflict scenarios.

The application of such a proactive approach to the real extreme drought episode of 2004/2005 to 2007/2008 has also been described, with successful results, as recognised by the stakeholders themselves (De Stefano et al., 2013).

However, even though the status and outcome of this case study in the Jucar River Basin are quite good with respect to drought planning and management, as acknowledged by many studies (e.g. Schwabe et al., 2013; De Stefano et al., 2013), improvements can, and must, still be made with respect to: (1) the applicability of the approach to other basins in the JRBD; (2) the consolidation of measures in the Jucar River Basin itself; (3) the refinement of the monitoring system of indicators (e.g. Ortega et al., 2015); and (4) the enhancement of institutional and legal aspects.

Acknowledgements

The authors would like to thank the Ministerio de Ciencia e Innovación de España (Comisión Interministerial de Ciencia y Tecnología, CICYT) for financing the projects 'SADIDMA' (contract CGL2005-07229/HI), 'INTEGRAME' (contract CGL2009-11798), 'NUTEGES' (contract CGL2012-34978), and 'SCARCE' (Consolider-Ingenio 2010 CSD2009-00065). Thanks are also due to the European Commission for financing the projects 'SIRIUS' (FP7-SPACE-2010), 'DROUGHT–R&SPI' (FP7 contract 282769), 'ENHANCE' (FP7 contract 308438), and 'IMPREX' (H2020 contract 641811). The authors would also like to express their gratitude to the Confederacion Hidrografica del Jucar for the data provided to develop the reported works.

References

Acácio, V., Andreu, J., Assimacopoulos, D., Bifulco, C., di Carli, A., Dias, S., Kampragou, E., Haro-Monteagudo, D., Rego, F., Seidl, I., and Vasiliou, E. (2013). Review of current drought monitoring systems and identification of (further) monitoring requirements. *DROUGHT-R&SPI Technical Report 6*.

AEMET (2008). *Generacion de escenarios regionalizados de cambio climatico para España*. Madrid.

Andreu, J. and Solera, A. (2006). Methodology for the analysis of drought mitigation measures in water resources systems. In: *Drought Management and Planning for Water Resources*, 1e (ed. J. Andreu, G. Rossi, F. Vagliasindi, and A. Vela), 133–168. Boca Raton: CRC Press.

Andreu, J., Capilla, J., and Sanchís, E. (1996). AQUATOOL, a generalized decision-support system for water-resources planning and operational management. *Journal of Hydrology* 177 (3–4): 269–291.

Andreu, J., Ferrer-Polo, J., Perez, M., Solera, A., and Paredes-Arquiola, J. (2013). Drought planning and management in the Jucar River Basin, Spain. In: *Drought in Arid and Semi-Arid Regions*, 1e (ed. K. Schwabe), 237–249. Dordrecht: Springer Science+Business Media.

Andreu, J., Perez, M., Ferrer, J., Villalobos, A., and Paredes, J. (2007). Drought management decision support system by means of risk analysis models. In: *Methods and Tools for Drought Analysis and Management*, 1e (ed. G. Rossi, T. Vega, and B. Bonnacorso), 195–216. Dordrecht: Springer.

Andreu, J., Perez, M., Paredes, J., and Solera, A. (2009). Participatory analysis of the Jucar-Vinalopo (Spain) water conflict using a decision support system. In: *18th World IMACS Congress and MODSIM09 International Congress on Modelling and Simulation* (ed. R.S. Anderssen, R.D. Braddock, and L.T.H. Newham), 3230–3236. [Online] Available at http://mssanz.org.au/modsim09/I3/andreu_b.pdf (accessed 11 March 2016).

Capilla, J.E., Andrew, J., Solera, A., and Quispe, S.S. (1998). Risk based water resources management. *WIT Transactions on Ecology and the Environment*, 26: 10.

CEDEX (2010). *Evaluacion del impacto del cambio climatico en los recursos hidricos en regimen natural*. Madrid.

CHJ (1998). *Plan Hidrologico de Cuenca del Jucar*. Valencia.

CHJ (2004). *Jucar Pilot River Basin, Provisional Article 5 Report Pursuant to the Water Framework Directive*. Valencia.

CHJ (2007). *Plan especial de alerta y eventual sequia en la Confederacion Hidrografica del Jucar*. Valencia.

CHJ (2014). *Propuesta de Proyecto de revision del plan hidrologico. Ciclo de planificacion hidrologica 2015–2021*. Valencia.

De Stefano, L., Urquijo, J., Krampagkou, E., and Assimacopoulos, D. (2013). Lessons learnt from the analysis of past drought management practices in selected European regions: experience to guide future policies. In: *13th International Conference on Environmental Science and Technology*. [Online] Available at http://environ.chemeng.ntua.gr/en/UserFiles/files/0255.pdf (accessed 11 March 2016).

Estrela, T., Fidalgo, A., Fullana, J., Maestu, J., Pérez, M.A., and Pujante, A.M. (2004). Júcar pilot river basin, provisional article 5 report pursuant to the water framework directive. Confederación Hidrográfica del Júcar, Valencia.

Estrela, T. and Vargas, E. (2012). Drought management plans in the European Union. The case of Spain. *Water Resources Management* 26 (6): 1537–1553.

Fernández, B., Andreu, J., and Sanchez-Quispe, S. (1998). Analisis de sequias en sistemas complejos sometidos a regulacion y gestión. In: *XVIII Congreso Latinoamericano de Hidráulica, Memorias, Avances en Hidráulica 1* (ed. A.A. Aldama, J. Aparicio, M. Berezowsky, C. Cruiskhank, R. Domínguez, R. Fuentes, J.A. Maza, C. Menéndez, D. Pérez Franco, and G. Sotelo). México: IAHR-AMH-IMTA.

Ferrer, J., Pérez-Martín, M., Jiménez, S., Estrela, T., and Andreu, J. (2012). GIS-based models for water quantity and quality assessment in the Júcar River Basin, Spain, including climate change effects. *Science of the Total Environment* 440: 42–59.

Haro, D., Paredes, J., Solera, A., and Andreu, J. (2012). A model for solving the optimal water allocation problem in river basins with network flow programming when introducing non-linearities. *Water Resources Management* 26 (14): 4059–4071.

Haro, D., Solera, A., Paredes, J., and Andreu, J. (2014). Methodology for drought risk assessment in within-year regulated reservoir systems. Application to the Orbigo River System (Spain). *Water Resources Management* 28 (11): 3801–3814.

Lavorel, S., Canadell, J., Rambal, S., and Terradas, J. (1998). Mediterranean terrestrial ecosystems: research priorities on global change effects. *Global Ecology and Biogeography Letters* 7(3): 157.

Lerma, N., Paredes-Arquiola, J., Andreu, J., and Solera, A. (2013). Development of operating rules for a complex multi-reservoir system by coupling genetic algorithms and network optimization. *Hydrological Sciences Journal* 58 (4): 797–812.

McKee, T.B., Doesken, N.J., and Kleist, J. (1993). The relationship of drought frequency and duration to time scales. In: *Proceedings of the 8th Conference on Applied Climatology* (Vol. 17, No. 22, pp. 179–183). Boston, MA, USA: American Meteorological Society.

Ochoa-Rivera, J.C., García-Bartual, R., and Andreu, J. (2002). Multivariate synthetic streamflow generation using a hybrid model based on artificial neural networks. *Hydrology and Earth System Sciences Discussions* 6(4): 641–654.

Ochoa-Rivera, J., Andreu, J., and García-Bartual, R. (2007). Influence of inflows modeling on management simulation of water resources system. *Journal of Water Resources Planning and Management* 133 (2): 106–116.

Ortega, T., Estrela, T., and Perez-Martin, M. (2015). The drought indicator system in the Jucar River Basin Authority. In: *Drought: Research and Science–Policy Interfacing*, 1e (ed. J. Andreu, A. Solera, J. Paredes-Arquiola, D. Haro-Monteagudo, and H. Van Lanen), 219–224. Leiden: CRC Press.

Paredes-Arquiola, J., Andreu-Álvarez, J., Martín-Monerris, M., and Solera, A. (2010). Water quantity and quality models applied to the Jucar River Basin, Spain. *Water Resources Management* 24 (11): 2759–2779.

Pérez-Martín, M., Estrela, T., Andreu, J., and Ferrer, J. (2014). Modeling water resources and river–aquifer interaction in the Júcar River Basin, Spain. *Water Resources Management* 28 (12): 4337–4358.

Pulido-Velazquez, M., Alvarez-Mendiola, E., and Andreu, J. (2013). Design of efficient water pricing policies integrating basinwide resource opportunity costs. *Journal of Water Resources Planning and Management* 139 (5): 583–592.

Sanchez-Quispe, S., Solera, A., and Andreu, J. (2000). Gestion de sistemas de recursos hidricos basado en la evaluacion del riesgo de sequia. In: *XIX Congreso Latinoamericano de Hidraulica*. Cordoba.

Schwabe, K. Albiac-Murillo, J., Connor, J.D., Hassan, R., and Meza González, L. ed. (2013). *Drought in Arid and Semi-Arid Regions*, 471–507. Dordrecht: Springer.

Van Lanen, H.A.J., Alderlieste, M.A.A., Acacio, V., Andreu, J., Garnier, E., Gudmundsson, L., Monteagudo, D.H., Lekkas, D., Paredes, J., Solera, A., Assimacopoulos, D., Rego, F., Seneviratne, S., Stahl, K., and Tallaksen, L.M. (2013a). *Quantitative Analysis of Historic Droughts in Selected European Case Study Areas* (No. 8, p. 61). Wageningen Universiteit.

Van Lanen, H.A.J., Alderlieste, M.A.A., Van Der Heijden, A., Assimacopoulos, D., Dias, S., Gudmundsson, L., Monteagudo, D.H., Andreu, J., Bifulco, C., Gero, F., Paredes, J., and Solera, A. (2013b). *Likelihood of Future Drought Hazards: Selected European Case Studies* (No. 11, p. 52). Wageningen Universiteit.

Vicente-Serrano, S., González-Hidalgo, J., De Luis, M., and Raventós, J. (2004). Drought patterns in the Mediterranean area: the Valencia region (eastern Spain). *Climate Research* 26: 5–15.

Villalobos de Alba, Ángel Alfonso. (2007). Análisis y seguimiento de distintos tipos de sequía en la cuenca del río Jucar. Universitat Politècnica de València.

Yevjevich, V. (1967). An objective approach to definitions and investigations of continental hydrologic droughts. *Hydrology Papers (Colorado State University)* 23: 1–18.

3.13
Drought Risk and Management in Syros, Greece

Dionysis Assimacopoulos[1] and Eleni Kampragou[2]

[1]*National Technical University of Athens, Athens, Greece*
[2]*Hellenic Ministry of Environment and Energy, Athens, Greece*

3.13.1 Introduction

Syros is an island located at the centre of the Cyclades complex in the Aegean Sea, in Greece (Figure 3.13.1). It is well known for two distinct reasons: (i) its physical features – that is, the presence of blueschist- and eclogite-facies rocks on the island (Keiter et al., 2004); and (ii) its socioeconomic features – that is, its industrial and administrative services character as compared to the rest of the touristic Cycladic islands. The port of Syros was among the most important in Greece during the nineteenth century, supporting the development of trade and industries and thus increasing the welfare of the locals. Today, Syros is the administrative centre of the South Aegean region, and many administrative services of the region as well as of the prefecture of Cyclades are located on the island.

The main economic activities are agriculture, services (employment in public services), trade/commerce, shipping, and tourism. There are numerous greenhouses producing fresh vegetables that are distributed in the local market, and also exported to other areas. The Neorion shipyard is an important economic source for the island, and there has also been an increase in tourism in the last few decades, owing to the proximity of the island to Athens and the port of Piraeus.

Syros is a semi-arid Mediterranean island (Massas et al., 2009) with low precipitation height (on average 293 mm for the 1970–2010 period, in the range 0–596 mm). Water scarcity is further aggravated by the lack of significant surface water bodies. As a result, Syros was one of the first islands that turned to seawater desalination for supporting the domestic water supply, benefiting from the technical capacity in the industrial and shipping sectors. The first unit was installed in 1989 in Hermoupolis, the capital of the island. As of 2015, there were 13 units in operation, with a total capacity 8340 m^3/day, covering the majority of domestic demand. Irrigation demand is mainly covered by

Drought: Science and Policy, First Edition.
Edited by Ana Iglesias, Dionysis Assimacopoulos, and Henny A.J. Van Lanen.

Main features
Area: 84 km²
Permanent population: 21 507 (2011 census)
Artificial surfaces: 3.1 km²
Agricultural areas: 43.5 km²
Forest and semi-natural areas: 38.2 km²

Figure 3.13.1 Location of Syros, Greece.

Table 3.13.1 Water supply and demand in Syros (Hellenic Ministry of Development, 2008).

Sector	Demand (m³)	Coverage (%)	
		Desalination	Groundwater
Urban use (includes domestic, cattle breeding, industrial, and public use)	1855053	49.4	21.5
Agriculture	1896552	—	54.2

groundwater. Overall, there is a negative water balance in Syros (Table 3.13.1), with the highest deficits reported in agriculture.

The primary goal of water management therefore was, and still is, to cope with water scarcity ensuring water supply. Even though the Cycladic region is one of the most drought-prone areas of Greece (according to Tigkas, 2008, the frequency of drought conditions equalled 45% of the years in the period 1955–2002), no particular emphasis has been placed on dealing with drought episodes in Syros. To date, a crisis-based management approach has been adopted, aimed at temporarily increasing water supply (e.g. water hauling from the mainland) or reducing demand by setting restrictions in use (both agricultural and domestic use).

The following text briefly describes the current drought management framework in Syros and discusses the approach followed for the assessment of options for future drought risk reduction.

3.13.2 Droughts in Syros

3.13.2.1 Past droughts

There are quite a few methods for analysing and characterising drought episodes (e.g. see the review by Keyantash and Dracup, 2002). Several studies on drought characterisation in the Cyclades have been undertaken using the reconnaissance drought index (e.g. Tigkas, 2008; Tsakiris and Vangelis, 2005), deciles (e.g. Kanellou et al., 2008), the Palmer drought severity index (e.g. Kanellou et al., 2008; Dalezios et al., 2000), and the standard precipitation index (e.g. Karavitis et al., 2011; Livada and Assimakopoulos, 2007). For Syros, the following three approaches have been selected for the analysis of past drought events:

1. *Standard precipitation index* (SPI; McKee et al., 1993), calculated for five different scales (1 month, 3 months, 6 months, 9 months, and 12 months) to analyse short-term and long-term effects of rainfall deficits.
2. *Threshold analysis* (Yevjevich, 1967; Hisdal et al., 2004), using precipitation data instead of runoff data. Typically, the Q80 or Q90 percentiles (values that are exceeded by 80% and 90% of the dataset, respectively) are used in drought analyses. However, for Syros, these percentiles are equal to zero, and would result in the identification of drought conditions for all months in the dataset. Two variable monthly thresholds have been selected for this study, corresponding to: (i) the 50th percentile (Q50) of monthly precipitation, and (ii) the monthly average precipitation.
3. Trend analysis of precipitation data using: (i) the Mann–Kendall test (S and Z statistics; Mann, 1945; Kendall, 1975); (ii) Sen's slope (Q; Sen, 1968); and (iii) the percentage change relative to the mean (T, %; Stahl et al., 2010, 2012).

Table 3.13.2 summarises the SPI results for the case of SPI-3 and SPI-12, corresponding respectively to the short-term and long-term effects of precipitation deficits. Drought conditions were estimated for almost 40% of the months in the 1970–2010 period, with severe and extreme events being calculated only in the case of SPI-12. This result is also confirmed by results from the threshold method, as drought events of less than 1 year duration are more frequent in the island (Figure 3.13.2).

The hypothesis of no trend (null hypothesis) or trend was tested at the 95% significance level. Trends found to be statistically significant are marked in bold. Trend analysis indicates a significant increasing trend in precipitation for most months of the year. It is important to note that, particularly for precipitation, the trend analysis result differs if the time period of analysis is divided in two sub-periods (1970–1990 and 1991–2010). There was an increase in average annual precipitation height in the second period (337 mm versus 244 mm), which explains the increasing trend found in precipitation height (Table 3.13.3).

Table 3.13.2 Frequency of drought and wet periods in Syros for 1970–2010, based on SPI.

SPI value	Category	Frequency = SPI-3	Frequency = SPI-12
≥0	Wet conditions	266 (58.8%)	268 (61.8%)
0 to −0.99	Mild drought	149 (33.0%)	100 (23.0%)
−1 to −1.49	Moderate drought	37 (8.2%)	20 (4.6%)
−1.50 to −1.99	Severe drought	0 (%)	36 (8.3%)
−2 or less	Extreme drought	0 (%)	10 (2.3%)

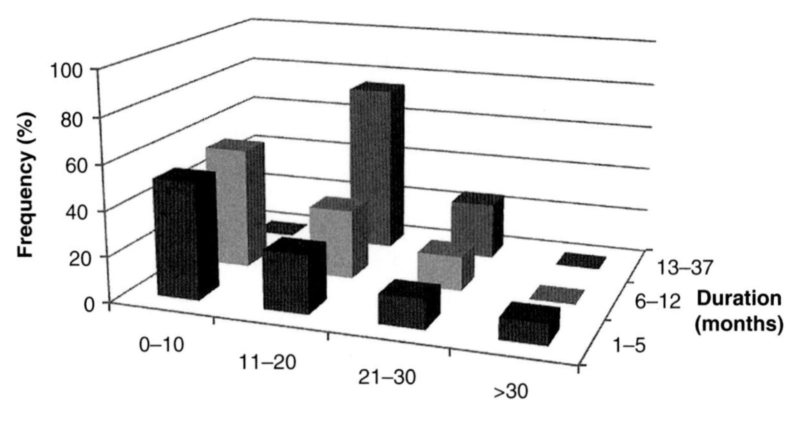

Figure 3.13.2 Drought intensity for different drought durations.

Table 3.13.3 Trend analysis results for Syros: Precipitation.

Variable	Z	Q	p-value	Trend	T (%)	Trend (1970–1990)	Trend (1991–2010)
P_{annual}	2.99	7.200	0.003	**Trend (↑)**	93.3	No trend	**Trend (↑)**
P_{Jan}	1.44	0.594	0.003	No trend	46.8	No trend	No trend
P_{Feb}	0.69	0.256	0.150	No trend	19.6	No trend	**Trend (↑)**
P_{Mar}	1.49	0.504	0.488	No trend	51.5	No trend	No trend
P_{Apr}	1.32	0.156	0.135	No trend	36.5	No trend	No trend
P_{May}	2.07	0.022	0.185	Trend (↑)	9.6	No trend	No trend
P_{Jun}	1.89	0.000	0.039	No trend	0.0	No trend	No trend
P_{Jul}	2.23	0.000	0.059	Trend (↑)	0.0	No trend	**Trend (↑)**
P_{Aug}	2.91	0.000	0.026	**Trend (↑)**	0.0	No trend	No trend
P_{Sep}	2.38	0.000	0.004	**Trend (↑)**	0.0	No trend	**Trend (↑)**
P_{Oct}	2.33	0.350	0.017	**Trend (↑)**	49.9	No trend	**Trend (↑)**
P_{Nov}	2.08	0.691	0.020	**Trend (↑)**	77.3	No trend	No trend
P_{Dec}	2.93	1.795	0.038	**Trend (↑)**	120.6	No trend	**Trend (↑)**

3.13.2.2 Future droughts

Future droughts in Syros have been a topic of research in the EU-funded FP7 DROUGHT–R&SPI project. Future drought characteristics (duration and deficit volume) were analysed for two SRES scenarios (A2, B1), using the simulated multi-model mean flow for three different GCMs and six large-scale hydrological models (van Lanen et al., 2013; Alderlieste et al., 2014). Results indicate an increase both in duration and deficit volume in the near (2021–2050) and far future (2071–2100), as a result of reduced precipitation height.

3.13.2.3 Key messages

The analysis of past and future droughts confirms the need for increasing awareness on the drought hazard. Drought events are (and are expected to be) frequent and of long duration. It is thus necessary to distinguish between water scarcity and drought; to improve preparedness for prolonged drought events; and to support drought monitoring, forecasting, and early warning.

3.13.3 Drought risk and mitigation

A national drought policy to guide drought management at the local level does not exist (Table 3.13.4). As a result, the authorities in charge of water management in Syros focus their efforts on coping with long-term water scarcity. The only drought-relevant management approach is the declaration of a 'state of emergency' (crisis management) for the island, which involves only the request for emergency funds (and for water transfer from the mainland in the past).

Drought vulnerability and management have been analysed for Syros using the DPSIR approach (Figure 3.13.3). Water scarcity, intensive abstraction of groundwater sources for use mainly in agriculture, limited drought awareness, and the lack of drought planning are the main factors that shape vulnerability to drought.

The selection of measures for coping with drought in the past (Tablc 3.13.5) was not associated with the type (e.g. meteorological, hydrological) or severity (e.g. mild, severe extreme) of drought; instead, it was mainly based on the availability of financial resources. These measures only partially addressed the underlying factors of vulnerability, and were mainly aimed at supply enhancement. They were assessed by local stakeholders in terms of effectiveness in drought impact mitigation, and their ranking in the order of efficiency is:

Table 3.13.4 Drought management framework in Syros.

Component	Description
Legal framework	• Law 1739/1987 'for the management of water resources' • Law 1650/1986 'for the protection of the environment', regarding the quality of water • European Framework Directive 2000/60/EU • Law 3199/2003 for the 'protection and management of water' • National Action Plan for Combating Desertification (NAPCD)
Competent authorities	• Ministry of Environment and Energy • Water Directorate of the Region of South Aegean • Municipality of Syros-Hermoupolis • Municipal Enterprise for Water Supply and Sewerage of Syros
Policy or regulatory gaps	• Lack of a reliable monitoring network • Lack of studies on vulnerability to drought • Limited assessment of the applicability/usefulness of established drought indices for insular areas • Fragmentation of responsibilities • Lack of capacity building in institutions regarding drought management • Lack of a coherent compensation policy, particularly for drought, as damage compensation schemes in agriculture address mostly other extreme events (e.g. heat waves, floods, etc.), which are easier to monitor and identify • Uneven allocation of funds (among users/municipalities)

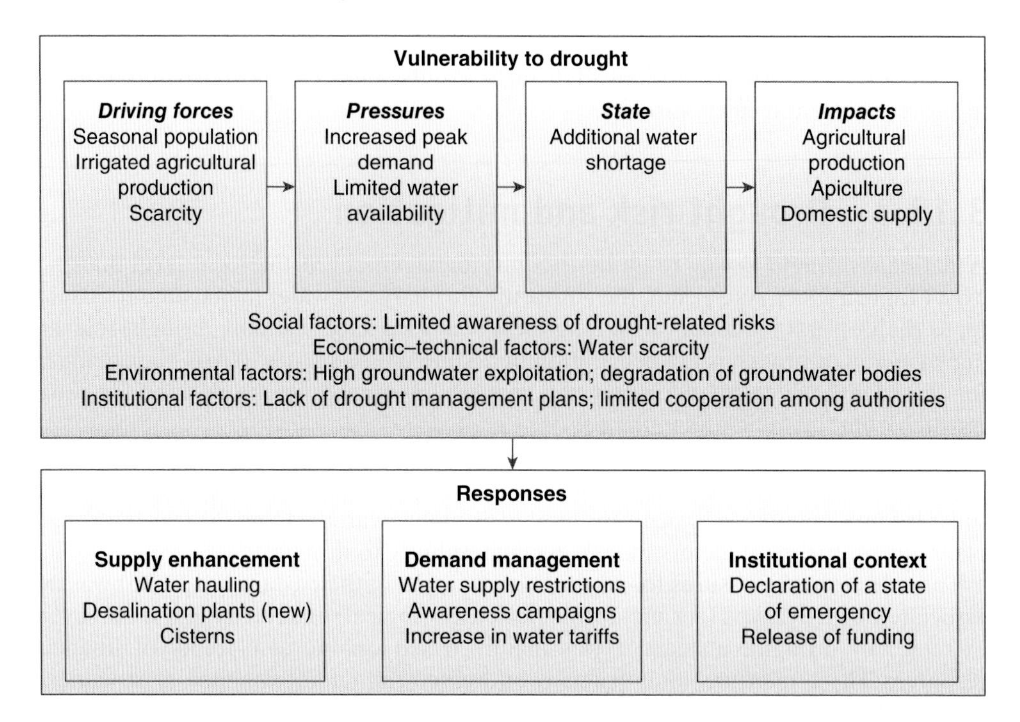

Figure 3.13.3 The DPSIR framework for analysing drought vulnerability and management in Syros. *Source*: Adapted from De Stefano et al. (2013).

1. Restrictions on water use
2. Installation of new desalination plants
3. Water storage in cisterns (rainwater harvesting)
4. Awareness campaigns
5. Compensation to farmers
6. Supplementary supply from desalination plants for irrigation

Options for future drought risk reduction have been evaluated by adopting a risk-based assessment approach. Options have been evaluated on the basis of risk of water deficit (and the corresponding economic losses) and vulnerability of the water system. The steps followed in the assessment are:

1. Water balance modeling to estimate water deficits – for baseline conditions (2010) and for the future (2011–2050), in three cases (baseline condition, best-case development scenario, and worst-case development scenario) – and for alternative drought mitigation options.
2. Estimation of economic losses (DEI_s, Equation 3.13.1) for the urban and agricultural sectors. Losses in the domestic sector are estimated as the substitution cost (for alternative water supply of adequate quality) for meeting water demand, and in the agriculture sector as the reduction in crop yield and income (no change in market prices assumed) owing to reduced water availability during drought events. The estimation of crop yield is based on Smith and Steduto (2012).

Table 3.13.5 List of measures for the past two episodes.

Drought episode	1999–2001 (severe drought)	2007 (drought and heat wave)
Impacts	• Agriculture: Reduction in the production of barley, vegetables, wine grapes • Cattle breeding: Reduced production of fodder • Apiculture: Reduction in honey production • Reduced water availability	• Some impacts on agriculture • Impacts on honey production
Responses	• Periodic restrictions in water use • Water transfer from the mainland • Construction of cisterns for rainwater harvesting • Awareness campaigns	• New desalination plants/ maintenance of existing ones • Supplementary supply from desalination plants for irrigation • Water pricing • Temporary transfer of water from one region to other • Increase of abstraction from groundwater bodies • Awareness campaigns

3. Estimation of the risk of economic losses (RL_s, Equation 3.13.2). Drought severity–duration–frequency curves have been developed in order to estimate the probability of occurrence of a drought event for a given return period T and duration D.
4. Calculation of the vulnerability of the water system (V_s, Equation 3.13.3), assuming that vulnerability for the baseline conditions equals one.

$$DEI_s = f\left(DI_{T,D}.WQ_sP_s\right) \tag{3.13.1}$$

$$RL_s = p_{T,D}.DEI_s \tag{3.13.2}$$

$$V_s = \frac{RL_s \text{with the implementation of options}}{RL_s\left(\text{current state}\right)} \tag{3.13.3}$$

Here, $DI_{T,D}$ is the drought intensity for a return period T and duration D; $p_{T,D}$ is the probability of occurrence of a drought event for a return period T and duration D; WQ_s is the water availability (or deficit) for the specific sector (s); P_s refers to parameters that specify the economic impacts in the specific sector (s); DEI_s is the drought economic impact for the specific sector (s); and RL_s is the risk of economic losses.

The list of options for future drought risk reduction was drafted on the basis of past practices to cope with water scarcity and drought, recommendations from stakeholders, and proposals in a previous study undertaken for the Ministry of Development (titled 'Development of systems and tools for water resources management in the South Aegean Water District', 2000–2008). Five options have been selected for further assessment: (i) water storage in cisterns for irrigation use (CI); (ii) direct wastewater reuse for irrigation (DWWI); (iii) rainwater harvesting in cisterns for domestic use (CD); (iv) wastewater reuse for groundwater recharge (WWG); and (v) desalination (DES). Figure 3.13.4 presents the contribution of each option to vulnerability reduction for a drought event of 5-year return period and 20–24-month duration,

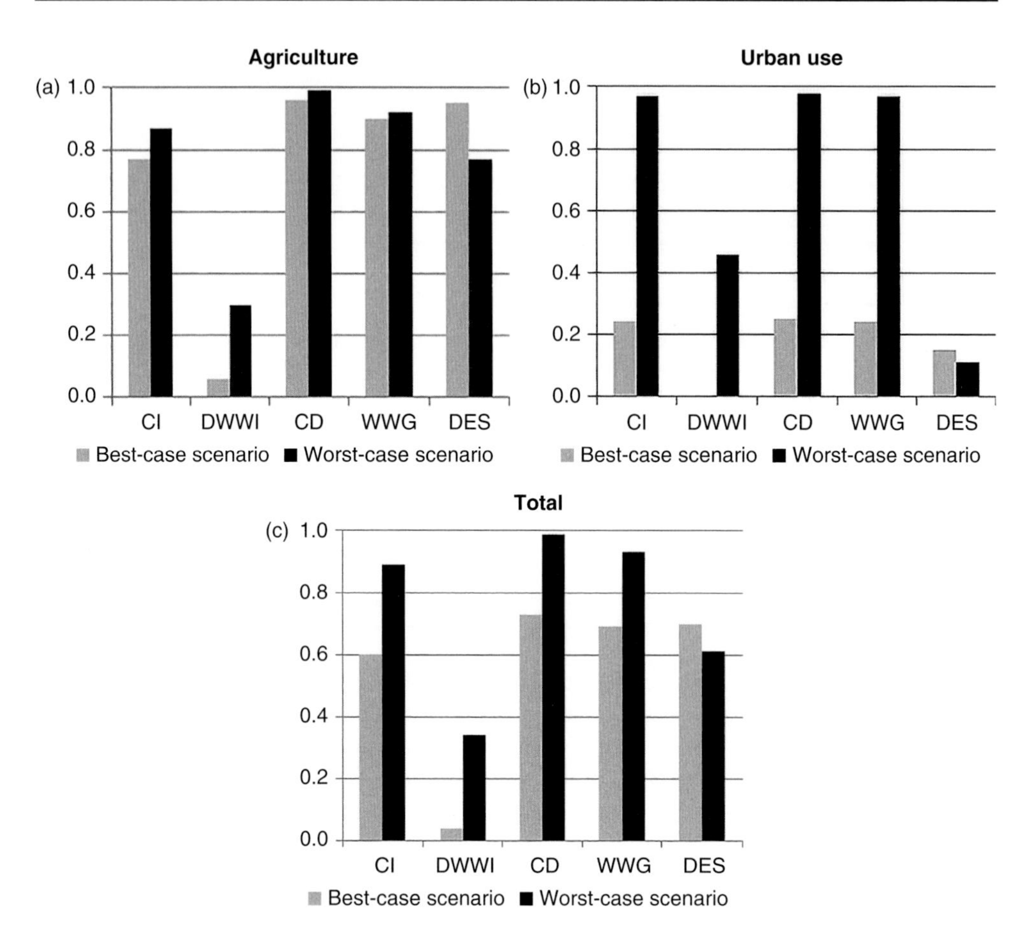

Figure 3.13.4 Change in vulnerability to drought after the implementation of options for future drought risk mitigation (for a drought event of 5-year return period and 20–24-month duration, 22% probability of occurrence).

since the analysis of past and future drought characteristics indicates that emphasis should be placed on prolonged events. Similar results have been obtained for drought events of different durations and return periods, indicating that a long-term strategy for integrated water and drought management in Syros should incorporate options for supply enhancement in agriculture and groundwater protection. Desalination, as an option, remains a reliable solution for the urban sector.

3.13.4 Lessons learnt – the need for participatory drought management

The analysis undertaken for Syros aimed at the identification of drought-related risks and the proposal of measures for future risk reduction. In order to better reflect the drought management framework in the island, local stakeholders were involved in the process through specific actions (Figure 3.13.5). Through a series of interviews (nine in

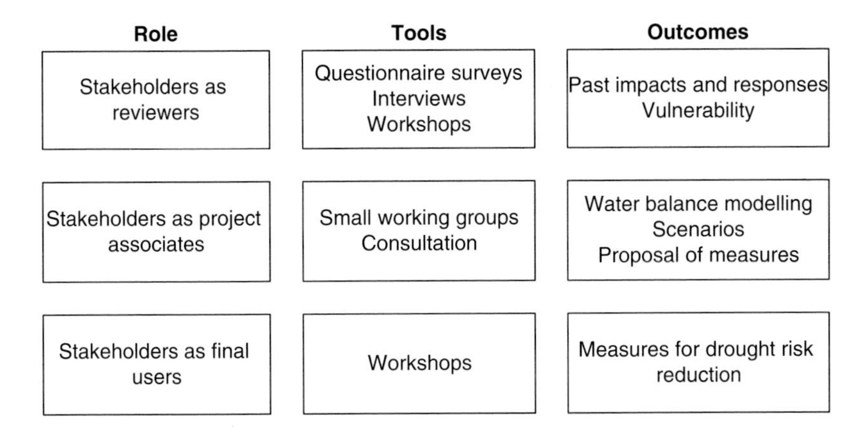

Figure 3.13.5 Stakeholder involvement.

total), a questionnaire survey, and workshops, information has been collected on drought awareness, past drought impacts, responses, and stakeholder involvement in water and drought management. In addition, stakeholders contributed to the development of the Syros water balance model and recommended measures for future drought risk reduction.

It was commonly pointed out by all stakeholders that addressing water scarcity is the main challenge and a priority for water managers. Measures taken in the past for coping with water scarcity had had a positive effect on drought management, and thus the need for separate management approaches was not widely acknowledged by local stakeholders. However, after presenting the research results on drought characteristics and impacts, the need for drought management has been accepted by locals as a means to protect water resources and maintain economic activities (i.e. agriculture) in the island.

Stakeholders identified three main practices that could improve water and drought management: (i) desalination; (ii) water storage in cisterns; and (iii) wastewater reuse (either for groundwater recharge or irrigation). These options were modelled, and their anticipated contribution to drought mitigation was quantified. The implementation of all these options would result in a vulnerability reduction, with the first and third options being the most promising ones.

The involvement of local stakeholders proved significant in formulating recommendations, as it allowed for incorporating the knowledge of local conditions and making proposals suitable to the local context. The current water management framework (Law 3199/2003) foresees stakeholder involvement processes, which have so far been limited to consultation on water management plans. More formal processes are needed for improving access to information and enabling the active involvement of stakeholders in planning.

Acknowledgements

This research was undertaken as part of the European Union (FP7)–funded DROUGHT–R&SPI Integrated project ('Fostering European Drought Research and Science–Policy Interfacing'), contract 282769.

References

Alderlieste, M.A.A., van Lanen, H.A.J., and Wanders, N. (2014). Future low flows and hydrological drought: how certain are these for Europe? In: *Hydrology in a Changing World: Environmental and Human Dimensions* (ed. T.M. Daniell, H.A.J. van Lanen, S. Demuth, G. Laaha, E. Servat, G. Mahe, J.-F. Boyer, J.-E. Paturel, A. Dezetter, and D. Ruelland), 60–65. Montpellier, France: IAHS Publ. No. 363.

Dalezios, N.R., Loukas, A., Vasiliades, L., and Liakopoulos, E. (2000). Severity–duration–frequency analysis of droughts and wets periods in Greece. *Hydrological Sciences Journal* 45 (5): 751–770. doi:10.1080/02626660009492375.

De Stefano, L., Urquijo, J., Kampragkou, E., and Assimacopoulos, D. (2013). Lessons learnt from the analysis of past drought management practices in selected European regions: experience to guide future policies. *13th International Conference on Environmental Science and Technology* (5–7 September 2013). Athens, Greece.

Hellenic Ministry of Development (2008). *Development of Systems and Tools for Water Resources Management in the Hydrological Department of the Aegean Islands, 2001–2008.*

Hisdal, H., Tallaksen, L.M., Clausen, B., Peters, E., and Gustard, A. (2004). Drought characteristics. In: *Hydrological Drought: Processes and Estimation Methods for Streamflow and Groundwater* (ed. L.M. Tallaksen and H. van Lanen), 139–198. Elsevier: Developments in Water Science 48.

Kanellou, E., Domenikiotis, C., Hondronikou, E., and Dalezios, N. (2008). Index-based drought assessment in semi-arid areas of Greece based on conventional data. *European Water* 23 (24): 87–98.

Karavitis, C., Alexandris, S., Tsesmelis, D., and Athanasopoulos, G. (2011). Application of the standardized precipitation index (SPI) in Greece. *Water* 3: 787–805. doi:10.3390/w3030787.

Keiter, M., Piepjohn, K., Ballhaus, C., Lagos, M., and Bode, M. (2004). Structural development of high-pressure metamorphic rocks on Syros island (Cyclades, Greece). *Journal of Structural Geology* 26: 1433–1445.

Kendall, M.G. (1975). *Rank Correlation Methods*, 4e., London, UK: Charles Griffin.

Keyantash, J. and Dracup, J.A. (2002). The quantification of drought: an evaluation of drought indices. *Bulletin of the American Meteorological Society* 83: 1167–1180.

Livada, I. and Assimakopoulos, V.D. (2007). Spatial and temporal analysis of drought in Greece using the standardized precipitation index (SPI). *Theoretical and Applied Climatology* 89: 143–153. doi:10.1007/s00704-005-0227-z.

Mann, H.B. (1945). Non-parametric test against trend. *Econometrica* 13: 245–259.

Massas, I., Ehaliotis, C., Gerontidis, S., and Sarris, E. (2009). Elevated heavy metal concentrations in top soils of an Aegean island town (Greece): total and available forms, origin and distribution. *Environmental Monitoring and Assessment* 151: 105–116. doi: 10.1007/s10661-008-0253-2.

McKee, T.B., Doesken, N.J., and Kleist, J. (1993). The relationship of drought frequency and duration to timescales. *8th Conference on Applied Climatology* (17–22 January 1993). Anaheim, California, USA.

Sen, P.K. (1968). Estimates of the regression coefficient based on Kendall's tau. *Journal of the American Statistical Association* 63: 1379–1389.

Stahl, K., Hisdal, H., Hannaford, J., Tallaksen, L.M., van Lanen, H.A.J., Sauquet, E., Demuth, S., Fendekova, M., and J. Jódar (2010). Streamflow trends in Europe: evidence from a dataset of near-natural catchments. *Hydrology and Earth System Sciences* 14: 2367–2382.

Stahl, K., Tallaksen, L.M., Hannaford, J., and van Lanen, H.A.J. (2012). Filling the white space on maps of European runoff trends: estimates from a multi-model ensemble. *Hydrology and Earth System Sciences Discussions* 9: 2005–2032.

Smith, M. and Steduto, P. (2012). Yield response to water: the original FAO water production function. In: *Crop yield response to water. FAO Irrigation and Drainage Paper No. 66* (ed. P. Steduto,

T.C. Hsiao, E. Fereres, and D. Raes), 6–13. Rome: Food and Agriculture Organisation of the United Nations.

Tigkas, D. (2008). Drought characterization and monitoring in regions of Greece. *European Water* 23 (24): 29–39.

Tsakiris, G. and Vangelis, H. (2005). Establishing a drought index incorporating evapotranspiration. *European Water* 9 (10): 3–11.

Yevjevich, V. (1967). An Objective Approach to Definitions and Investigations of Continental Hydrologic Droughts. *Hydrology Papers 23.* Colorado State University, Fort Collins, USA.

van Lanen, H.A.J. Alderlieste, M.A.A., Van Der Heijden, A., Assimacopoulos, D., Dias, S., Gudmundsson, L., Haro Monteagudo, D., Andreu, J., Bifulco, C., Gero, F., Paredes, J., and Solera, A. (2013). Likelihood of future drought hazards: selected European case studies. *DROUGHT–R&SPI Technical Report No. 11.* Available from http://www.eu-drought.org/technicalreports (last accessed on 8/02/2016).

Index

Page references in *italics* refer to Figures; those in **bold** refer to Tables

actual evapotranspiration (ETa), 11, 12
agricultural insurance, 148
  failure of market, 148
agronomic drought indicators, 151, **152**
ANalysis Of VAriance (ANOVA) technique, 84, 85
Aquatool DSS, 224, 227, 228, 231
  SIMRISK MODULE, 227
area-yield (revenue) index insurance, 150
artificial neural network approaches, 222
atmospheric blocking, 7
atmospheric synoptic circulation changes, 9
Australia
  Millennium Drought, 20
  New South Wales, cumulative rainfall insurance contract, 155
autoregressive–moving average (ARMA) modelling, 221–2

Barker, Thomas: meterological journal, *46*
Base Flow Index (BFI), 17
Bedford Level Company, 47
benchmark networks, 31

Cabina di Regia, 208, 210, 211, 212
Canal du Midi, 47
Central American Dry Corridor (CADC), 124
CGCM2, 223
Cherwell, River, 53
CHJ, 231
classical rainfall deficit drought, 20
climate change
  drought risk and, 149
  human influence and, 9–11, 13
  impact on drought and low flows, 76, 96–7
  impact on river flow deficits, 74
climate drivers of drought, 7–11
  atmospheric and oceanic drives, 7–8, *8*
  human influence and climate change, 9–11
  summer drought (2015), 8–9, *10*
Climate Model Intercomparison Project Phase 5 (CMIP5), 13, 32, 75, 86
climate water deficit (P-PET), 6
climate-induced drought, 6
CNRM, 71, 84
CNRM GCM data, 169
CNRM-CM3, 76
cold snow season drought, 20
composite droughts, 20
critical soil moisture (SMc), 11
crop insurance, 148
crop moisture index (CMI), **152**
CRUTS database, 130
culumative precipitation indices, **152**

data envelopment analysis (DEA), 121–2
deciles, **152**
decision support systems (DSSs), 223–4, 225, 233, 235
deficit of soil moisture drought (DSMD), 12, 13f
desalination(DES), 245
diagnosis of drought-generating processes, 3–23
direct wastewater reuse for irrigation (DWWI), 245
DPSIR approach, 243, *244*
drought duration (DD), 12, 13f
Drought Early Warning System for the Po River Basin (DEWS-Po), 208, 209

Drought: Science and Policy, First Edition.
Edited by Ana Iglesias, Dionysis Assimacopoulos and Henny A.J. Van Lanen.

drought hygrometer, 209
drought indemnity-based (or index-based) insurance, 150
drought index-based insurance, 150–4
  advantages of, 151
  basis risk, 151–3
  drought indicators, 151, **152**
  selection of, 151
drought indices
  standardised approaches, 26
  threshold approaches, 6–7
drought insurance, 147–57
  difficulties and challenges in developing, 148–9
  types of, 149–56, **149**
drought management plan (DMP), 98, 101
drought occurrence measure, 75
drought risk assessment in Latin America, 123–4, *123*
drought severity measure, 75
  drought vulnerability, 111–24
  composite indicator (weighting and aggregation), 118–22
  definition, 111, 112
  normalisation of variables to common baseline, 117–18
  policy-relevant variables, 115–17
    adult literacy rate, 116
    agricultural value added per unit of GDP, 116
    agricultural water use, 115
    average precipitation, 115
    energy use, 116
    fertiliser consumption, 117
    government effectiveness, 116
    gross domestic product (GDP) per capita, 116
    institutional capacity, 116
    irrigated area, 115
    life expectancy at birth, 116
    population density, 115
    population living below poverty line, 116
    population without access to improved water, 116
    reasonable access, 116
    total water use, 115
    water infrastructure, 117
  sensitivity analysis and model validation, 122–3
  theoretical framework, 112–15, *113*, **113–14**
drought vulnerability index *see under* DVI
DVI, 112–15, **113**, 116
  agricultural value added per unit of GDP, 116
  energy use, 116
  gross domestic product (GDP) per capita, 116
  normalisation of variables, 117–18
  population living below poverty line, 116
DVI_{Kim}, 119–20
$DVI_{Naumann}$, 120
  equal weights (*EW*), 120
  proportional weights (*PW*), 120
  random weights (*RW*), 120
drought wells, 233
DROUGHT–R&SPI project, 83
Dutch Association of Regional Water Authorities, 175

ECHAM, 71
ECHAM GCM data, 169
ECHAM4, 223
ECHAM5, 84
ECHAM5/MPIOM, 76
Economy First (EcF) scenario, 80
emergence of signal, 76
estimation of economic losses (*DEIs*), 244
estimation of risk of economic losses (*RL*), 245
EURO-FRIEND, 32
European Cohesion Fund, 195
European Fire Database, 191
European Regional Development Fund, 195
European Union
  EU-WATCH project, 71, 84
  Water Framework Directive, 29, 105, 229
European Water Archive (EWA), 31
EUROSTAT, 191
Evelyn, John, 51, *52*
executive-level institutions, 99

FCDD indexes, **152**
FIC, 223
fire weather index (FWI), 187
FP7 DROUGHT–R&SPI project, 190, 194, 242
Frasers, chronicle of, 53
future drought, 69–87
  assessment of hydrological drought, 71–8, **73**
    future drought across climate regions, 71–2
    future low flows, 76–8
    future streamflow drought in Europe, 72–5, *74*
    hotspots of future drought, 75–6
  future needs, 87
  human influences on, 78–82, *79*, *81*
    impact of gradual change of hydrological regime on, 81–2
    impact of reservoirs on global future drought, 78–80
    impact of water use on streamflow drought in Europe, 80–1
  overview of studies, 70–1
  uncertainties in, 82–6
    impact sources of uncertainty, 84–6
    uncertainty assessments, 82–4

General Circulation Models (GCMs), 40, 71
geographic information system (GIS), 131
glaciermelt drought, 20
global circulation models, 223

Global Climate Models (GCMs), 29, 75
global drought index, 75
Global Hydrological Models (GHMs), 71, 75
Global Impact Models (GIMs), 75
global warming, 9
graphical user interface (GUI), 140, *141*
Great Fire of London (1666), 50, 51, *52*
Great Winter (1709), 49
GRI, 6
groundwater (hydrological drought processes), 14–17
human influences, 15–17
processes, 16–17
groundwater drought, 6
GWAVA, 32

H08, 75
HadAM3, 223
HadCM3, 223
historical drought, recent trends, 29–41
future needs, 41
future needs, 41
trend analysis and data, 30–2, *33*
data, 31–2
methodology, 30
trends in river flow across Europe, 32–8, *35*
influence of decadal-scale variability on trends in long streamflow records, 36–8
trends in modelled runoff, 34–6
trends in observed river flow, 32–4
historic drought records, 45–65
France: Ile-de-France, 55–9, *56*, *57*
historical material, 45–7
methodology, 45–8
Rhenish droughts, 59, *62*
United Kingdom, 48–55
Valley of the Upper Rhine (Germany, Switzerland, France), 59, *62*
HSDS, 47–8, **48**, 50
HTESSEL, 32
Human Development Index(HDI), 116, 119
human-induced drought, 6
human-modified drought, 6
hydrogeology, 15
hydrological drought, 6
hydrological drought, 69, 70
hydrological drought indicators, 151, **152**
hydrological drought index insurance, 151, 155–6, **156**
hydrological drought processes (groundwater and streamflow), 14–18
groundwater, 14–17
human influences, 15–17
processes, 16–17
streamflow, 17–18
human influences, 18
processes, 17–18

Ile-de-France historic drought records, 55–9, *56*, *57*
chronological variation and severity of, 55–7, *56*, *57*
very severe droughts, 58
1578 drought, 58, *58*
eighteenth-century, 58–9, *59*
index-based insurance, 150
India: Varsha Bima rainfall index insurance scheme, 154
INPE, 129
institutional framework for drought planning and early action, 95–108
drought management in Mediterranean countries, 106–8, **107**
drought planning and water resources planning, 96–8
early action and risk management plans, 98–9
institutional analysis of drought planning, 99–100
institutions involved in drought planning, 99–108
legal framework and complexity, 100–2
Intergovernmental Panel on Climate Change (IPCC), 127
assessment reports, 32
emission scenarios, 70
SRES A1B emission scenario, 80
SRES A2 emission scenario, 71
Inter-Sectoral Impact Model Intercomparison Project (ISI-MIP), 70, 75, 79, 87
intertemporal adverse selection, 149
IPSL, 71, 84
IPSL GCM data, 169
IPSL-CM4, 77

Jucar River Basin (Spain):drought planning and management, 217–35
12-month standardised precipitation index (SPI12), 219, 220
droughts characterisation, 219–23
future droughts, 223
methods for drought vulnerability and risk assessment, 223–8, *228*
during the management phase (real time), 224–6, *225*
during the planning phase, 224
use of DSSs, 226–8
use of SODMIs, 224, 225–6, *227*
past droughts, 219–23
annual precipitation, *219*
drought impacts, 222
hydrological characterisation of droughts, 220–1, *221*
meterological characterisation, 219–20, *219*
operational drought (water resources drought), 222
proactive and participatory drought management, 228–30
culture of adaptation to droughts, 228–30

Jucar River Basin (Spain):drought planning and management (*Continued*)
institutional, legal, and normative framework, 230
measures included in the SDP, 232
responses to past drought events and drought impact mitigation, 232–4
stakeholder involvement, 231–2
Jucar River Basin District (JRBD), 217–18
location, *218*
Plan of, 1998 226, 230
population, 217
Jucar River Basin Partnership (Confederacion Hidrografica del Jucar; CHJ), 229
JULES, 32, 75, 76, 83, 85

Kendall–Theil robust line, 30
KNMI, 176
Köppen–Geiger climate type acronyms, 12, 14, 77

La Plata Basin:drought vulnerability under climate change (case study), 127–44
conclusions, 140–4
emergency measures, 143, 144
strategic measures, 143–4
tactical measures, 143, 144
methods, 128–33
data analysis, 130–2
data and regional extent, 129, *130*, **130**
framework, 128–9, *129*
limitations of methodology, 132–3
variables for drought characterisation, 132–3, **132**
results and discussion, 133–40
drought characterisation: changes in rainfall, 133
drought characterisation: SPEI drought indicator, 134–40, *135*, **137**, *138–9*
graphical user interface (GUI), 140, *141*
Land Surface Models (LSMs), 75
Latin America, drought risk assessment in, 123–4, *123*
LISFLOOD, 62, 80, 83
Little Ice Age, 49
Locke, John, 53, *52*, 53
LOESS, 30, 36, *37*
LPJmL, 32, 75, 76

Mac-PDM0.9, 75
Mann–Kendall test, 30, 241
MATSIRO, 32, 75
Maunder Minimum, 60
Meteorological Station of Montsouris, 57
meteorological-based index insurance, 154–5
meterological drought, 6, 69, 147
frequency and severity, 9–11
meterological drought index insurance, 151
meterological drought indicators, 151, **152**
Millennium Development Goals, 116
Millennium Drought, Australia, 20
MME, 83
MME mean runoff (MME-R), 84
model intercomparison projects (MIPs), 32
Monte Carlo experiments, 122
Montsouris, Meteorological Station of, 57
MPCI (yield) insurance, 150
MPI-HM, 32, 75, 76
multidimensional poverty index (MPI), 119
multi-model ensembles (MMEs), 71

NAO Index, 31
Netherlands, The
Association of Regional Water Authorities (water boards), 173
Bureau for Economic Policy Analysis (CPB), 167
Directorate General of Public Works and Water Management (Rijkswaterstaat), 173
Environmental Assessment Agency (MNP), 167
Helpdesk Water Service, 175
Interprovincial Consultation Committee, 173
National Coordination Centre Crises Management, 173
National Coordination Committee for Environment Pollution, 174
National Coordination Committee for Flooding Threat, 174
National Institute for Spatial Research (RPB), 167
National Water Distribution Committee (LCW), 171, 174, 175
Water Management Centre for the Netherlands (WMCN), 171, 174–5
Netherlands, The: drought and water management in, 165–80
2015 drought monitoring and management, 175–7, *176*
2015 drought vs 2003 drought, 177–8
current management framework, 169–70
Delta Plan on Freshwater Supply, 169
Delta Programme, 167, 169
drought characteristics: frequency and severity, 168–9
drought management framework, 171–4, *172*, *173*, *174*
drought policy, 171
drought studies, 171
drought risk and mitigation, 170–1
location-related water users, 170
network-related water users, 170
vulnerability to drought, 170
free-draining areas, 166
freshwater sources, 165
international dimension, 180

long-term average precipitation, 165
maximum potential precipitation deficit, 165, *166*
past drought events and drought impact mitigation, 178, **179**
physical system, 165–6
scenarios, 167–8, *168*
service levels, 180
socioeconomic systems, 166
stakeholder involvement, 174–5
New South Wales, Australia, cumulative rainfall insurance contract, 155
non-governmental organisations (NGOs), 98, 99, 196, 229, 230
normalised difference vegetation index (NDVI), **152**
North Atlantic Oscillation (NAO), 8, 37, 39
atmospheric circulation patterns, 7

Orchidee, 32, 76

Palmer drought severity index (PDSI), 13, 38, **152**, 186, 187, 241
Paris Observatory, 56, 57, 58
Pathe News: 'Peril of the Drought, The' (1921), 54, *55*
PCR-GLOBWB, 75, 79
Penman–Monteith concept (PDSI_PM), 13
Pepys, Samuel, 51
percentage change relative to the mean (*T*, %), 241
PET, 11
plague epidemics, UK, 50
Po River Basin (Po-DMP) (Italy), drought management, 201–14
drought characteristics and water availability, 203–6, *203*
2003 and 2005/2007 drought events, 204
2012 and 2015 drought events, 204–5, *205–6*
drought risk and mitigation, 206–13
annual water demand volume by use, **207**
framework for drought management, 208–10
policy responses to the, 2003 drought event 210–13, *211*, **212**
vulnerability to drought, 206–8
physical and socioeconomic system, 201–3, *202*
Po Water Balance Plan, 208, 209
policy-level institutions, 99
Portugal
Drought Commission, 188, 190
Drought Commission for Reservoir Management, 189, 194
Drought Monitoring and Early Warning (SVAS), 189
Drought Monitoring and Impact Mitigating Program, 190
Drought Observatory, 189
Environment Agency (APA), 188
General Directorate of Agriculture and Rural Development (DGADR), 189
Institute for Nature Conservation and Forests (ICNF), 189
National Authority for Civil Protection (ANPC), 189
National Information System for Water Resources (SNIRH), 189, 193
National Program for the Efficient Use of Water (PNUEA), 186
Regulatory Entity for Water and Waste (ERSAR), 189
Sea and Atmosphere Institute (IPMA), 187
Portugal: improving drought preparedness in, 183–95
2004–2006 drought event, 188, **188**
climate and land use, 183–4, *184*
drought management framework, 189–90
drought monitoring systems, 189
national drought information systems, 193
past drought events, impacts, and forecasted trends, 186–8, *187*
policy gaps and improvement measures, 193–5, **194**
SPIs as indicators of future drought impacts, 191–2, **192**
stakeholder involvement in drought management, 190–1
water resources: use and consumption, 184–6, *185*
potential evapotranspiration (PET), 4, 128, 129, 131, *131*
precipitation (P), 128, 129, 131, *131*
probability distribution functions (PDFs), 122
PROMES, 223
propagation, drought, 19–21, *19*
climate–hydrology links, 19
hydrological drought typology, 20
human influences, 20–1, *21*
Protocollo d'Intesa, 212

rainfall distribution index (RDI), 154
rainfall index insurance programmes, 154
rain-to-snow-season drought, 20
rainwater harvesting in cisterns for domestic use (CD), 245
RCAO, 223
reactive policies, 98–9
reference networks, 31
Regional Climate Models (RCMs), 29
regionalisation approaches, 223
remote sensing–based index insurance, 155
remote sensing drought index insurance, 151
remote sensing drought indicators, 151, **152**
reservoirs, 6

Rhenish droughts, 59, *62*
  1540–41 drought, 63–4
  chronological variation and severity, 60–1, *60*, *61*
  Laufenstein, 61–4, *62*, *63*
Rogers, Samuel, 51
Roman Catholic Church as source of historical information, 46–7
Royal Academy of Sciences of Paris, 47, 56
Royal Netherlands Meteorological Institute (KNMI), 173, 174
Royal Society of London, 47

Sacromento Valley index, **152**
SDSM, 223
sea surface temperatures (SSTs), 8
Sen's slope (*Q*), 241
sensitivity analysis, 122–3
SGI (Standardised Groundwater level Index), 173
Sicc-Idrometro (Drought-hydrometer), 209
SIMRISK module of Aquatool, 227
snowmelt drought, 20
Societas Meteorologica Palatina of Mannheim, 47, *48*
soil moisture anomaly (SMA), 13
soil moisture change (SM'), 11–12
soil moisture drought processes, 11–12
  human influences, 13–14
  processes, 11
soil moisture supply capacity (SMSC), 11
soil water drought, 6
soil water storage (SM), 11
sowing failure (SF), 154
Spain
  Agricultural Insurance Law, 103
  CEDEX (Spanish Governmental Research Institute)., 223
  Centre of Hydrographical Studies (CEH), 223
  drought management legislation, 101
  drought management plans, 98, 101
  Geological Institute, 230
  Law of the National Hydrological Plans, 103
  legal provision in, 103–6, **104–5**
  Ministers' Council, 101
  National Drought Monitoring System for Basin Water Resources Systems, 226
  permanent drought committee (PDC), 230–1
  State Agency for Meteorology (AEMET), 223
  Water Law, 101, 103
  Water Laws (1879 and 1985), 229
special drought plan (SDP), 224
standardised operative drought monitoring indicators (SODMIs), 224, 225–6, 232, 233, 234
standardised precipitation index (SPI), 6, 9, 11, 38, **152**, 173, *187*, 241
  SPI-6, 7
  SPI-12, 219, 220
Standardised Precipitation–Evapotranspiration Index (SPEI), 6, 9, 38, 128, 129, 131–2
  drought indicator, 134–40, *135*, **137**, *138–9*
  SPEI index, 127
  SPEI-6, 7
Standardized Runoff Index (SRI), 7, 173
status indicators, **152**
streamflow (hydrological drought processes), 17–18
  human influences, 18
  processes, 17–18
streamflow drought, 6
Subtropical High-Pressure Belt, 4, 7, 22
summer drought (2015), 8–9, *10*
Syros (Greece) Drought Risk and Management in, 239–47, *240*, **245**
  drought management framework, **243**
  droughts in, 240
    drought risk and mitigation, 243–6
    future droughts, 242
    key messages, 242–3
    past droughts, 240–2
  frequency of drought and wet periods (1970–2010), **241**
  need for participatory drought management, 246–7
  stakeholder involvement, *247*
  water supply and demand, **256**

Thames, River, 53
Thornthwaite approach (PDSI_Th), 13
threshold analysis, 75, 80, 241
transient threshold approach, 81
trend analysis, 30–2, *33*
  of precipitation data, Syros, 241, **242**
Tunisia Drought National Commission, 106

United Kingdom historic drought records, 48–55, *51*
  1634 drought, 51
  1666 (Great Drought), 50, 51–3, *52*, *53*
  1785 drought, 53–4, *54*
  1921 drought, 54, *55*
  chronological variation and severity of, 49–50, *50*
  most extreme events in 500-year period, 50
uncertainty analysis, 122
UNESCO-FRIEND-Water program
  European Water Archive (EWA), 31
United Nations Convention to Combat Desertification (UNCCD), 102, 103
United Nations Development Programme, 127
United Nations High Commissioner for Refugees (UNHCR), 116
United Nations International Strategy for Disaster Reduction (UNISDR), 102, 111
  Hyogo Framework for Action, 127
Upper Rhine Valley (Germany, Switzerland, France) historic drought records, 59, *62*

1540–41 drought, 63–4
chronological variation and severity, 60–1, *60*, *61*
Laufenstein, 61–4, *62*, *63*
user-level institutions, 99

Valencia Water Tribunal, 229
variable *Q80* threshold method, 71
Varsha Bima (India) rainfall index insurance scheme, 154
vegetation index-based index insurance, 155
VIC, 75
Vietnam rainfall-based drought index insurance, 154
vulnerability of water system (V_s), 245

warm snow season drought, 20
wastewater reuse for groundwater recharge (WWG), 245
WATCH Forcing Dataset (WATCH-WFD), 7, 32
ERA-Interim (WFDEI), 9
WATCH project, 32
water balance modeling, 244
water storage in cisterns for irrigation use (CI), 245
water stores and fluxes, 4–6, *5*
WaterGAP, 32, 75, 76, 80
WaterMIP, 32, 70, 87
weather index insurance, 148, 150–4
advantages of, 151
basis risk, 151–3
drought indicators, 151, **152**
liability, 153
selection of, 151
strike, triggure or upper threshold, 153
structure of index contract, *153*
tick size, 153
wet-to-dry-season drought, 20
White, Gilbert, 53
within-type changes, 9
WMO Global Runoff Data Centre (GRDC), 31
World Climate Research Programme (WCRP)
Climate Model Intercomparison Project Phase 5 (CMIP5), 13, 32, 75, 86